George H. Howison

A Treatise on Analytic Geometry

Salzwasser

George H. Howison

A Treatise on Analytic Geometry

1. Auflage | ISBN: 978-3-75250-008-0

Erscheinungsort: Frankfurt, Deutschland

Erscheinungsjahr: 2020

Salzwasser Verlag GmbH

Reprint of the original, first published in 1869.

ECLECTIC EDUCATIONAL SERIES.

A TREATISE

ON

ANALYTIC GEOMETRY

ESPECIALLY AS APPLIED TO

THE PROPERTIES OF CONICS:

INCLUDING THE MODERN METHODS OF ABRIDGED NOTATION.

WRITTEN FOR THE MATHEMATICAL COURSE OF
JOSEPH RAY, M.D.,

BY
GEORGE H. HOWISON, M.A.,
PROFESSOR IN WASHINGTON UNIVERSITY.

CINCINNATI:
WILSON, HINKLE & CO.
PHIL'A: CLAXTON, REMSEN & HAFFELFINGER.
NEW YORK: CLARK & MAYNARD.

PREFACE.

In preparing the present treatise on Analytic Geometry, I have had in view two principal objects: to furnish an adequate introduction to the writings of the great masters; and to produce a book from which the topics of first importance may readily be selected by those who can not spare the time required for reading the whole work. I have therefore presented a somewhat extended account of the science in its latest form, as applied to Loci of the First and Second Orders; and have endeavored to perfect in the subject-matter that natural and scientific arrangement which alone can facilitate a judicious selection.

Accordingly, not only have the equations to the Right Line, the Conics, the Plane, and the Quadrics been given in a greater variety of forms than usual, but the properties of Conics have been discussed with fullness; and the Abridged Notation has been introduced, with its cognate systems of Trilinear and Tangential Co-ordinates. On the other hand, to facilitate selection, these modern methods have been treated in separate chapters; and, in the discussion of properties, distinct statement, as well as natural grouping, has been constantly kept in view.

It is to be hoped, however, that omissions will be avoided rather than sought, and that the modern methods, which are here for the first time presented to the American student, may awaken a fresh interest in the subject, and lead to a wider study of it, in the remarkable properties and elegant forms with which

it has been enriched in the last fifty years. The labors of PON-
CELET, STEINER, MÖBIUS, and PLÜCKER have well-nigh wrought
a revolution in the science; and though the new properties which
they and their followers have brought to light, have not yet received any sufficient application, nevertheless, in connection with
the elegant and powerful methods of notation belonging to them,
they constitute the chief beauties of the subject, and have very
much heightened its value as an instrument of liberal culture.

To render the book useful as a work of reference, has also been
an object. In the Table of Contents, a very full synopsis of
properties and constructions will be found, which it is hoped
will meet the wants, not only of the student in reviewing, but
of the practical workman as well.

In the demonstrations, convenience and elegance have been
aimed at, rather than novelty. When it has seemed preferable
to do so, I have followed the lines of proof already indicated by
the leading writers, instead of striking out upon fresh ones. My
chief indebtedness in this respect, is to the admirable works of
Dr. GEORGE SALMON. The treatise of Mr. Todhunter has furnished some important hints; while those of O'Brien and Hymers
have been often referred to. For examples, I have drawn upon
the collections of Walton, Todhunter, and Salmon. Of American
works, those of Peirce and Church have been consulted with
advantage.

To Professor William Chauvenet, Chancellor of Washington
University, formerly Head of the Department of Mathematics
in the United States Naval Academy, I am indebted for many
valuable suggestions.

H.

WASHINGTON UNIVERSITY, }
ST. LOUIS, Sept., 1869. }

NOTE TO SECOND EDITION.

At the suggestion of several instructors, I place here an OUTLINE OF THE COURSES OF STUDY which seem to me most judicious in using the present treatise.

MINIMUM COURSE.

BOOK I, PART I. ARTS. 1—6; 13—28; 46—64; 74—85; 95, 96, 98, 99; 101—103; 106; 133—138; 145—152; 165—172; 179—184.

BOOK II, PART I. ARTS. 293—305; 310—317; 351—357; 359—372; 376; 379—385; 389—392; 402—406; 411—413; 416—418; 421, 422; 427—429; 442—444; 446—454; 456—469; 473; 476—481; 485—488; 497—501; 506, 507, 510, 511, 514, 515; 520—522; 535—543; 546—550; 553—557; 559—576; 579—586; 594—609; 622—634.

BOOK II. ARTS. 674—690, including the general doctrine of Space-coördinates and of the Plane.

To these articles there should be added such a selection from the Examples as the Course implies. The Course will thus include about 210 pages.

INTERMEDIATE COURSE.

This Course is what I suppose the leading Colleges will be most likely to pursue, and should therefore include

THE INTRODUCTION. ARTS. 1—6; 13—45.

BOOK FIRST, PART I. Chapter I. PART II, Chapter I to Art. 274. Chapter II to Art. 332, omitting, however, Arts. 307, 324—327. Chapters III—V, omitting Articles in fine type. Chapter VI to Art. 670.

BOOK SECOND. Chapter I. Chapter II, Arts. 713, 714; 731—741.

´ THE FULL COURSE

Is intended for such students as desire to make Mathematics a specialty; and students in Schools of Technology will naturally read the whole of Book Second, even when they omit large portions of Book First.

THE AUTHOR.

ERRATA.

Page 63, line 20: for $sin - \omega$ read $sin\, \omega$.

" 103, " 21: for $+$ read $=$.

" 103, " 22: *dele* the period at the close.

" 221, " 24: for $k''=A:sin C$ read $k''=sin A:sin C.$

" 247, " 9: put the ^{2} *outside* of the brace.

" 269, " 28: for *of* read *to*, and *dele* the period.

CONTENTS.

INTRODUCTION:—THE NATURE, DIVISIONS, AND METHOD OF THE SCIENCE.

PAGE.

I. DETERMINATE GEOMETRY:

Principles of Notation, ... 5
Examples, ... 8
Principles of Construction, ... 8
Examples, ... 16

DETERMINATE PROBLEMS:

In a given triangle, to inscribe a square, ... 16
" " " " a rectangle with sides in given ratio, ... 17
To construct a common tangent to two given circles, . . 18
" a rectangle, given area and difference of sides, . 21
Examples, ... 23

II. INDETERMINATE GEOMETRY:

I°. *Development of its Fundamental Principle:*
The Convention of Co-ordinates, ... 26
Distinction between Variables and Constants; definition of a Function, ... 28
Equations between co-ordinates: their geometric meaning, ... 29
The Locus defined and illustrated, ... 33

II°. *Its Method outlined; in what sense it is Analytic:*
Manner of employing geometric equations to establish properties, ... 35
Special analytic character of the Algebraic Calculus, 38
Elements of analysis added by the Convention of Co-ordinates, ... 39

III°. *Its Divisions and Subdivisions:*
Algebraic and Transcendental Geometry, . . . 41
Orders of algebraic loci: Elementary and Higher Geometry, ... 42
Loci in a Plane and in Space: Geometry of Two and of Three Dimensions, ... 42

(v)

BOOK FIRST:—PLANE CO-ORDINATES.

Part I. On the Representation of Form by Analytic Symbols.

CHAPTER FIRST.

THE OLDER GEOMETRY: BILINEAR AND POLAR CO-ORDINATES.

Section I.—The Point.

PAGE.

BILINEAR OR CARTESIAN SYSTEM OF CO-ORDINATES: Explanation in
detail, ... 48
Expressions for Point on either Axis;—for the Origin,. . 50
POLAR SYSTEM OF CO-ORDINATES: 52
Expression for the Pole;—for Point on Initial Line, . . 53
Distance, in both systems, between any Two Points in a plane, . 55
Co-ordinates of Point cutting this distance in a given ratio, . . 56
TRANSFORMATION OF CO-ORDINATES:
I. To change the Origin, Axes remaining parallel to their
first position, ... 59
II. To change the Inclination of the Axes, Origin remaining
the same, ... 59
Particular Cases:—1. From Rectangular Axes to
Oblique, . . . 60
2. From Oblique to Rectangular, 61
3. From Rectangular to Rect-
angular, ... 61
III. To change System — from Bilinears to Polars, and con-
versely, ... 62
IV. To change the Origin, and make either previous Trans-
formation at the same time, ... 63
GENERAL PRINCIPLES OF INTERPRETATION:
I. Any single equation between co-ordinates represents a
Locus, ... 65
II. Any two simultaneous equations represent Determinate
Points, ... 67
III. Any equation lacking absolute term, represents Locus
passing through Origin, ... 69
IV. Transformation of Co-ordinates does not affect Locus, nor
change the Degree of its Equation, . . . 69
SPECIAL INTERPRETATION OF EQUATIONS: Tracing their Loci by
means of Points, ... 71
Definitions and illustrations, ... 72
Examples:—Equations to some of the Higher Plane Curves, 75

Section II.—The Right Line.

A. THE RIGHT LINE UNDER GENERAL CONDITIONS.

I. Geometric Point of View:—*Equation to Right Line is always of the First Degree.*

PAGE.

Equation in terms of angle made by Line with axis X, and of its
intercept on axis Y, ... 79
" " its intercepts on the two axes, . . 81
" " its perpendicular from the Origin, and angle
of perp'r with axis X, ... 83
Polar equation, deduced geometrically, ... 84

II. Analytic Point of View:—*Every equation of First Degree in two variables represents a Right Line.*

Proof of the theorem by Algebraic Transformation of the general
equation of First Degree, ... 87
Proof by means of the Trigonometric Function implied in the equation, 87
Proof by Transformation of Co-ordinates, ... 89
Analytic deduction of the Three Forms of the equation, . . 92
Reduction of $Ax + By + C = 0$ to the form $x \cos a + y \cos \beta - p = 0$, . 95
Polar Equation obtained by Transformation of Co-ordinates, . . 97

B. THE RIGHT LINE UNDER SPECIAL CONDITIONS.

Equation to Right Line passing through Two Fixed Points, . 98
Angle between two Right Lines: condition that they shall be
parallel or perpendicular, ... 100
Equation to Right Line parallel to given Line;—perpendicular to
given Line, ... 101
Equation to Right Line passing through given Point, and parallel
to given Line, ... 103
Equation to any Right Line through a Fixed Point, . . 103
Equation to Right Line through a given Point, and cutting a given
line at given angle, ... 105
Equation to Perpendicular through a given Point, . . . 106
Length of Perpendicular from (x, y) on $x \cos a + y \cos \beta - p = 0$; also
on $Ax + By + C = 0$, ... 107
Equation to any Right Line through the intersection of two given ones, 109
Meaning of equation $L + kL' = 0$, ... 110
Equation to Bisector of angle between any two Right Lines, . . 113
Equation to the Right Line situated at Infinity, ... 116
Equations of Condition:
Condition that Three Points shall lie on one Right Line, . . 117
" " Three Right Lines shall meet in One Point, 118
" " Movable Right Line shall pass through a Fixed
Point, ... 118

C. EXAMPLES ON THE RIGHT LINE.

PAGE.

Examples in Notation and Conditions, ... 120
Examples of Rectilinear Loci, ... 125

SECTION III.—PAIRS OF RIGHT LINES.

I. Geometric Point of View :—*Equation to a Pair of Right Lines is always of Second Degree.*

Formation of equations in the type of $LM.V.\ldots = 0$: their consequent meaning, ... 130
Interpretation of equation $LL' = 0$, ... 130
Equation to Pair of Right Lines passing through a Fixed Point, . 131
   Meaning of the equation $Ax^2 + 2Hxy + By^2 = 0$, . . 132
Angle between the Pair $Ax^2 + 2Hxy + By^2 = 0$, . . . 133
   Condition that they shall cut at right angles, ... 134
Equation to Bisectors of angles between $Ax^2 + 2Hxy + By^2 = 0$, . 134
   Case of Two Imaginary Lines having Real Bisectors of their angles, ... 134

II. Analytic Point of View:—*The Equation of Second Degree in two variables, upon a Determinate Condition, represents Two Right Lines.*

Proof of the theorem by the mode of forming $LL' = 0$, . . . 135
Condition on which $Ax^2 + 2Hxy + By^2 + 2Gx + 2Fy + C = 0$ represents Two Right Lines, ... 136

SECTION IV.—THE CIRCLE.

I. Geometric Point of View:—*Equation to Circle is always of Second Degree.*

Equation to the Circle, referred to any Rectangular Axes, deduced
   from geometric definition, . . 138
   "   "   "   " Oblique Axes, ... 139
   "   "   "   " Rectangular Axes with Origin at Center, . . . 139
   "   "   "   " Diameter and Tangent at its extremity, ... 140
Polar Equation to the Circle, ... 140

II. Analytic Point of View: — *The Equation of Second Degree in two variables, upon a Determinate Condition, represents a Circle.*

Proof of theorem by comparison of the General Equation with that
   to Circle, ... 142

CONTENTS.

PAGE.

Condition that $Ax^2 + 2Hxy + By^2 + 2Gx + 2Fy + C = 0$ shall represent a Circle, ... 143

To determine Magnitude and Position of Circle, given its equation, ... 144

Condition that a Circle shall touch the Axes, ... 145

Examples, ... 146

SECTION V.—THE ELLIPSE.

I. Geometric Point of View:—*Equation to Ellipse is always of Second Degree.*

Equation to the Ellipse deduced from geometric definitions, . 149

Its general Form, referred to Axes of Curve and Focal Center, . 150

Center of a Curve defined: proof that Focal Center is center of the Ellipse, ... 151

Polar Equation to Ellipse, Center being Pole, ... 151

"   "   "   Focus   "   "   153

II. Analytic Point of View:—*The Equation of Second Degree in two variables, upon a Determinate Condition, represents an Ellipse.*

Reduction of $Ax^2 + 2Hxy + By^2 + 2Gx + 2Fy + C = 0$ to Center of its Locus, ... 155

Condition that it represent an Ellipse is $H^2 - AB < 0$, . . 159

The Point, as intersection of Two Imaginary Right Lines, a particular case of the Ellipse, ... 161

Examples, ... 165

SECTION VI.—THE HYPERBOLA.

I. Geometric Point of View:—*Equation to Hyperbola is always of Second Degree.*

Equation to the Hyperbola deduced from geometric definitions, 169

Its general Form, referred to the Axes and Focal Center, . 170

Proof that the Focal Center is the center of the Hyperbola, . 172

Polar Equation to Hyperbola, Center being Pole, . . . 172

"   "   "   "   Focus   "   "   . . . 174

II. Analytic Point of View:—*The Equation of Second Degree in two variables, upon a Determinate Condition, represents an Hyperbola.*

Equation to Hyperbola compared with Reduced Equation of Second Degree, ... 175

Condition that the latter shall represent an Hyperbola is $H^2 - AB > 0$, ... 176

Two Right Lines intersecting, a particular case of the Hyperbola, . 177

                                                                    PAGE.
Examples on the Hyperbola, ... 178

               SECTION VII.—THE PARABOLA.

I. Geometric Point of View:—*Equation to Parabola is always of
        Second Degree.*

Equation to Parabola deduced from geometric definitions,   .   181
    Its general Form, referred to Axis and Directrix,   .   .   . 182
Polar Equation to Parabola, Focus being Pole, ... 183

II. Analytic Point of View:—*The Equation of Second Degree in two
        variables, upon a Determinate Condition, represents a Parabola.*

Additional transformation of General Equation, under condition of
        non-centrality, ... 185
Condition that it represent a Parabola is $H^2 - AB = 0$,   .   .   . 188
    The Right Line as Center of Two Parallels,   .   .   . 189
        "   "   " as Limit of Two Parallels, a particular case of
                the Parabola, ... 190
    Examples, ... 191

         SECTION VIII.—LOCUS OF SECOND ORDER IN GENERAL.

Summary of Conditions already imposed upon the General Equa-
        tion, ... 195
Proof that these exhaust the varieties of its Locus,   .   .   . 196
Conics defined: Classification into Three Species, ... 197
Résumé of argument for the theorem: Every Equation of the Second
        Degree in two variables, represents a Conic, ... 197

                        CHAPTER SECOND.

    THE MODERN GEOMETRY:—TRILINEAR AND TANGENTIAL
                        CO-ORDINATES.

            SECTION I.—TRILINEAR CO-ORDINATES.

Trilinear Method of representing a Point, ... 199
Origin of the Method: the Abridged Notation, ... 200
Geometric meaning of the constant k in $a + k\beta = 0$, ... 201
Interpretation of the equations $a \pm k\beta = 0$, $a \pm \beta = 0$,   .   . 201
The Notation extended to equations in the form $Ax + By + C = 0$, . 202
Meaning of the equation $la + m\beta + n\gamma = 0$: condition that it repre-
        sent any Right Line, ... 203
    Examples: Any line of Quadrilateral, in terms of any Three, . 206

CONTENTS.

PAGE.

The symbols a, β, γ may be considered as Co-ordinates, . . . 207

Peculiar Nature of Trilinear Co-ordinates : each a Determined Function of the other two, ... 208

Equation expressing this Condition is $a\alpha + b\beta + c\gamma = M$, . . 209

General trilinear symbol for a Constant ; namely, $k(a \sin A + \beta \sin B + \gamma \sin C)$, ... 210

To render homogeneous any given equation in Trilinears, . . 210

Trilinear Equation to Right Line, ... 212

"   "   "   " parallel to a given one, . . 212

"   "   "   " situated at infinity, . . . 212

Condition, in Trilinears, that two Right Lines shall be at right angles to each other, ... 213

Trilinear Equation to Right Line joining Two Fixed Points, . . 214

"   " any Conic, referred to Inscribed Triangle, . 215

"   " Circle,   "   "   " . 216

Same for any Concentric Circle, . . 216

"   "   " Triangle of Reference having any situation, ... 217

General Equation of Second Degree in Trilinears : i. e., Trilinear Equation to any Conic, Triangle of Reference having any situation, ... 218

Trilinear Equations to Chord and Tangent of any Conic, . . . 219

Examples of Trilinear Notation and Conditions, . . . 220

SECTION II.—TANGENTIAL CO-ORDINATES.

In Tangential system, Lines are represented by *co-ordinates*, and Points by *equations*, ... 225

Cartesian *Co-efficients* are Tangential *Co-ordinates : * Tangential *Equations* are Cartesian *Equations of Condition* — namely, that a Line shall pass through two Consecutive Points on a given Curve, ... 225

Geometric interpretation of Tangentials : how they represent a *Locus*. Reason for Name, ... 226

The Right Line in the Tangential System, ... 227

ENVELOPES defined : Condition that a Right Line shall touch a Curve is the Tangential equation to the Curve ; or simply the Equation to the Envelope of the Line, ... 228

Development of the Tangential Equation to a Conic, referred to Inscribed Triangle, ... 228

Reciprocal relation between Points and Lines : the Principle of Duality, ... 235

Description of the Method of Reciprocal Polars : its relation to the Modern Geometry, ... 238

Examples illustrating Tangentials, ... 242

PART II. ON THE PROPERTIES OF CONICS.

CHAPTER FIRST.

THE RIGHT LINE.

                                                   PAGE.

Area of a Triangle in terms of the Co-ordinates of its vertices,   . 246

"   "   "   given the Right Lines which inclose it,   .   . 246

Compound ratio of segments of the Three Sides by any Transversal $= -1$, ... 248

Compound ratio of segments of the Three Sides by any three Convergents $= +1$, ... 248

Various cases of Three Convergents occurring in any Triangle, solved by the Abridged Notation, ... 249

Further application of Trilinears : The property of Homology ; Axis and Center of Homology, ... 250

Quadrilaterals — when Complete : Centers of their three Diagonals lie on one Right Line, ... 251

Harmonic and Anharmonic Properties : 253

    Constant ratio among Segments of Transversals to any Linear Pencil, ... 253

    Harmonic and Anharmonic Pencils, ... 254

    Anharmonic of a, β, $a + k\beta$, $a + k'\beta = \dfrac{k}{k'}$, ... 255

    a, β, $a + k\beta$, $a - k\beta$, form a Harmonic Pencil, . . . 255

    Anharmonic of any Pencil $a + k\beta$, $a + l\beta$, $a + m\beta$, $a + n\beta = \dfrac{(n-k)(m-l)}{(n-m)(l-k)}$, ... 255

    Definition of Homographic Systems of lines, . . . 257

Examples involving properties of the Right Line :

    Triangles, ... 257

    Harmonies of a Complete Quadrilateral, ... 258

CHAPTER SECOND.

THE CIRCLE.

I. THE AXIS OF X :

    Every Ordinate a mean proportional between the corresponding segments, ... 259

    Every Right Line meets the Curve in Two Points, . . 260

    Discrimination between Real, Coincident, and Imaginary points, ... 260

    Chords defined. Equation to any Chord, . . . 262

II. DIAMETERS :

    Definition : Locus of middle points of Parallel Chords. Equation, ... 263

CONTENTS.

xiii

PAGE.

Every diameter passes through Center, and is perpendicular
to bisected Chords, ... 264

CONJUGATE DIAMETERS defined — each bisects Chords parallel
to the other, ... 264

Conjugates of the Circle are at right angles, . . . 265

III. TANGENT:

Definition : Chord meeting Curve in Two Coincident Points, 265

Equation, ... 266

Condition that a Right Line shall touch Circle. Auxiliary
Angle, ... 267

Analytic Construction of Tangent through (x', y'): *Two* Tangents, real, coincident, or imaginary, ... 268

Length of Tangent from $(x, y) = 1 \; \bar{S}.$ 269

Subtangent — its definition and value, ... 270, 271

IV. NORMAL:

Definition of Normal. Equation, ... 270

The Normal to Circle passes through Center. Length constant, ... 271

Subnormal — its definition and value, ... 271

V. SUPPLEMENTAL CHORDS:

Definition : Equation of Condition, ... 271, 272

In the Circle, they are always at right angles, . . . 272

VI. POLE AND POLAR:

Development of the conception of the Polar, ... 273

I. Chord of Contact to Tangents from (x', y'). . . 273

II. Locus of intersection of Tangents at extremities of
convergent chords, ... 273

III. Tangent brought under this conception, . . . 274

Construction of Polar from its definition, . . . 276

Polar is perpendicular to Diameter through Pole : — its distance from Center, ... 277

Simplified geometric construction, ... 277

Distances of any two points from Center are proportional to
distance of each from Polar of the other, . . . 278

Conjugate and Self-conjugate Triangles defined : — they are
homologous, ... 278, 279

SYSTEMS OF CIRCLES.

I. SYSTEM WITH COMMON RADICAL AXIS:

Radical Axis defined : its Equation, $S - S' = 0$, . . . 280

It is perpendicular to the line of the centers, . . . 281

Construction : Combine $S - S' = 0$ with $y = 0$ and observe
the foregoing, ... 281

The three Radical Axes belonging to any three Circles meet
in one point : *Radical Center*, ... 281

PAGE.

To construct the Radical Axis by means of the Radical
 Center, ... 281
Radical Axis of Point and Circle;—of Two Points, . . 282
Definition of System of Circles with Common Radical Axis;—
 Their Centers lie on one Right Line. Their Equa-
 tion: $x^2 + y^2 - 2kx \pm \delta^2 = 0$, ... 282
To trace the System from the equation, ... 283
Locus of Contact of Tangents from any point in the C. R. A.
 is Orthogonal Circle, ... 283
 Geometric construction of the System: Limiting Points, 284
 Analytic proof of the existence of the Limiting Points, . 285

II. Two Circles with Common Tangent :
To determine the Chords of Contact, ... 287
The Tangents intersect on Line of Centers, and cut it in ratio
 of Radii, ... 289
Every Right Line through these Points of Section is cut sim-
 ilarly by the two Circles, ... 289
The Centers of Similitude, ... 289
The three homologous Centers of Similitude belonging to any
 three Circles, lie on one Right Line. The Axis of
 Similitude, ... 290

THE CIRCLE IN THE ABRIDGED NOTATION.

If a Triangle be inscribed in a Circle, and Perpendiculars be dropped
 from any point in the Circle upon the three sides, their feet
 will lie on one Right Line, ... 291
Angle between Tangent and Chord = angle inscribed under corre-
 sponding arc, ... 292
Trilinear Equation to the Tangent, referred to Inscribed Triangle, . 292
Tangents at Vertices of Inscribed Triangle cut Opposite Sides
 in points lying on one Right Line, ... 292
Lines joining Vertices of Inscribed Triangle to those of Tri-
 angle formed by Tangents meet in one Point, . . 292
Radical Axis in Trilinears, ... 293
Examples on the Circle, ... 293

CHAPTER THIRD.

THE ELLIPSE.

I. The Curve referred to its Axes.

THE AXES.

Theorem I. Focal Center bisects the Axes. Corresponding interpretation of $\frac{x^2}{a^2} + \frac{y^2}{b^2} = 1$, ... 297

CONTENTS.
XV

PAGE.

THEOREM   II. Foci fall *within* the Curve, ... 298
THEOREM III. Vertices of Curve equidistant from Foci, . . 298
THEOREM IV. Sum of Focal Radii = length of Transverse Axis. To
           construct Curve by Points, ... 298
THEOREM   V. Semi-conjugate Axis a geometric mean between
           Focal Segments of Transverse, ... 299
           Cor. Distance from Focus to Vertex of Conjugate =
           Semi-transverse. To construct Foci, . . . 299
THEOREM VI. Squares of Ordinates to Axes are proportional to
           Rectangles of corresponding Segments, . . 300
           Cor. The *Latus Rectum*, and its value, . . . 300
THEOREM VII. Squares of Axes are to each other as Rectangle of
           any two segments to Square of Ordinate, . . 301
THEOREM VIII. Ordinate of Ellipse : Corresponding Ordinate of
           Circumscribed Circle :: Semi-conjugate : Semi-
           transverse, ... 301
           Cor. 1. Analogous relation to Inscribed Circle. Con-
           structions for the Curve, ... 302
           Cor. 2. Interpretation of $\dfrac{a^2 - b^2}{a^2} = e^2$: e defined as
           the *Eccentricity*, ... 303
THEOREM IX. The Focal Radius of the Curve is a Linear Function
           of corresponding Abscissa, ... 304
           LINEAR EQUATION to Ellipse: $\rho = a \pm ex$, . . 304
Verification of the Figure of the Curve by means of its equation, . 304

DIAMETERS.

Diameters: Equation to Locus of middle points of Parallel
       Chords, ... 305
THEOREM   X. Every Diameter is a Right Line passing through
           the Center, ... 306
           Cor. Every Right Line through Center is a Diam-
           eter, ... 306
THEOREM XI. Every Diameter of an Ellipse cuts Curve in Two
           Real Points, ... 306
Length of Diameter, in the Ellipse, ... 306
THEOREM XII. Transverse Axis the *maximum*, and Conjugate the
           minimum Diameter, ... 306
THEOREM XIII. Diameters making supplemental angles with Axis
           Major are equal, ... 307
           Cor. Given the Curve, to construct the Axes, . . 307
THEOREM XIV. If a Diameter bisects Chords parallel to a second,
           second bisects Chords parallel to first, . . 307
CONJUGATE DIAMETERS defined : Ordinates to any Diameter, . . 308
           To construct a pair of Conjugates, ... 308

PAOE.

Equation of Condition to Conjugates, in the Ellipse, is
$\tan\theta.\tan\theta' = -\frac{b^2}{a^2}$, ... 308

THEOREM XV. Conjugates in the Ellipse lie on *opposite* sides of
Axis Minor, ... 308
Equation to Diameter Conjugate to that through any given point, . 309
Cor. The Axes are a case of Conjugates, . . 309
Given the co-ordinates to extremity of any Diameter, to find those
to extremity of its Conjugate, ... 309
THEOREM XVI. Abscissa to extremity of Diameter : Ordinate to
extremity of its Conjugate :: Axis Major :
Axis Minor, ... 310
THEOREM XVII. Sum of squares on Ordinates to extremities of
Conjugates, constant and = b^2, . . 310
Length of any Diameter in terms of Abscissa to extremity of its
Conjugate, ... 310
THEOREM XVIII. Square on any Semi-diameter = Rectangle Focal
Radii to extremity of its Conjugate, . . 311
THEOREM XIX. Distance, measured on a Focal Chord, from ex-
tremity of any Diameter to its Conjugate, is
constant, and equals the Semi-Major, . 311
THEOREM XX. *Sum* of squares on any two Conjugates is constant
and = sum squares on Axes, . . . 312
Angle between any two Conjugates, . . . 312
THEOREM XXI. Parallelogram under any two Conjugates, constant
and = Rectangular under Axes, . . . 313
Cor. I. Curve has but one set of Rectangular Con-
jugates, ... 313
Cor. 2. Inclination of Conjugates is *maximum*
when $a' = b'$, ... 314
THEOREM XXII. Equi-conjugates : they are the Diagonals of Cir-
cumscribed Rectangle, ... 314
Cor. Curve has but one pair of Equi-conju-
gates, ... 314
Anticipation of the Asymptotes in the Hyperbola, . . . 315

THE TANGENT.

Equation to the Tangent, referred to the Axes, . . . 315
Condition that a Right Line touch an Ellipse. *Eccentric Angle*, . 316
Analytic construction of the Tangent through any point: *Two*,
real, coincident, or imaginary, ... 317
THEOREM XXIII. Tangent at extremity of any Diameter is parallel
to its Conjugate, ... 318
Cor. Tangents at extremities of any Diameter are
parallel. Circumscribed Parallelogram, 318

PAGE.

THEOREM XXIV. Tangent bisects the *External* Angle between
Focal Radii of Contact, ... 319
Cor. 1. To construct a Tangent to the Ellipse at
a given point, ... 319
Cor. 2. Derivation of the term *Focus*, . . 319
Intercept by Tangent on Axis Major: Constructions for Tangent
by means of it, ... 320
The SUBTANGENT defined: Distinction between Subtangent of the
Curve and of a Diameter, ... 320
THEOREM XXV. Subtangent of Curve is Fourth Proportional to
Abscissa of Contact and the corresponding
segments of Axis Major, ... 321
Cor. Construction of Tangent by means of Cir-
cumscr. Circle: Subtan. not function of b, . 321
THEOREM XXVI. Perpendicular from Center on Tangent is Fourth
Proportional to the corresponding Semi-
conjugate and the Semi-axes: $p = \dfrac{ab}{b'}$, . 322
Length of the Central Perp'r in terms of its angles with the Axes, . 322
THEOREM XXVII. Locus of Intersection of Tangents cutting at
right angles is Concentric Circle, . . 323
Focal Perpendiculars upon Tangent: their length, ... 323
THEOREM XXVIII. Focal Perpendiculars on Tangent are propor-
tional to adjacent Focal Radii, . . 324
THEOREM XXIX. Rectangle under Focal Perpendiculars, constant
and $= b^2$, ... 324
THEOREM XXX. Locus of foot of Focal Perpendicular is the Cir-
cumscribed Circle, ... 324
Cor. Method of drawing the Tangent, common
to all Conics, ... 325
THEOREM XXXI. If from any Point *within* a circle a Chord be
drawn, and a perpendicular to it at the
point of section, the Perpendicular is Tangent to an Ellipse, ... 326
Cor. The Ellipse as Envelope. . . . 326
THEOREM XXXII. Diameters through feet of Focal Perpendiculars,
parallel to Focal Radii of Contact, . 327
Cor. Diameters parallel to Focal Radii of Contact, meet Tangent at the feet of the Focal
Perpendiculars, and are of the constant
length $= 2a$, ... 327

THE NORMAL.

Equation to the Normal, referred to the Axes, ... 328
THEOREM XXXIII. Normal bisects the *Internal* Angle between Focal
Radii of Contact, ... 328

An. Ge. 2.

PAGE.

Cor. 1. To construct a Normal at given point
on the Ellipse, ... 329

Cor. 2. To construct a Normal through any
point on Axis Minor, ... 329

Intercept of Normal on Axis Major: Constructions by means of it, 329

THEOREM XXXIV. Normal cuts distance between Foci in segments
proportional to adjacent Focal Radii, . 330

The SUBNORMAL defined: Subnormal of the Curve — its length, . 330

THEOREM XXXV. Normal cuts its Abscissa in constant ratio =
$$\frac{a^2 - b^2}{b^2},$$
. . . . 330

Length of Normal from Point of Contact to either Axis, . 330

THEOREM XXXVI. Rectangle under Segments of Normal by Axes = .
Square on Conjugate Semi-diameter, . 331

Cor. Equal, also, to Rectangle under corresponding Focal Radii, . . . 331

THEOREM XXXVII. Rectangle under Normal and Central Perpendicular, constant and = a^2, . . . 331

SUPPLEMENTAL AND FOCAL CHORDS.

Equation of Condition to Supplemental Chords, ... 332

THEOREM XXXVIII. Diameters parallel to Supplemental Chords are
Conjugate, ... 332

Cor. 1. To construct Conjugates at a given inclination. Caution, ... 333

Cor. 2. To construct a Tangent parallel to
given Right Line, ... 333

Cor. 3. To construct the Axes in the empty
Curve, ... 333

Focal Chords — Special Properties, ... 334

THEOREM XXXIX. Focal Chord parallel to any Diameter, a third
proportional to Axis Major and the Di-
ameter, ... 335

II. THE CURVE REFERRED TO ANY TWO CONJUGATES.

DIAMETRAL PROPERTIES.

Equation to Ellipse, referred to Conjugates, ... 336

THEOREM XL. Squares on Ordinates to any Diameter, proportional
to Rectangles under corresponding Segments, 337

THEOREM XLI. Square on Diameter : Square on Conjugate :: Rectangle under Segments : Square on Ordinate, 337

THEOREM XLII. Ordinate to Ellipse : Corresponding Ordinate to
Circle on Diameter :: b' : a', . . . 338

Cor. 1. Given a pair of Conjugates, to construct
the Curve, ... 338

PAGE.

Cor. 2. General interp'n of $x^2 + y^2 = a'^2$: Ellipse, referred to Equi-conjugates, . . . 339

Figure of the Ellipse with respect to any two Conjugates, . 339

CONJUGATE PROPERTIES OF THE TANGENT.

Equation to Tangent, referred to Conjugates, ... 339

THEOREM XLIII. Intercept of Tangent on any Diameter, third proportional to Abscissa of Contact and the Semi-diameter: $x = \dfrac{a'^2}{x'}$, ... 340

THEOREM XLIV. Rectangle under Intercepts *by* Variable Tangent on Two Fixed Parallel Tangents, *constant* and = Square on parallel Semi-diameter, . 341

THEOREM XLV. Rectangle under Intercepts *on* Variable Tangent *by* Two Fixed Parallel Tangents, *variable* and = Square on Semi-diameter parallel to Variable, ... 341

THEOREM XLVI. Rectangle under Intercepts on Variable Tangent by any two Conjugates equals Square on Semi-diameter parallel to Tangent, . . 342

Cor. 1. Diameters through intersections of Variable Tangent with Two Fixed Parallel ones, are *Conjugate*, ... 342

Cor. 2. Given two Conjugates in position and magnitude, construct the Axes, . . . 342

THE SUBTANGENT TO ANY DIAMETER: its length = $\dfrac{a'^2 - x'^2}{x'}$, . . 343

Cor. Construction of Tangent by means of Auxiliary Circle, ... 343

THEOREM XLVII. Rectangle under Subtangent and Abscissa of Contact : Square on Ordinate :: $a'^2 : b'^2$, . 344

THEOREM XLVIII. Tangents at extremities of any Chord meet on its bisecting Diameter, ... 345

PARAMETERS.

Parameter to any Diameter defined: Third proportional to Diameter and Conjugate, ... 345

Parameter of the Curve: identical with Latus Rectum, . . . 345

THEOREM XLIX. In the Ellipse, no Parameter except the Principal is equal in value to the Focal Double Ordinate, ... 346

THE POLE AND THE POLAR.

Development of the Equation to the Polar: Definition, . . 346—349

Cor. Construction of Pole or Polar from its definition, ... 349

PAGE.

THEOREM   L. Polar of any Point, parallel to Diameter Conjugate to the Point, ... 350

Special Properties: Polar of Center; — of *any* point on axis of *x*; — on Axis Major, ... 350

Cor. Second geometric construction for Polar, . 351

POLAR OF FOCUS: its distance from center, and its direction, . . 351

THEOREM   LI. Ratio between Focal and Polar distances of any point on Ellipse, constant and = *e*, . . . 352

Cor. 1. On the construction of the Ellipse according to this theorem, ... 352

Cor. 2. Polar of Focus hence called the *Directrix*, . 353

Cor. 3. Second basis for the name *Ellipse*, . . 353

THEOREM LII. Line from Focus to Pole of any Chord, bisects focal angle which the Chord subtends, . . . 354

Cor. Line from Focus to Pole of Focal Chord, perpendicular to Chord, ... 354

III. THE CURVE REFERRED TO ITS FOCI.

Interpretation of the Polar Equations to the Ellipse, . . . 355

Development of the Polar Equation to a Tangent, ... 355

Polar proof of Theorem XIX compared with former proof, and with that by pure Geometry, ... 356

IV. AREA OF THE ELLIPSE.

THEOREM LIII. Area of Ellipse = π times the Rectangle under its Semi-axes, ... 358

V. EXAMPLES ON THE ELLIPSE.

Loci, Transformations, and Properties, ... 358

CHAPTER FOURTH.

THE HYPERBOLA.

I. THE CURVE REFERRED TO ITS AXES.

THE AXES.

THEOREM   I. Focal Center bisects the Axes. Corresponding interpretation of $\frac{x^2}{a^2} - \frac{y^2}{b^2} = 1$, ... 363

The Axis Conjugate, conventional: Equation to the Conjugate Hyperbola, ... 364

THEOREM II. Foci full *without* the Curve, . . . 365

THEOREM III. Vertices of Curve, equidistant from Foci, . . 365

THEOREM IV. Difference Focal Radii = Transverse. The Curve by Points, ... 365

PAGE.

THEOREM   V. Conventional Semi-conjugate, geometric mean be-
               tween Focal Segments of Transverse,   .   . 366
           Cor. Dist. from Center to Focus = dist. between ex-
               tremities of Axes.   To construct Foci,   .   . 366
THEOREM   VI. Squares on Ordinates to Axes, proportional to Rect-
               angles under corresponding Segments,   .   367
           Cor. The *Latus Rectum*, and its value,   .   . 367
THEOREM VII. Squares on Axes are as Rectangle under any two
               Segments to Square on their Ordinate,   .   . 367
Analogy of Hyperbola to Ellipse, with respect to Circle on Trans-
    verse, defective.   Circle replaced by the *Equilateral Hyperbola*, 368
THEOREM VIII. Ordinate Hyperbola   Corresponding Ordinate of its
               Equilateral :: b : a, ... 368

           Cor. Interpretation of $\dfrac{a^2 + b^2}{a^2} = e^2$. e defined as the

               Eccentricity, ... 369
THEOREM   IX. Focal Radius of Curve, a Linear Function of corre-
               sponding Abscissa, ... 370
           LINEAR EQUATION to Hyperbola: $\rho = ex \pm a$,   .   . 371
Verification of Figure of Curve by means of its equation,   .   . 371

DIAMETERS.

Diameters : Equation to Locus of middle points of Parallel Chords,   371
THEOREM   X. Every Diameter a Right Line passing through the
               Center, ... 371
           Cor. Every Right Line through center is a Diam-
               eter, ... 372
THEOREM   XI. "Every Diameter cuts Curve in Two Real Points"
               untrue for Hyperbola, ... 372
           Cor. 1. Limit of those diameters having real intersec-

               tions : $\theta = \tan^{-1}\dfrac{b}{a}$, ... 372

           Cor. 2. All diameters cutting Hyperbola in Imagi-
               nary Points, cut its Conjugate in Two Real ones,   373
Length of Diameter, in the Hyperbola, ... 373
THEOREM XII. Each Axis the *minimum* diameter for its own curve, . 373
THEOREM XIII. Diameters making supplemental angles with Trans-
               verse arc equal, ... 374
           Cor. Given the Curve, to construct the Axes,   .   . 374
THEOREM XIV. If a Diameter bisects Chords parallel to a second,
               second bisects those parallel to first,   .   . 374
CONJUGATE DIAMETERS :   Ordinates, ... 374
               To construct a pair of Conjugates, .   .   . 374

Equation of Condition to Conjugates, in Hyperbola : $\tan \theta . \tan \theta' = \dfrac{b^2}{a^2}$,   375

PAGE.

THEOREM XV. Conjugates in the Hyperbola lie on *same* side of
Conjugate Axis, . . . 375

Equation to Diameter Conjugate to that through any given point, 375

Cor. The Axes are a case of Conjugates, . . 376

Given co-ordinates to extremity of Diameter, to find those to extremity of its Conjugate, ... 376

THEOREM XVI. Abscissa ext'y of any Diameter . Ordinate ext'y
of its Conjugate :: Transverse : Conjugate, 376

THEOREM XVII. Diff. squares on Ordinates to extremities of Conjugates, constant and = b^2, ... 376

Length of any Diameter in terms of Abscissa to extremity of its
Conjugate, ... 377

THEOREM XVIII. Square on any Semi-diameter = Rectangle Focal
Radii to extremity of its Conjugate, . 378

THEOREM XIX. Distance, measured on Focal Chord, from extremity of any Diameter to its Conjugate, constant
and equal to Semi-Transverse, . . . 378

THEOREM XX. *Difference* of squares on any two Conjugates, constant and = difference of squares on Axes, 378

Angle between any two Conjugates, . . . 379

THEOREM XXI. Parallelogram under any two Conjugates, constant
and = Rectangular under Axes, . . . 379

Cor. 1. Curve has but one set of Rectangular
Conjugates, . . . 380

Cor. 2. Inclination of Conjugates diminishes without limit : the conception of Equi-conjugates
replaced by that of *Self-conjugates*, . 380

THEOREM XXII. The Self-conjugates in the Hyperbola are Diagonals of the Inscribed Rectangle, 381

Cor. Curve has two, and but two, Self-conjugates, 381

Analogy of the Self-conjugates to the Equi-conjugates of the Ellipse, 382

THE TANGENT.

Equation to Tangent, referred to the Axes, ... 382

Condition that a Right Line touch an Hyperbola. *Eccentric Angle*, . 383

Analytic construction of the Tangent through any point : *Two*, real,
coincident, or imaginary, ... 384

THEOREM XXIII. Tangent at extremity of any Diameter, parallel
to its Conjugate, . . . 385

Cor. Tangents at extremities of any Diameter are
parallel. To circumscribe Parallelogram, . 385

THEOREM XXIV. Tangent bisects the *Internal* Angle between Focal
Radii of Contact, . . . 386

Cor. 1. To construct Tangent to Hyperbola, at a
given point, ... 386

Cor. 2. Derivation of the term *Focus*, . . . 386

CONTENTS.

None

Intercept by Tangent on the Transverse Axis: Constructions by means of it, ... 387

The SUBTANGENT to the Hyperbola 387

THEOREM XXV. Subtangent of Curve a Fourth Proportional to Abscissa of Contact and corresponding segments of Transverse, ... 387

Cor. 1. Defect supplied in the analogy between Hyperbola and Ellipse, respecting Circle on 2a, ... 388

Cor. 2. Construction of Tangent by means of Inscribed Circle, ... 388

THEOREM XXVI. Central Perpendicular on Tangent, a Fourth Proportional to corresponding Semi-conjugate and the Semi-axes: $p = \frac{ab}{b'}$, . . . 389

Length of Central Perpendicular in terms of its angles with Axes, . 389

THEOREM XXVII. Locus of Intersection of Tangents cutting at right angles is Concentric Circle, . 390

Focal Perpendiculars on Tangent: their length, . . . 390

THEOREM XXVIII. Focal Perpendiculars proportional to adjacent Focal Radii, ... 390

THEOREM XXIX. Rectangle under Focal Perpendiculars, constant and = b^2, ... 391

THEOREM XXX. Locus of foot of Focal Perpendicular is Inscribed Circle, ... 391

Cor. To draw Tangent by the method common to all Conics, ... 391

THEOREM XXXI. If, from any point *without* a Circle, a Chord be drawn, and a perpendicular to it at the point of section, the Perpendicular is Tangent to an Hyperbola, ... 392

Cor. The Hyperbola as Envelope, . . . 392

THEOREM XXXII. Diameters through feet of Focal Perpendiculars are parallel to Focal Radii of Contact, . 393

Cor. Diameters parallel to Focal Radii of Contact, meet Tangent at the feet of Focal Perpendiculars, and are of the constant length = 2a, ... 393

THE NORMAL.

Equation to the Normal, referred to the Axes, ... 393

THEOREM XXXIII. Normal bisects the *External* Angle between Focal Radii of Contact, ... 394

Cor. 1. If Ellipse and Hyperbola are confocal, Normal to one is Tangent to other at intersection, ... 394

PAGE.

Cor. 2. To construct Normal at any point on
Hyperbola, ... 394
Cor. 3. To construct Normal through any
point on Conjugate Axis, . . . 394
Intercept of Normal on Transverse Axis: Constructions by means
of it, ... 395
THEOREM XXXIV. Normal cuts distance (produced) between Foci
in segments proport'l to Focal Radii, . 395
The SUBNORMAL: Subnormal of the Hyperbola—its length, . . 396
THEOREM XXXV. Normal cuts its Abscissa in the constant
ratio $= \dfrac{a^2 + b^2}{b^2}$, ... 396
Length of Normal from Point of Contact to either Axis, . . 396
THEOREM XXXVI. Rectangle under Segments of Normal by
Axes = Square on Conjugate Semi-diam-
eter, ... 396
Cor. Equal, also, to Rectangle under corro-
sponding Focal Radii, ... 396
THEOREM XXXVII. Rectangle under Normal and Central Perp'r
on Tangent, constant and $= a^2$, . . 397

SUPPLEMENTAL AND FOCAL CHORDS.

Equation of Condition to Supplemental Chords, ... 397
THEOREM XXXVIII. Diameters parallel to Supplemental Chords are
Conjugate, ... 397
Cor. 1. To construct Conjugates at a given
inclination, ... 397
Cor. 2. To construct a Tangent parallel to
given Right Line, ... 398
Cor. 3. To construct the Axes in the empty
Curve, ... 398
Focal Chords—Properties analogous to those for the Ellipse, . . 398
THEOREM XXXIX. Focal Chord parallel to any Diameter, a third
proportional to the Transverse and the
Diameter, ... 399

II. THE CURVE REFERRED TO ANY TWO CONJUGATES.

DIAMETRAL PROPERTIES.

Equation to Hyperbola, referred to Conjugates. Conjugate and
Equilateral Hyperbola, ... 400
THEOREM XL. Squares on Ordinates to any Diameter, proportional
to Rectangles under corresponding Segments, . 401
THEOREM XLI. Square on Diameter . Square on its Conjugate ::
Rectangle under Segments : Square on their
Ordinate, ... 401

PAGE.

THEOREM XLII. Ordinate to Hyperbola : Corresponding Ordinate
to Equilateral on Axis of x :: b' : a',   . 401
  Rem. Failure of analogy to Ellipse in respect to
    Diametral Circle, ... 401
Figure of the Hyperbola, with respect to any pair of Conjugates,   . 402

CONJUGATE PROPERTIES OF TANGENT.

Equation to Tangent, referred to Conjugates, ... 402
THEOREM XLIII. Intercept of Tangent on any Diameter, third pro-
portional to Abscissa of Contact and the
Semi-diameter: $x = \dfrac{a'^2}{x'}$, ... 402

THEOREM XLIV. Rectangle under Intercepts *by* Variable Tangent
on Two Fixed Parallel Tangents, *constant*
and = Square on parallel Semi-diameter, . 403

THEOREM XLV. Rectangle under Intercepts *on* Variable Tangent
by Two Fixed Parallel Tangents, *variable*
and = Square on Semi-diameter parallel to
Variable, ... 403

THEOREM XLVI. Rectangle under Intercepts *on* Variable Tangent
by any two Conjugates = Square on Semi-
diameter parallel to Tangent, . . . 403
  Cor. 1. Diameters through intersection of Vari-
    able Tangent with Two Fixed Parallel Tan-
    gents are *Conjugate*, ... 403
  Cor. 2. Given two Conjugates in position and
    magnitude, to construct Axes, . . . 403
THE SUBTANGENT TO ANY DIAMETER: its length $= \dfrac{x'^2 - a'^2}{x'}$, . 404
  Cor. Construction of Tangent by means of Aux-
    iliary Circle, ... 404
THEOREM XLVII. Rectangle under Subtangent and Abscissa of
Contact Square on Ordinate :: a'^2 : b'^2, . 405
THEOREM XLVIII. Tangents at extremities of any Chord meet on its
bisecting Diameter, ... 405

PARAMETERS.

Definitions. Parameter of Hyperbola identical with its Latus Rectum, 406
THEOREM XLIX. In the Hyperbola, no Parameter except the
Principal equal in value to the Focal
Double Ordinate, ... 406

THE POLE AND THE POLAR.

Development of the Equation to the Polar: Definition, .   . 406—408
  Cor. Construction of Pole or Polar from its
    definition, ... 409

An. Ge. 3.

PAGE,

THEOREM   L. Polar of any Point, parallel to Diameter Conjugate
to the Point, ... 409

Polar of Center ;—of *any* point on Axis of x ;—on Transverse Axis, 410

     Cor. Second geometric construction for Polar, . 410

POLAR OF FOCUS : its distance from center, and its direction, . . 410

THEOREM   LI. Ratio between Focal and Polar distances of any
point on Hyperbola, constant and $= c$, . . 411

     Cor. 1. Curve described by continuous Motion, . 411

     Cor. 2. Polar of Focus hence called the *Directrix*, 412

     Cor. 3. Second basis for name *Hyperbola*, . . 412

THEOREM   LII. Line from Focus to Pole of any Chord, bisects
focal angle which the Chord subtends, . . 413

     Cor. Line from Focus to Pole of Focal Chord, per-
pendicular to Chord, ... 413

III. THE CURVE REFERRED TO ITS FOCI.

Interpretation of the Polar Equations to the Hyperbola, . . . 413

Development of the Polar Equation to a Tangent, ... 414

IV. THE CURVE REFERRED TO ITS ASYMPTOTES.

ASYMPTOTES defined : Derivation of the name, ... 414

THEOREM   LIII. Self-conjugates of Hyperbola are Asymptotes, . 416

Angle between the Asymptotes ; — its value in the Eq. Hyperb., . 416

Equations to the Asymptotes, ... 416

THEOREM   LIV. Asymptotes par. to Diag'ls of Semi-conjugates, . 417

THEOREM   LV. Asymptotes limits of Tangents, ... 418

THEOREM   LVI. Perpendicular from Focus on Asymptote $=$ Conjugate Semi-axis, ... 419

THEOREM   LVII. Focal distance of any point on Hyperbola $=$ distance to Directrix on parallel to Asymptote, . 419

Equation to Hyperbola, referred to its Asymptotes, ... 420

     Equation to Conjugate Hyperbola, . . . 420

THEOREM LVIII. Parallelogram under Asymptotic Co-ordinates,
constant and $= \dfrac{ab}{2}$, ... 421

THEOREM   LIX. Right Lines joining two Fixed Points on Curve to
a Variable one, make a constant intercept on
Asymptote, ... 422

Equation to the Tangent, referred to Asymptotes, ... 422

     "   Diameter through any given point,   . . . 422

     "   " Conjugate to $x'y'$. Equations to the Axes, . 422

Co-ordinates of extremity of Diameter conjugate to $x'y'$,   . . 423

THEOREM   LX. Segment of Tangent by Asymptotes, bisected at
Contact, ... 423

     Cor. The Segment $=$ Semi-diameter conjugate to
point of contact, ... 423

PAGE.

THEOREM LXI. Rectangle intercepts by Tangent on Asymptotes, constant and $= a^2 + b^2$, ... 423

THEOREM LXII. Triangle included between Tangent and Asymptotes, constant and $= ab$, ... 424

THEOREM LXIII. Tangents at extremities of Conjugates meet on Asymptotes, ... 424

THEOREM LXIV. Asymptotes bisect the Ordinates to any Diameter, ... 425

    Cor. 1. Intercepts on any Chord between Curve and Asymptotes are equal, ... 425

    Cor. 2. Given Asymptotes and Point, to construct the Curve, ... 425

THEOREM LXV. Rectangle under Segments of Parallel Chords by Curve and Asymptote, constant and $= b'^2$, . 426

V. AREA OF THE HYPERBOLA.

THEOREM LXVI. Area of Hyperbolic Segment equals log. Abscissa extreme point, in system whose modulus $=$ $\sin \phi$: or, $A = \sin \phi . lx'$, ... 428

VI. EXAMPLES ON THE HYPERBOLA.

Loci, Transformations, and Properties, ... 428

CHAPTER FIFTH.

THE PARABOLA.

I. THE CURVE REFERRED TO ITS AXIS AND VERTEX.

THE AXIS.

THEOREM I. Vertex of Curve bisects distance from Focus to Directrix, ... 431

    Interp'n of symbol p in $y^2 = 4p(x - p)$, . . . 431

THEOREM II. Focus falls *within* the Curve, ... 431

    Transformation to $y^2 = 4px$, ... 432

    Cor. To construct the Curve by points, . . . 432

THEOREM III. Square on any Ordinate $=$ Rectangle under Abscissa and four times Focal distance of Vertex, . . 432

    Cor. Squares on Ordinates vary as the Abscissas, . 432

The *Latus Rectum* defined. Its value $= 4p$, ... 433

Relation between the Parabola and the Ellipse: proof that Ellipse becomes Parabola when a increases without limit, . . . 433

    Cor. 1. Analogue, in Parabola, of Circumscribed Circle in Ellipse, ... 434

    Cor. 2. Interpretation of $\left(\dfrac{a^2 - b^2}{a^2}\right)_{a=\infty} = 1 = e^2$: e defined as the *Eccentricity*, ... 435

                                                                 PAGE.

THEOREM   IV. Focal Radius of Curve, a linear function of corre-
                sponding Abscissa, ... 436
              LINEAR EQUATION to Parabola: $p = p + x$, . 436
Verification of Figure of Curve by means of its equation, . . 437
              Nature of its infinite branch as distinguished from
                that of Hyperbola, ... 437

                          DIAMETERS.

Diameters: Equation to Locus of middle points Parallel Chords, . 438
THEOREM   V. Every Diameter is a Right Line parallel to the Axis, 439
              Cor. 1. All Diameters are parallel, . . 439
              Cor. 2. Every Right Line parallel to Axis, i. e., per-
                pendicular to Directrix, is Diameter, . 439
THEOREM   VI. Every Diameter meets Curve in Two Points — one
                finite, the other at infinity, ... 440
CONJUGATE DIAMETERS — in case of Parabola, vanish in the paral-
lelism of all Diameters, ... 440

                        THE TANGENT.

Equation to the Tangent referred to Axis and Vertex, . . . 441

Condition that a Right Line touch Parabola: $y = mx + \dfrac{p}{m}$, . . 441

Analytic construction of Tangent through $x'y'$: Two, real, coinci-
dent, or imaginary, ... 441
THEOREM   VII. Tangent at extremity of any Diameter is parallel to
                its Ordinates, ... 442
              Cor. Vertical Tangent is the Axis of y, . . 443
THEOREM VIII. Tangent bisects the Internal Angle of Diameter
                and Focal Radius to its Vertex, . . . 443
              Cor. 1. To construct Tangent at any point on Curve, 443
              Cor. 2. Derivation of Term *Focus*, . . . 443
Intercept by Tangent on Axis: its length $= x'$, ... 444
THEOREM   IX. Foot of Tangent and Point of Contact equally dis-
                tant from Focus, ... 444
              Cor. To construct Tangent at any point on Curve, or
                from any on Axis, ... 444
SUBTANGENT TO THE CURVE: its length $= 2x'$, ... 445
THEOREM   X. Subtangent to Curve is bisected in Vertex, . . 445
              Cor. 1. To construct Tangent at any point on Curve,
                or from any on Axis, ... 445
              Cor. 2. Envelope of lines in Isosceles Triangle, . 446
Focal Perpendicular on Tangent: to determine its length, . . 446
THEOREM   XI. Focal perpendicular varies in subduplicate ratio to
                Focal distance of Contact, ... 447
Length of Focal Perpendicular in terms of its angle with Axis, . 447

PAGE.

THEOREM XII. Locus of foot of Focal Perp'r is the Vertical Tangent, 447
   Cor. 1. Construction of Tangent by general Conic Method, ... 448
   Cor. 2. Circle to radius infinity is the Right Line, 448
THEOREM XIII. If from any Point a right line be drawn to a fixed Right Line, and a perpendicular to it through the point of section, the Perpendicular will touch a Parabola, ... 449
   Cor. The Parabola as Envelope, ... 449
THEOREM XIV. Locus of intersection of Tangent with Focal Chord at any fixed angle is Tangent of same inclination to Axis, ... 450
THEOREM XV. The angle between any two Tangents to a Parabola = half the focal angle subtended by their Chord of Contact, ... 451
THEOREM XVI. Locus of intersection of Tangents cutting at right angles is the Directrix, ... 451
   Cor. New illustration of Right Line as Circle with infinite radius, ... 451

THE NORMAL.

Equation to the Normal, referred to Axis and Vertex, . . . 451
THEOREM XVII. Normal bisects External Angle of Diameter and Focal Radius to its Vertex, ... 452
   Cor. To construct Normal at any point, . . 452
Intercept by Normal on Axis: its length in terms of the Abscissa of Contact, ... 452
Constructions for the Normal by means of its Intercept, . . . 453
THEOREM XVIII. Foot of Normal equidistant from Focus with Foot of Tangent and Point of Contact, . . 453
   Cor. Corresp'g constructions for Normal or Tang., 453
SUBNORMAL TO THE PARABOLA, ... 453
THEOREM XIX. Subnormal to the Parabola, constant and = 2p, . 453
Length of the Normal determined, ... 453
THEOREM XX. Normal double corresp'g Focal Perpendicular, . 454

II. THE CURVE IN TERMS OF ANY DIAMETER.
DIAMETRAL PROPERTIES.

Equation to Parabola, referred to any Diameter and Vert'l Tangent, 454
THEOREM XXI. Vertex of any Diameter bisects distance between Directrix and the point in which the Diameter is cut by its Focal Ordinate, . . . 455
THEOREM XXII. Focal distance Vertex of any Diameter = Focal distance Principal Vertex divided by the square of the Sine of Angle between Diameter and its Vertical Tangent, . . . 456

PAGE.

Theorem XXIII. Square on Ordinate to any Diameter = Rectangle
under Abscissa and four times Focal distance
of its Vertex, ... 456
Cor. Squares on Ordinates to any Diameter vary
as the corresponding Abscissas, . . . 456
Theorem XXIV. Focal Bi-ordinate to any Diameter = four times
Focal distance of its Vertex, . . . 457
Rem. Analogy of this Double Ordinate to the
Latus Rectum peculiar to Parabola, . . 457
Figure of the Curve with reference to any Diameter, . . . 457

GENERAL DIAMETRAL PROPERTIES OF THE TANGENT.

Equation to Tangent, referred to any Diameter and Vert'l Tangent, 457
Intercept by Tangent on any Diameter: its length $= x'$, . . 458
Theorem XXV. Subtangent to any Diameter is bisected in its
Vertex, ... 458
Cor. 1. To construct a Tangent to a Parabola
from any point whatever, ... 458
Cor. 2. To construct an Ordinate to any Diameter, 458
Theorem XXVI. Tangents at extremities of any Chord meet on its
bisecting Diameter, ... 459

THE POLE AND THE POLAR.

Development of the Equation to the Polar, ... 459—461
Cor. Construction of Pole and Polar from their
definitions, ... 461
Theorem XXVII. Polar of any Point, parallel to Ordinates of corre-
sponding Diameter, ... 462
Polar of any point on Axis of x; — on principal Axis, . . 462
Polar of Focus: its identity with the *Directrix*, ... 462
Theorem XXVIII. Ratio between Focal and Polar distances of any
point on Parabola, constant and $= e$, . . 463
Rem. 1. Vindication of original definition and
construction of Curve, . . . 463
Rem. 2. Second basis for the name *Parabola*, . 464
Theorem XXIX. Line from Focus to Pole of any Chord bisects
focal angle which Chord subtends, . . 464
Cor. Line from Focus to Pole of Focal Chord
perpendicular to Chord, ... 464
Examples: Intercept on Axis between any two Polars = that be-
tween perp'rs from their Poles, ... 465
Circle about Triangle of any three Tangents passes
through Focus, ... 465

PARAMETERS.

Parameter defined as Third Prop'l to Abscissa and its Ordinate, . 465

PAGE.

THEOREM XXX. Parameter of any Diameter = four times Focal
distance of its Vertex, ... 466
Cor. Parameter of the Curve = four times Focal
distance of the Vertex, ... 466
Rem. New interpretation of various Theorems
and Equations, ... 466
THEOREM XXXI. Parameter of any Diameter = its Focal Bi-
ordinate, ... 466
Cor. Parameter of Curve = Latus Rectum, . 466
Rem. The Theorem holds in the Parabola alone
of all the Conics, ... 466
Parameter of any Diameter in terms of Abscissa of its Vertex, . 467
" " " " Principal, ... 467
THEOREM XXXII. Parameter inversely proportional to sin² of Ver-
tical Tangency, ... 467

III. THE CURVE REFERRED TO ITS FOCUS.

Interpretation of the Polar Equation to Parabola, ... 467
Development of the Polar Equation to Tangent, ... 468

IV. AREA OF THE PARABOLA.

THEOREM XXXIII. Area of any Parabolic Segment = Two-thirds
the Circumscribing Rectangle, . . . 470

V. EXAMPLES ON THE PARABOLA.

Loci, Transformations, and Properties, ... 470

CHAPTER SIXTH.

THE CONIC IN GENERAL.

I. THE THREE CURVES AS SECTIONS OF THE CONE.

Definitions, ... 473
Conditions of the several Sections, and their Geometric order, . 474

II. VARIOUS FORMS OF EQUATION TO THE CONIC IN GENERAL.

General Equation in Rectangular Co-ordinates at the Vertex, . . 475
Equation to the Conic, in terms of the Focus and its Polar, . . 478
Linear Equation to the Conic, ... 479
Equation to the Conic, referred to any two Tangents, . . . 480
The Conic as Locus of the Second Order in General, . . . 482

III. THE CURVES IN SYSTEM AS SUCCESSIVE PHASES OF ONE FORMAL LAW.

Order of the Curves, as given by Analytic Conditions, . . . 483
Classification of the Conics, ... 485

PAGE.

IV. Discussion of the Properties of the Conic in General.

The Polar Relation, ... 486
Diameters: Development of the Center, ... 495
Development of the conception of Conjugates and of the Axes, . 498
Development of the Asymptotes in general symbols, . . . 501
Similar Conics defined, ... 506

V. The Conic in the Abridged Notation.

Fundamental Anharmonic Property of Conics, ... 507
Development of Pascal's and Brianchon's Theorems, . . . 508

BOOK SECOND:—CO-ORDINATES IN SPACE.

CHAPTER FIRST.

THE POINT.

Rectangular Co-ordinates in Space explained, ... 514
    Expressions for Point on either Reference-plane ;—for Point
    on either Axis ;—for Origin, ... 515
Polar Co-ordinates in Space, ... 516
The doctrine of Projections, ... 517
Distance between any Two Points in Space, . . 520
    Relation between the Direction-cosines of any Right Line, . 521
Co-ordinates of Point dividing this distance in Given Ratio, . 522
Transformation of Co-ordinates : . . 522
    I. To change Origin, Reference-planes remaining parallel to
        first position, ... 522
    II. To change Inclination of Reference-planes, . 522
    III. To change System —from Planars to Polars, and con-
        versely, ... 523
General Principles of Interpretation : 524
    I. Single equation represents a Surface, ... 524
    II. Two equations represent Line of Section, ... 524
    III. Three equations represent mnp Points, . . . 524
    IV. Eq. wanting abs. term represents Surface through Origin, . 524
    V. Transf'n of Co-ordinates does not affect Space-Locus, . 524

CHAPTER SECOND.

LOCUS OF FIRST ORDER IN SPACE.

Equation of First Degree in Three Variables represents a Plane, . 525
General Form of Equation to any Plane, ... 526
Equation to Plane in terms of its Intercepts on Axes, . . . 527

PAOE.

Equation to Plane in terms of its Perpendicular from Origin and
Direction-cosines, ... 527
Transformation of $Ax + By + Cz + D = 0$ to the form last obtained, 528
THE PLANE UNDER SPECIAL CONDITIONS, ... 529
   Equation to Plane through Three Fixed Points, . . . 529
   Angle between two Planes : conditions that they be par-
   allel or perpendicular, ... 529
   Equation to Plane parallel to given Plane ; — perpendicular to
   given Plane, ... 531
   Length of Perp'r from (xyz) on x cos $a + y$ cos $\beta + z$ cos $\gamma = 0$, 532
     "   "   "   " on $Ax + By + Cz + D = 0$, . . 532
   Equation to Plane through Common Section of two given
   ones, $P + kP' = 0$, ... 532
   Equation to Planar Bisector of angle between any two Planes, 533
   Condition that Four Points lie on one Plane, ... 533
     "   " Three Planes pass through one Right Line, . 533
     "   " Four Planes meet in one Point, . . . 534
QUADRIPLANAR CO-ORDINATES: Abridged Notation in Space, . . 534
$la + m\beta + n\gamma + r\delta = 0$ represents a Plane in Quadriplanars, . 535
LINEAR LOCI IN SPACE: solved as Common Sections of Surfaces, . 535
   The Right Line in Space as common Section of Two Planes, . 535
   Equation to Right Line in terms of Two Projections, . . 536
   Symmetrical Equations to the Right Line in Space, . . 537
   To find the Direction-cosines of a Right Line, . . . 537
   Angle between Two Right Lines in Space, ... 538
   Conditions as to Parallelism and Perpendicularity, . . . 539
   Equation to Right Line perpendicular to given Plane, . . 539
   Angle between a Right Line and a Plane, ... 540
   Condition that a Right Line lie wholly in given Plane, . 540
   Condition that Two Right Lines in Space shall intersect, . . 541
Examples involving Equations of the First Degree, ... 541

CHAPTER THIRD.

LOCUS OF SECOND ORDER IN SPACE.

General Equation of Second Degree in Three Variables : — its gen-
eral interpretation, ... 542
SURFACES OF SECOND ORDER IN GENERAL :
   Criterion of the Form of any Surface furnished by its Sections
   with Plane, ... 543
   Every Plane Section of Second Order Surface is a Curve of Sec-
   ond Order, ... 543
   Properties common to all Quadrics, ... 544—552
   Classification of QUADRICS, or Surfaces of Second Order, . . 553
   Summary of Analogies between Quadrics and Conics, . . 561

CONTENTS.

	PAGE.
SURFACES OF REVOLUTION OF THE SECOND ORDER:	562
General Method of Revolutions explained,	563
Equation to the Cone,	564
Demonstration that all Curves of Second Order are Conics,	565
Equation to the Cylinder,	567
" " Sphere,	567
Equations to the Ellipsoids of Revolution,	568
" " Hyperboloids "	568
The Ellipse of the Gorge, and its Equation,	569
" to the Paraboloid,	569
The Tangent and Normal Planes to the Quadrics,	570
EXAMPLES,	573

NOTE.

The references in the present treatise are to RAY's *New Higher Algebra*, and RAY's *Geometry and Trigonometry*.

INTRODUCTION:
THE NATURE, DIVISIONS, AND METHOD OF
ANALYTIC GEOMETRY.

ANALYTIC GEOMETRY:

ITS NATURE, DIVISIONS, AND METHOD.

1. By **Analytic Geometry** is meant, speaking generally, *Geometry treated by means of algebra.* That is to say, in this branch of mathematics, the properties of Figures, instead of being established by the aid of diagrams, are investigated by means of the symbols and processes of algebra. In short, *analytic* is taken as equivalent to *algebraic.*

2. Accordingly, and within recent years especially, the science has sometimes been called *Algebraic Geometry.* It is preferable, however, for reasons which will appear farther on, to retain the older and more usual name. *Why* algebraic treatment should be considered analytic,—in what precise sense geometry is called analytic if treated by algebra, when it is not called so if treated by the ordinary method,—will appear as we proceed. But, for the present, the attention needs to be fixed upon the simple fact, that, in connection with geometry, *analytic* means *algebraic.*

3. The properties of Figures are of two principal classes : they either refer to *magnitude,* or else to *position and form.* Thus, *The areas of circles are to each other as the squares upon their radii,* is a property of the first

(3)

kind; of the second is, *Through any three points, not in the same right line, one circle, and but one, can be passed.*

4. Accordingly, geometric problems are either *Problems of Dimension* or *Problems of Form.*

5. Corresponding to these two classes of problems, there are, in Analytic Geometry, two main divisions; namely, DETERMINATE GEOMETRY and INDETERMINATE GEOMETRY.

The methods of these two divisions we now proceed to sketch.

I. DETERMINATE GEOMETRY.

6. The geometry of Dimension is called DETERMINATE, because the conditions given in any problem in which a dimension is sought must be sufficient to determine the values of the required magnitudes; or, to speak from the algebraic point of view, these conditions are always such as give rise to a group of independent equations, equal in number to the unknown quantities involved, and therefore determinate.*

7. In completing a problem of Determinate Geometry, there are two distinct operations: the SOLUTION and the CONSTRUCTION.

8. The **Solution** for the required parts, consists in representing the known and unknown parts of the figure in question by proper algebraic symbols, and finding the roots of the equations which express the given relations of those parts.

The **Construction** of the parts when found, consists in drawing, according to geometric principles, the geometric equivalents of the determined roots.

---

* Algebra, Art. 159, compared with 168.

The principles underlying each of these operations will now be developed.

PRINCIPLES OF NOTATION.

9. These are all derived from the algebraic convention that a single letter, unaffected with exponent or index, shall stand for a single dimension; or, as it is commonly put, that each of the literal factors in a term is called a *dimension* of the term. From this, it follows that the degree of a term is fixed by the number of its dimensions. Our principles therefore are:

1st. *Any term of the first degree denotes a* LINE, *of determinate length.* For, by the convention just stated, it denotes a quantity *of one dimension*. When applied to geometry, therefore, it must denote a *magnitude* of one dimension. But this is the definition of a line of fixed length. Accordingly, a, x denote lines whose lengths have the same ratio to their unit of measure that a and x respectively have to 1.

2d. *Any term of the second degree denotes a* SURFACE, *of determinate area.* For it denotes a magnitude of *two* dimensions, that is, a surface; and since each of its dimensions denotes a line of fixed length, the term, as their product, must denote a surface of equal area with the rectangle under those lines. In fact, it is usually cited as their rectangle. Thus, ab denotes the rectangle under the lines whose lengths are a and b. Similarly, x^2 denotes the square upon a side whose length is x.

3d. *Any term of the third degree denotes a* SOLID, *of determinate volume.* For it is the product of the lengths of three lines, and hence denotes a volume equal to that of the right parallelopiped between those lines, and is so cited. Thus, abc is the right parallelopiped whose edges have severally the lengths a, b, c; and x^3 denotes the cube on the edge whose length is x.

4th. *An abstract number, or any other term of the zero degree, denotes* some TRIGONOMETRIC FUNCTION, *to the radius* 1. The general symbol for a term of the zero degree may be written $\dfrac{a}{b}$, since the number of dimensions in a quotient equals the number in the dividend less that in the divisor. If, now, we lay off any right line $AB = b$, describe a semicircle upon it, and, taking the chord $BC = a$, join AC: we shall have (Geom., Art. 225) the triangle ABC right-angled at C. Hence, (Trig., Art. 818,)

$$\frac{a}{b} = \sin{\llcorner} 1.$$

If the base, instead of the hypotenuse, were taken $= b$, we should have

$$\frac{a}{b} = \tan A.$$

If the hypotenuse were taken $= a$, and the base $= b$, we should have

$$\frac{a}{b} = \sec A \, ;$$

and so on.

5th. *A polynomial, in geometric use, is always* HOMOGENEOUS, *and denotes (according to its degree) a length, an area, or a volume, equal to* THE ALGEBRAIC SUM *of the magnitudes denoted by its terms.* By the ordinary convention of signs, it must denote the sum mentioned; it is therefore necessarily homogeneous, since the summation of magnitudes of unlike orders is impossible. We can not add a length to an area, nor an area to a volume.

Corollary.—Hence, if a given polynomial be apparently not homogeneous, it is because one or more of the linear dimensions in certain of its terms are equal to the unit of measure, and consequently represented by the implicit factor 1. *When, therefore, such a polynomial occurs, before constructing it, render its homogeneity apparent by supplying the suppressed factors.* Thus,

$$a^3 + a^2b - c - fg = a^3 + a^2b - c \times 1 \times 1 - fg \times 1.$$

Similarly, for

$$x = \sqrt{a},$$

we may write

$$x = \sqrt{a \times 1};$$

and so on.

6th. *Terms of higher degrees than the third have no geometric equivalents.* For no magnitude can have more than three dimensions.

Corollary.—*If expressions apparently of such higher degrees occur, they are to be explained by assuming 1 as a suppressed divisor,* and constructed accordingly. Thus,

$$a^5 = \frac{a^5}{1 \times 1} = \frac{a^5}{1 \times 1 \times 1};$$

and so on.

Remark.—These six principles enable us to represent by proper symbols the several parts of any geometric problem, and to interpret the result of its solution, as indicating a line, a rectangle, a parallelopiped, etc. We then construct the magnitude thus indicated, according to the principles to be explained in the next article.

An. Ge. 4.

EXAMPLES.

1. Render homogeneous $a^2b + c - d^2$.

2. In what different ways may the degree of $2a$ be reckoned? Of $5xy$? Of $\sqrt{5(x^2 + y^2)}$? State their geometric meaning for each way.

3. Interpret geometrically $\sqrt{2ab}$; $\sqrt{3}$; $\sqrt{a}$; and $\sqrt{3a}$.

4. Adapt ab to represent a line; also, $\sqrt{abc}$.

5. What does $\sqrt{a^2 + b^2}$ represent? What $\sqrt{m^2 + n^2 - l^2 - r^2}$?

6. $\sqrt{1}$ being given as denoting a surface, render its form consistent with its meaning.

7. Render $\dfrac{a + d^3 + rh}{l^2 + m^2}$ homogeneous of the second degree.

8. Render $\dfrac{abc + d + e^2}{f}$ homogeneous of the first degree.

9. Render $\dfrac{a^2b + c - d^2}{f + g^2}$ homogeneous of the zero degree.

10. Adapt a^5b^2 to represent a solid; — a surface; — a line. What is the geometric meaning of a^4b^{-2}? of a^5b^{-1}? of a^{-2}?

PRINCIPLES OF CONSTRUCTION.

10. In these, we shall confine ourselves to constructing the roots of Simple Equations and Quadratics.

I. The Root of the Simple Equation.—This may assume certain forms, the construction of which can be generalized. The following are the most important:

1st. Let $x = a \pm b$. Here, (Art. 9, 5) x denotes a line whose length is the algebraic sum of those denoted by a and b. Therefore, on any right line, take a point A as the starting-point, or *Origin*, and lay off (say to the right) the unit of measure till $AB = a$. Then, if b is *positive*, by laying off, in the same direction and on the same line as before, $BC = b$, we obtain AC as the required line; for it is evidently the sum of the given lengths. But, if b is *negative,* its effect will be to *diminish* the departure from A; hence, in that case, it must be laid off as $BD = BC = b$. We thus obtain AD for the required line; and it is obviously equal to the difference of the given lengths.

If $b > a$ and negative, then x is wholly negative. Our construction answers to this condition. For then the extremity of b will fall to the left of the origin A, say at E, and the line x will therefore be represented by AE, and measured wholly to the *left* of A.

This brings into view the important principle that *the signs* $+$ *and* $-$ *are the symbols of measurement in opposite directions.* Hence, if we have a linear polynomial, its negative terms are to be constructed by retracing such a portion of the distance made from the origin, corresponding to its positive terms, as their length requires. If we have two monomials with contrary signs, they must be laid off in opposite directions from the origin.

2d. Let $x = \dfrac{ab}{c}$. In this case, x denotes a line whose length is a fourth proportional to c, a, and b. Therefore, draw two right 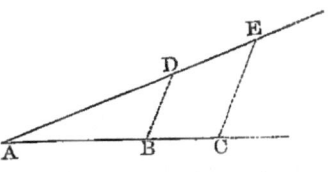 lines, AC and AE, making any angle with each other. On the one, lay off $AB = c$, and $AC = a$; on the other, $AD = b$. Join BD, and draw CE parallel to BD; then will AE be the line required. For, (Geom., 307) the triangles ABD, ACE being similar, we have

$$AB : AC :: AD : AE; \quad \text{or,} \quad c : a :: b : AE.$$

$$\therefore \quad AE = \frac{ab}{c} = x.$$

3d. Let $x = \dfrac{abc}{fg}$. Putting $\dfrac{ab}{f} = k$, this may be written

$$x = \frac{kc}{g}.$$

Therefore, construct $k = \dfrac{ab}{f}$, as in the preceding case, and, with the line thus found, apply the same construction to x.

In like manner, $\dfrac{abcd}{fgh} = \dfrac{k'd}{h}$, by putting $k' = \dfrac{abc}{fg} = \dfrac{kc}{g}$.

And, in the same way, we may construct *any* quotient of the first degree.

II. The Roots of the Pure Quadratic.—Three or four cases deserve attention:

1st. Let $x = \sqrt{ab}$. Here we have a line whose length is the geometric mean of a and b. There are several constructions, but the following is as elegant as any. On any right line, lay off $AB = a$, and $BC = b$. Upon $AC = a + b$, describe a semicircle. At B erect a perpendicular meeting the curve in D: BD is the line sought. For (Geom., 325)

$$BD = \sqrt{AB \times BC} = \sqrt{ab} = x.$$

Strictly speaking, since the radical $\sqrt{ab}$ has a double sign, there are two lines answering to x, equal in length, but measured in opposite directions. And for this, in fact, the construction provides: since there is a semicircle *below*, as well as above AC, to which the perpendicular dropped from B has the same length as BD, but is drawn in a direction exactly opposite.

2d. Let $x = \sqrt{a^2 + ac}$. Writing this in the form
$$x = \sqrt{a(a + c)},$$
we perceive that we have to construct the geometric mean of a and $a + c$. On any right line lay off $AB = a$, and $BC = c$: then $AC = a + c$. Describe a semicircle about AC, erect the perpendicular BD, and join AD: AD is the line required. (Geom., 324, 2.) In this

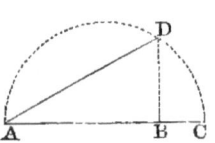

case, to satisfy the negative value of x, we must lay off a and c from A to the *left*, and throw our semicircle *below* the $-(a + c)$ thus formed. The geometric equivalent of x is the chord joining A and the point where the perpendicular from the extremity of $-a$ meets this downward semicircle. And this chord is obviously drawn from A in exactly the opposite direction to AD.

3d. Let $x = \sqrt{a^2 + b^2}$. This gives us the side of a square whose area equals the sum of two given squares. Accordingly, lay off $AB = a$; at B erect a perpendicular, and upon it take $BC = b$; join AC: then (Geom., 408) AC is the line required.

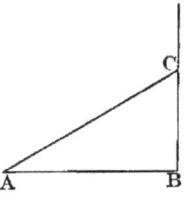

4th. Let $x = \sqrt{c^2 - b^2}$. In this case, we are to construct the side of a square whose area equals the difference of two given squares. Lay off $AB = c$, and erect upon it a semicircle. From A as a center, with a radius $AC = b$, describe an arc cutting the semicircle in C. Join CB,

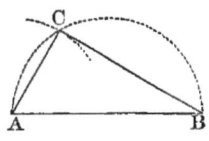

and (Geom., 409) BC will be the required line.

We leave the discussion of the negative values of x, in the last two cases, to the student.

III. Roots of the Complete Quadratic. — Let us consider the Four Forms separately. (See Alg., 231.)

1st. The First Form of the complete Quadratic is

$$x^2 + 2px = q^2 \quad \text{*}$$

Its roots are given in the formula

$$x = -p \pm \sqrt{p^2 + q^2}.$$

To construct these: Lay off $AB = q$. At B erect the perpendicular $BC = p$. From C as center, with the radius BC describe a circle. Join AC, and produce the line to meet the circle in E: AD and EA are the required roots. For, by the con-

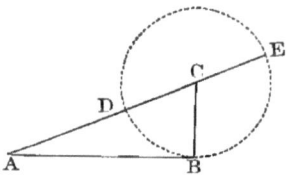

struction, $AC = \sqrt{p^2 + q^2}$; and, the first of the above roots having the radical *positive*, we must lay it off to the *right*, from A to C. The negative p must now be laid off to the *left*, from C to D. Hence,

$$AD = -p + \sqrt{p^2 + q^2} = x' \qquad (1).$$

In the second of the above roots, the radical is *negative*. Both that and p must therefore be measured to the *left* from their origin. We begin, then, at E and lay off $EC = -p$; continuing in the same direction, $CA = -\sqrt{p^2 + q^2}$. Whence

$$EA = -p - \sqrt{p^2 + q^2} = x'' \qquad (2).$$

2d. The Second Form is written

$$x^2 - 2px = q^2.$$

---

* By writing the absolute term as of the 2d degree, since (Art. 9, 5) the equation is homogeneous.

Solving for x, we obtain

$$x = p \pm \sqrt{p^2 + q^2}.$$

The construction is the same as in the First Form, except that the roots are laid off differently. In the first root, p and the radical are both positive; accordingly, we begin at A, and take $\sqrt{p^2 + q^2} = AC$; p must be taken in the same direction $= CE$: hence we obtain

$$AE = p + \sqrt{p^2 + q^2} = x' \qquad (3).$$

In the second root, p being positive and the radical negative, we begin at D, and take $p = DC$; we then retrace our steps, taking $-\sqrt{p^2 + q^2} = CA$. Whence

$$DA = p - \sqrt{p^2 + q^2} = x'' \qquad (4).$$

3d. The Third Form is

$$x^2 + 2px = -q^2:$$

whence

$$x = -p \pm \sqrt{p^2 - q^2}.$$

We construct these roots as follows: On any right line, lay off $AB = p$; erect at B a perpendicular, and take upon it $BC = q$; from C as a center, with a radius equal to p, describe an arc intersecting AF in D and E: DA and EA are the roots required. For, by the construction, $DB = BE = \sqrt{p^2 - q^2}$. From D as origin, take $DB = \sqrt{p^2 - q^2}$; from B, measure backwards $BA = -p$: and we obtain

$$DA = -p + \sqrt{p^2 - q^2} = x' \qquad (5).$$

If from E as origin we measure to the left, $EB = -\sqrt{p^2 - q^2}$, and $BA = -p$. Whence we have

$$EA = -p - \sqrt{p^2 - q^2} = x'' \qquad (6).$$

4th. The Fourth Form is

$$x^2 - 2px = -q^2.$$

Its roots are

$$x = p \pm \sqrt{p^2 - q^2}.$$

Using the same general construction as in the Third Form, we find the linear equivalent of the first of these by assuming A as origin, taking $AB = p$ to the right, and, in the same direction, $BE = \sqrt{p^2 - q^2}$. Whence,

$$AE = p + \sqrt{p^2 - q^2} = x' \qquad (7).$$

For the second root, still making A the origin, we have $AB = p$, and $BD = -\sqrt{p^2 - q^2}$. Therefore,

$$AD = p - \sqrt{p^2 - q^2} = x'' \qquad (8).$$

Remarks.—The constructions just explained furnish a good example of the clearness and completeness with which algebraic and geometric properties reflect each other, when the necessary conventions are established. Thus,

First: The construction in the First Form reflects the algebraic property (Alg. 234, Prop. 4th) that the absolute term of a complete quadratic is equal to the product of its roots.* For, by the construction, $AB = q$ is tangent to the circle DBE at B. Hence (Geom., 333) $AB^2 = AD \times AE$. That is, $q^2 = x'x''$.

On the other hand, we may see that the algebraic condition, $q^2 = x'x''$, gives us the geometric property that *the square on the tangent to a circle, from any point without the curve, is equal to the rectangle under the segments*

---

* We assume here, as we shall generally throughout the book, that the absolute term is written in the first member of the equation.

of the corresponding secant. For, multiplying together equations (1) and (2), we have,

$$x'x'' = AD \times EA.$$

But our condition gives us $x'x'' = q^2$; and, by the construction, $q^2 = AB^2$. Hence,

$$AB^2 = AD \times EA.$$

Now, AB and AE are respectively the tangent and secant from A to the circle DBE.

Second: If we compare equations (1) and (4), (2) and (3), we observe that the linear equivalents of x' in (1) and x'' in (4), of x'' in (2) and x' in (3), are identical lengths, *measured in opposite directions.* In other words, the *positive* root of the First Form is the *negative* root of the Second, and *vice versâ.* This is as it should be: for, obviously, the First Form becomes the Second, if we put $-x$ for $+x$.

Third: If, in equations (5), (6), (7), (8), we suppose $p > q$, the roots are real and unequal. The construction also indicates this. For, so long as the hypotenuse $CD = p$ is greater than the perpendicular $CB = q$, it will intersect AF in two real points.

If we suppose $p = q$, the roots are real, but equal. This, too, is involved in the construction. For, when the radius $CD = p$ becomes equal to $CB = q$, the circle touches AF; that is, its two points of section with AF become coincident in B, and $AD = AB = AE$.

Fourth: If, in (5), (6), (7), (8), we suppose $q > p$, the roots become imaginary. With this, again, the construction perfectly agrees. For, if $q > p$, the radius CD is less than the perpendicular CB, and the circle cuts AF in imaginary points. The supposition, moreover, requires us to construct a right triangle whose hypotenuse

An. Ge. 5.

shall be less than its perpendicular — a geometric impossibility. This agrees with the well-known algebraic principle, that *imaginary roots arise out of some incongruity in the conditions upon which the equation is founded.*

EXAMPLES.

1. Construct $x = \sqrt{l^2 + m^2 - n^2}$, first, by placing $m^2 - n^2 = k^2$; secondly, by placing $l^2 + m^2 = k^2$; thirdly, by placing $l^2 - n^2 = k^2$. Show that the three constructions give the same line.

2. Construct $x = \dfrac{lmn + k^2 h}{ng}$

3. Construct $\sqrt{5}$, $\sqrt{3abc}$, and $\sqrt{a^2 + b^2 - c + de}$.

4. Construct $x = \dfrac{lmn - k^2 h}{ng}$. *

5. Construct $x = \sqrt{\dfrac{l^2 m}{n}} \pm \sqrt{\dfrac{m}{n}(l^2 - mn)}$.

DETERMINATE PROBLEMS.

11. The mode of applying the foregoing principles to the solution of these problems, may be best exhibited in a few examples.

EXAMPLES.

1. *In a given triangle, to inscribe a square.*—A triangle is given when its base and altitude are given; we are therefore here required to find the side of the inscribed square in terms of the base and altitude of the given triangle. If we draw the annexed diagram, representing the problem as if solved, and designate the base of the triangle by b, its altitude by h, and the side of the inscribed square by x: then, since the triangles CAB, CEF are similar, we have (Geom., 310)

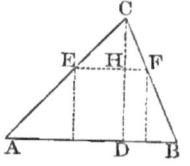

---

* The student should pay strict attention to the geometric meaning of the signs + and −, as explained above. He should also *see that the given algebraic expressions are put into the most convenient forms*, before constructing.

$$AB : EF :: CD \cdot CH; \quad \text{or,} \quad b : x :: h \,.\, h - x.$$

$$\therefore \quad x = \frac{bh}{b + h}.$$

That is, the side of the inscribed square is a fourth proportional to the base, the altitude, and their sum. We therefore construct it as in Art. 10, I. 2d. Or it may be more conveniently done as follows:

Produce the base of the given triangle until $BL = h$; through L draw LM, parallel and equal to BC; join MA, and from N, the point where MA cuts BC, drop a perpendicular upon AB: then is NO the side required. For, letting 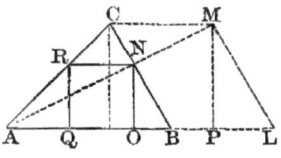 fall MP perpendicular to AL, we have, by the similar triangles MAL, NAB,

$$AL : AB :: MP : NO; \quad \text{that is,} \quad b + h : b :: h \quad NO.$$

$$\therefore \quad NO = \frac{bh}{b + h} = x.$$

Note.—The student should consider what several positions the side of the square may assume according as the triangle is acute-angled, right-angled, or obtuse-angled.

2. *In a given triangle, to inscribe a rectangle whose sides are in a given ratio.*—Let x and y represent the two sides, and r their constant ratio. Then we shall have

$$\frac{y}{x} = r \quad \therefore \quad y = rx \qquad (1).$$

And, as in the previous example, (the other 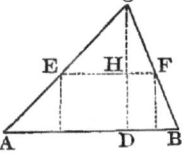 symbols remaining the same,) we obtain the proportion $b \quad y :: h : h - x$.

$$\therefore \quad hy = bh - bx \qquad (2).$$

Eliminating y between (1) and (2),

$$x = \frac{bh}{b + rh}.$$

This value we construct in the same manner as the side of the inscribed square. In fact, as is obvious, the first problem is merely a particular case of the present one; for a square is a rectangle, the ratio between the sides of which is equal to 1. The solution, too, shows this; for the value of x in the present example becomes that obtained in the former, when $r = 1$.

Produce, then, the base of the given triangle until BL equals rh; and complete the drawing exactly as in the case of the inscribed square: the point N, in which the diagonal AM cuts the side of the triangle, is a vertex of the required rectangle. Let the student prove this.

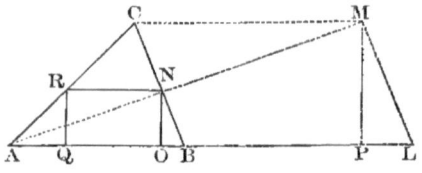

3. *To draw a common tangent to two given circles.*—Here our data are the radii of the circles, and the distance between their centers. Let r denote the radius of the circle *on the left*, and r' that of the other. Let $d =$ the distance between the centers.

The problem may be otherwise stated: *Required a point, from which, if a tangent be drawn to one of two given circles, it will also touch the other.* From the method of constructing a tangent, (see Geom., 230,) it follows that this point is somewhere on the line joining the centers. Hence, drawing the diagram as annexed, it is evident that our unknown quantity is the intercept made by the tangent on this line; that is, we let

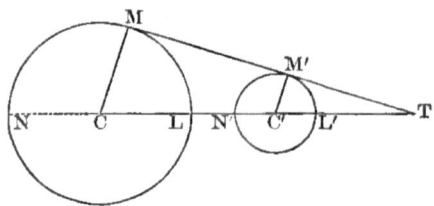

$$x = CT.$$

Now we have (Geom., 333)

$$MT^2 = NT \times LT; \quad \text{or,} \quad MT^2 = (x + r)(x - r) \tag{1}.$$

In like manner,

$$M'T^2 = N'T \times L'T; \quad \text{or,} \quad M'T^2 = (x - d + r')(x - d - r') \tag{2}.$$

Expanding, and dividing (1) by (2),

$$\frac{MT^2}{M'T^2} = \frac{x^2 - r^2}{(x - d)^2 - r'^2} \tag{3}.$$

But, by similar triangles,

$$MT : M'T :: r \quad r' \qquad \therefore \quad \frac{MT^2}{M'T^2} = \frac{r^2}{r'^2}.$$

Substituting in (3), and reducing,

$$(r^2 - r'^2) x^2 - 2r^2dx + r^2d^2 = 0:$$

$$\therefore \quad x = \frac{rd}{r \pm r'}.$$

Before constructing this result, let us interpret it. We observe that our problem involves the solution of a quadratic, and that we thus obtain a double value for $x = CT$. The required tangent, therefore, cuts the line of the centers in *two* points; that is, there are *two* points from which a common tangent to the two circles may be drawn.*

Let us now consider the two values of x more minutely. We shall find all the geometric facts of the problem perfectly represented in them.

First take the value numerically the greater, namely,

$$x = \frac{rd}{r - r'}.$$

Since this must be numerically greater than d, and since the definition of a tangent renders it impossible that the point sought should fall *within* either of the circles, the point determined by this value of x is *beyond both*. If $r > r'$, x is positive, and the point T falls to the *right* of both circles, as in the diagram; if $r < r'$, x is negative, and the point then falls to the *left* of both.

Secondly, the value numerically the less,

$$x = \frac{rd}{r + r'}.$$

This is numerically *less* than d, and, in connection with the definition of the tangent, indicates that the corresponding point lies *between* the two circles — a fact with which the *sign* of x agrees: for the present value being necessarily positive places the point to the right of C.

We learn, then, from this analysis, that (1) there are two points which satisfy the conditions of the problem; that (2) one lies beyond both circles, and the other between the two; that (3) the former falls to the *right* of both circles, or to the *left* of both, according as the circle whose center is taken as the origin has a

---

* Of course, there are *four* common tangents — two tangents from any given point to a circle being always possible. But as these exist *in pairs*, all the analytic conditions will be exhausted in *two*. Hence, we have a quadratic to solve, rather than an equation of the fourth degree.

greater or less radius than the other. How perfectly all this agrees with the geometric conditions, is manifest. By merely inspecting the diagram, we can see that two common tangents can be drawn, one passing without both circles, and intersecting the line of the centers beyond the smaller circle; the other passing between the two.

Resuming now the general expression,

$$x = \frac{rd}{r \pm r'}:$$

if we suppose $r = r'$ we obtain $x = \infty$ and $x = \frac{d}{2}$; from which we learn that, in the case of two equal circles, the external tangent is parallel to the line of the centers, while the internal tangent bisects the distance between the centers. This, again, obviously accords with the geometric conditions.

If $r = 0$, x vanishes for both its forms: hence, in this case, the two tangents are drawn from the center which was assumed as origin, and are coincident. This should be so, since, if $r = 0$, the corresponding circle is reduced to a point, and we have the ordinary problem of the tangent to a circle from a given point without.

If $r' = 0$, the two values of x again coincide, and $x = d$. In this case, therefore, the problem is reduced, as before, to that of the tangent from a given point; but the point is now the vanished *second* circle.

If $r = 0 = r'$, we have $x = \frac{0}{0}$; that is, the required tangent may be drawn from *any* point in the line of the centers. This, too, is as it should be; for, when both circles are reduced to points, the two tangents coincide with each other and with the line of the centers; and a line coincident with a given line may always be drawn from any point in the latter.

Passing now to the construction of the intercept represented by $x = \frac{rd}{r \pm r'}$, we see that we are to find a fourth proportional to $r \pm r'$, r, and d. We shall obtain this most simply as follows: Draw any set of parallel radii, as CK, $C'K'$, producing the latter to meet its circle in K''. Through (K, K') and (K, K'') draw

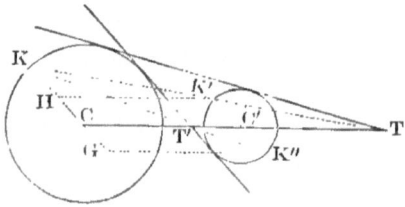

right lines: the points T and T', in which these intersect the line of the centers, are the extremities of the required intercepts. For, drawing $K'H$ and $K''G$ parallel to CT, we have, by similar triangles,

$$\left. \begin{array}{c} KG \\ KH \end{array} \right\} . \ CK :: CC' \ \left\{ \begin{array}{c} CT' \\ CT \end{array} ; \quad \text{or,} \ r \pm r' : r :: d \ \left\{ \begin{array}{c} CT' \\ CT \end{array} \right. \right.$$

Whence,

$$CT = \frac{rd}{r-r'}; \qquad CT' = \frac{rd}{r+r'}.$$

Therefore, draw through T, or T', a tangent to either of the given circles, and the construction is complete. *

4. *To construct a rectangle, given its area and the difference between its sides.*—This problem is of .importance, as illustrating the fact that we are not always to interpret the presence of a quadratic in our investigations as indicating a double solution of the problem in hand. On the contrary, a quadratic not unfrequently arises when but one solution is possible. One of its most important interpretations in that case, will appear in solving the present example.

Let (Art. 9, 2) $a^2 =$ the given area of the rectangle, and $d =$ the difference between its sides. Let $x =$ the less side; then will $x + d =$ the greater.

By the data (Geom., 379) we have

$$x(x+d) = a^2; \quad \text{or,} \ x^2 + dx = a^2.$$

$$\therefore \ x = -\frac{d}{2} \pm \sqrt{a^2 + \frac{d^2}{4}} = \text{the } \textit{less} \text{ side.}$$

$$\therefore \ x + d = \ \ \frac{d}{2} \pm \sqrt{a^2 + \frac{d^2}{4}} = \text{the } \textit{greater} \text{ side.}$$

---

* This construction, so well adapted for analytic discussion, sometimes fails in practice, as the extremity of the outer intercept may not fall upon the paper. The following elegant construction is practicable in all cases:

From the center of the larger circle, with a radius equal to the difference between the given radii, describe a circle, to which draw a tangent from the center of the smaller circle. A tangent to either given circle, parallel to this, is tangent to both.

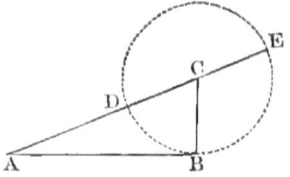

This is a case of the roots of a quadratic in the First Form. We therefore construct x as in Art. 10, III, 1. We then have, taking the upper sign in the values of both sides,

$$AD = -\frac{d}{2} + \sqrt{a^2 + \frac{d^2}{4}} = x, \text{ the } \textit{less } \text{side};$$

$$AE = \frac{d}{2} + \sqrt{a^2 + \frac{d^2}{4}} = x + d, \text{ the } \textit{greater}.$$

If we take the lower sign in the values of the two sides, we obtain

$$x = -\frac{d}{2} - \sqrt{a^2 + \frac{d^2}{4}} = -\left(\frac{d}{2} + \sqrt{a^2 + \frac{d^2}{4}}\right);$$

$$x + d = \frac{d}{2} - \sqrt{a^2 + \frac{d^2}{4}} = -\left(-\frac{d}{2} + \sqrt{a^2 + \frac{d^2}{4}}\right).$$

Comparing with the former values, we see that the present *less* side is the negative of the former *greater;* and the present *greater*, the negative of the former *less.* Hence, in this case,

$$- AE = \text{the } \textit{less } \text{side}.$$
$$- AD = \text{the } \textit{greater}.$$

It is obvious that the expressions *less* and *greater* are here used in their *algebraic* sense; for AE is still numerically greater than AD.

Now, by the construction, (Geom., 333) the rectangle of the parts gives us

$$AD \times AE = - AE \times - AD = AB^2 = a^2.$$

Hence, in both cases, the rectangle is positive, and absolutely the same. The quadratic, therefore, does not here indicate two solutions. It merely signifies that the required rectangle may be obtained either by representing its sides by x and $x + d$, or by $- x$ and $- (x + d)$. That is, it points not to two *rectangles* answering the given conditions, but merely to two correlated *modes of expressing the conditions* of one and the same rectangle.

We learn, then, that the algebraic discussion of a problem not only possesses the greatest generality — indicating by the equations to which the problem gives rise every possible solution; but that, if there are various modes of expressing conditions, which still lead to the same equation, the equation formed on the basis of any *one* of these modes will include all of them in the form of its roots.

12. From the foregoing examples, we gather the following rule for solving Determinate Problems:

Draw a diagram representing the problem as if solved, inserting any auxiliary lines needed to develop the relations between the known and unknown parts. By means of the geometric properties which the diagram involves, form equations between these parts, taking care that they be independent and equal in number to the unknown quantities. Construct upon a single figure the roots of these equations.

A few exercises are added, which the beginner should carefully perform. In each, let the problem be discussed, as to the number of its solutions, their various meanings, etc. In the construction, select that method which is neatest and most convenient.

EXAMPLES.

1. To construct a square of equal area with a given rectangle.

2. In a given triangle, to inscribe a rectangle of a given area.

3. In a given semicircle, to inscribe a square.

4. To draw, parallel to the base of a triangle, a line which shall divide it into two parts equal in area.

5. Through a given point without a circle to draw a secant whose internal segment shall be equal to a given line.

6. To describe a circle equal in area to two given ones.

7. To draw, from a given line to a given circle, a tangent of a given length.

8. To draw, from a given line to a given circle, the tangent of the least length.

9. Through two given points, to describe a circle touching a given right line.

10. In a given circle, to inscribe three equal circles touching each other externally.

II. INDETERMINATE GEOMETRY.

13. The geometry of Form is called INDETERMINATE, because all Forms are conceived to arise out of the relative positions of points; that is, out of a point's being so far indeterminate as to be capable of assuming any one of a series of positions which define a Form : or, from the algebraic point of view, because the equations which express the conditions under which a point may vary its position, are always found to be less in number than the unknown quantities they contain, and hence, admitting of an infinite number of values for these, are indeterminate.

14. It is in this second main division of the subject, that we come upon the proper province of Analytic Geometry. In fact, as the student has doubtless already noticed, the method of Determinate Geometry is rather that of ordinary geometry than of algebra: the reasoning is based mainly on the diagram, and the only use of the algebraic symbol is to abbreviate the terms of ordinary language. But, in the geometry of Form, as we shall soon discover, the method is really analytic: the reasoning is strictly algebraic, while the symbol has assumed a meaning and power entirely new. In the articles immediately following, we will endeavor, *first*, to show and establish the fundamental principle of this Geometry of Form, or of Analytic Geometry strictly so called ; *secondly*, to explain in outline its method and

---

* Alg., 168.

the reasons for calling it *analytic*; and *thirdly*, to un-
fold its several subdivisions, especially those discussed
in the body of the present work.

I. DEVELOPMENT OF THE FUNDAMENTAL PRINCIPLE.

15. The principle upon which the whole method of
Indeterminate Geometry is founded is this: *The algebraic symbol of geometric form is the Equation.*

16. The figure of any magnitude is obviously determined by that of its boundaries. Hence, all Forms are
either *surfaces* or *lines*. If, then, we can show that an
equation, geometrically interpreted, represents either
some line, or else some surface, the fundamental principle will be established.

17. There is, of course, no *necessary* connection between the symbols of algebra and the conceptions of
geometry: the former are merely conventional *marks*,
denoting *magnitudes* and *operations;* while the latter are
forms, which can be imagined and pictured, and which
are necessarily the same to every mind.

18. The truth of the proposition in Art. 15 is accordingly not *necessary*, but must depend upon certain
arbitrary assumptions. In other words, if the symbols
of algebra are to be applied to represent lines and
surfaces — if symbols of magnitude are to be converted
into symbols of form — we must introduce some convention as to their meaning in the new connection. This
convention, summarily stated, consists in making *the
algebraic symbols of magnitude denote the distances, linear
or angular, of a point from certain assumed limits.* One
form of it, the most important and characteristic, we will
now illustrate, confining ourselves, for the sake of simplicity, to points in a given plane.

19. The Convention of Co-ordinates.—Let $X'X$, $Y'Y$, be any two intersecting right lines, having any extent we please. From P, any point in their plane, draw PN parallel to the first, and PM parallel to the second. If, now, we assume $X'X$ and $Y'Y$ as the fixed limits to which all positions in their plane shall be referred, it is obvious that we know the position

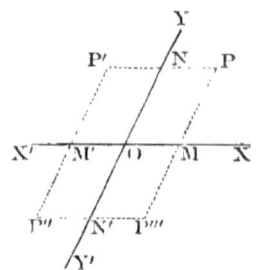

of P, so soon as we know the distances NP (or its equal OM) and MP. Hence, if we know the *distances* of any point from the two fixed limits, and the *directions* in which they are measured from these, we know the position of the point. Accordingly, if we can represent those distances and directions algebraically, we can represent the point.

This simple apparatus therefore enables us at once to convert the algebraic symbols of magnitude and direction into symbols of position. For we have only to represent the *lengths* corresponding to OM and MP by letters, and their *directions*, upward or downward from $X'X$, to the right or to the left from $Y'Y$, by the signs $+$ and $-$. The letters are applied according to the conventions for notation given in Art. 9; and the signs, according to the usage, familiar in trigonometry, that $+$ shall denote measurement *upward* from $X'X$ or to the *right* from $Y'Y$, and $-$ measurement *downward* from $X'X$ or to the *left* from $Y'Y$. Thus,

$+a = OM$, with $+b = MP$, represents P;

$-a = OM'$, with $+b = M'P'$, represents P';

$-a = OM'$, with $-b = M'P''$, represents P'';

$+a = OM$, with $-b = MP'''$, represents P'''.

The lines $X'X$, $Y'Y$ are called *axes;* their intersection O is called the *origin; OM, MP,* etc., are called *co-ordinates.*

If, now, instead of the particular lines *OM, MP,* etc., we take x and y as general symbols for the co-ordinates of a point, and denote by a and b the values they assume for any particular point, we obtain *the algebraic expression for a determinate point in a given plane,* namely,

$$\left. \begin{array}{l} x = a \\ y = b \end{array} \right\} :$$

in which a and b may have any value from 0 to ∞, and be either positive or negative.

As already hinted, the foregoing is only one form of the convention upon which rests the whole structure of the geometry of Form. Several others are used, differing from the present in the nature of the assumed limits and of the means by which the point is referred to them: in some, as in the present one, the co-ordinates are linear; in others, one of them is angular. The present form, moreover, applies only to points in a given plane; forms suitable for representing a point anywhere in space, are obtained by assuming for Fixed Limits *planes* instead of *lines.* But whether the forms of the convention apply to points in space or to points in a given plane, and however they may differ in their details, they all agree in this: that a point shall be determined by referring it to certain Fixed Limits by means of certain Elements of Reference, called Co-ordinates.

Definition.—The **Co-ordinates** of a point are its *linear or angular distances from certain assumed limits.*

Corollary.—By the **Convention of Co-ordinates** we therefore mean, *The agreement that the algebraic symbols of magnitude shall denote the co-ordinates of a point.*

20. This convention once established, the connection between an equation and a geometric form will readily become apparent. The discovery of this connection, in its universal bearing, and the first exhaustive application of it to the discussion of curves, was the work of the French philosopher DESCARTES. His method was first published in 1637, in his treatise *De la Géométrie.* We shall now show that the connection alluded to really exists; but must first define certain conceptions on which it depends.

21. Variables and Constants.—In analytic investigations, the quantities considered are of two classes: *variables* and *constants.*

Definition.—A **Variable** is a quantity susceptible, in a given connection, of an infinite number of values.

Definition.—A **Constant** is a quantity susceptible of but one value in any given connection.

Remark.—In problems of analysis, constants *impose* the conditions; variables *are subject to* them. Constants are represented by the *first* letters of the alphabet; variables by the *last*. At times, both are designated by such Greek letters as may be convenient.

22. Functions.—In the investigations belonging to Indeterminate Geometry, the variables are so connected by the conditions of the problem in hand, that any change in the value of one produces a corresponding change in that of the others.

Definition.—A **Function** is a variable so connected with others, that its value, in every phase of its changes, is derived from theirs in a uniform manner. Thus, in $y = ax + b$, y is a function of x; in $z^2 = mx^3 + ny^2 + l$, z is a function of x and y.

Remark.—Functions are classed, according to the number of the variables on which they depend, as *functions of one variable, functions of two variables,* etc.

23. With these definitions in view, we may state our Fundamental Principle with greater exactness, thus: *Every equation between variables that denote the co-ordinates of a point, represents, in general, a geometric form.* The proof of this now follows.

24. Equations between Co-ordinates: their Geometric Meaning.—Every equation is the expression of a constant relation between the variables which enter it. Further, if we solve any equation for one of its variables in terms of the others, it becomes apparent that such variable is a function of the rest. Accordingly, by varying either, we may cause all of them to vary together, by differences as great or as small as we please; but, so long as the constants that express the manner in which each is derived from the others remain unchanged, all the changes must comply with one uniform law. That is, whatever be the *absolute* value of either variable, its *relative* value, as compared with the others, is always the same. If either changes by infinitely small differences, the others must change by corresponding infinitesimals.

If, then, we assume that the variables in an equation denote co-ordinates, the equation itself must represent a number of points, as many as we please, all of which have co-ordinates *of the same relative* values. Now, since these co-ordinates vary by differences *as small as we please*, the equation really represents an infinite number of points, lying infinitely near to each other, and thus forming *a continuous series.* This continuous series of positions, moreover, has a definable *form*, of the same nature in all its parts; since, from the definition of an equation, every point in the infinite succession must comply with a *law* of position, the same for all:—a law expressed by the constants in the equation, which subject the variable co-ordinates to an inflexible relation in value. Every equation between variables that denote co-ordinates must therefore, in general, represent a geometric form.

Remark.—It will represent a line or a surface, according as the co-ordinates are taken in a plane or in space.

25. A few illustrations will render the principle just proved still more apparent. For the sake of variety, we will take these from the converse point of view, from which it will appear that every attempt to state a law of form in algebraic symbols results in an equation between co-ordinates. To simplify, let us confine ourselves to rectilinear co-ordinates in a given plane, and (since the axes of reference may be *any* two intersecting right lines) suppose $X'X$, $Y'Y$ to intersect at right angles: the co-ordinates OM, MP will then be at right angles to each other.

First: Let it be required to represent in algebraic symbols a right line parallel to the axis $Y'Y$. The law of this form plainly permits the variable point P of the line to be at any distance above or below $X'X$, but restricts it to being at a

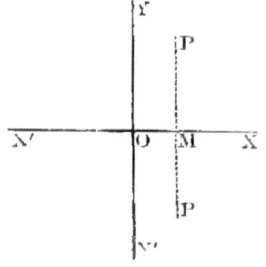

constant distance from $Y'Y$: a condition imposed by

assuming that $y = MP$ varies without limit, while, at the same time, $x = OM$ remains unchanged. Thus we see that, in a right line parallel to the axis $Y'Y$, the co-ordinate y has no determinate value, but the co-ordinate x has a fixed and unchangeable value. Hence the algebraic expression for such a line is the *equation*

$$x = \text{constant.}$$

Similarly, a right line parallel to the axis $X'X$ is represented by the *equation*

$$y = \text{constant.}$$

Second: Let it be required to represent algebraically a circle whose center is at the origin O. The law of this form is, that the variable point P shall maintain a constant distance from O. But it is obvious, upon inspecting the diagram, that the distance of *any* point from the origin is equal to $\sqrt{x^2 + y^2}$. Hence, the condition that the point shall be upon a circle whose center is

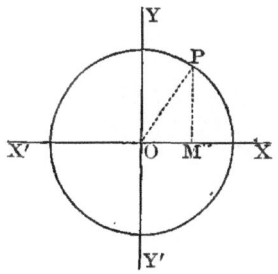

at O, gives us $\sqrt{x^2 + y^2} = \text{constant}$; and, squaring, we represent the circle by the *equation*

$$x^2 + y^2 = \text{constant} = r^2.$$

26. Let us now return to a more exact consideration of the Fundamental Principle. The student will have noticed, that, in both forms of stating it hitherto, — the forms used in Arts. 23 and 24, — we have been careful to say that it is true *in general*. This restriction is necessary, and of great importance; for there are certain

An. Ge. 6.

equations which no real values of the variables will satisfy : and such, of course, can denote only imaginary, or
impossible, forms.   Others can only be satisfied by infinite values of the variables, and consequently denote a
series of points situated at infinity : a conception as impossible, geometrically, as that corresponding to the
previous class of equations.   Others, again, can be satisfied by only one set of real values for the variables,
and therefore represent a single point; while others,
which can be satisfied by a fixed, finite number of
values, but by no others, represent a finite number of
separate points.   Others, still, are satisfied by distinct
sets of values, each set being capable of an infinite
number of values within itself, and having a distinct
relation among the variables which belong to it; and
such equations represent a group of distinct, though
related forms.

All this makes it clear that, to hold universally, our
Fundamental Principle must be stated in more abstract
terms.   We should be obliged to say, merely, that *every
equation between co-ordinates represents some conception
relating to form or position,* were it not that the happy
expedient of a technical term saves us from this cum-
brous circumlocution.   It being established that every
equation between co-ordinates has *some* equivalent in the
province of geometry, it only remains to assign a name
to that equivalent — a name generic enough to include
not only surfaces and lines, but all the exceptional
cases, real, imaginary, or at infinity, that have been
mentioned above.

27. Loci.—To include all the cases that may arise
under the conception that an equation has geometric
meaning, the term *locus* is used.

Definition.—A **Locus** is the series of positions, real or imaginary, to which a point is restricted by given conditions of form.

Corollary.—Since the locus is the geometric equivalent of the equation, we may state the Fundamental Principle of Indeterminate Geometry universally as follows: *Every equation between variables which denote the co-ordinates of a point, represents a locus.*

28. The locus being the fundamental conception of purely analytic geometry, it is of the utmost importance that correct views of it be secured at the outset. The beginner is liable to conceive of it loosely, or else too narrowly. To guard against these errors, let us illustrate what has been said or implied above somewhat more at length.

I. Classification.—Loci are either *Geometric* or merely *Analytic:* the former, when they can be represented in a diagram; the latter, when they can not.

Geometric Loci include real *surfaces, lines, points,* and related *groups* of either.

Merely **Analytic Loci** include *imaginary* loci and loci at *infinity,* and loci to be explained hereafter under the conception of a *locus in general.* The first have no existence whatever, except in the equations which symbolize them; the last two exist to abstract thought, but can neither be drawn nor imagined. The value of considering these merely analytic loci, lies in their important bearings upon some of the higher problems of the science.

II. Conformity to Law.—It is essential to the conception of a locus, that it shall conform to some definite law. No form that comes within the scope of analytic geometry can be generated at hazard; no locus is the least capricious. For it is always the counterpart of an equation; and every equation, by means of its constants, maintains among its variables, throughout their infinite changes, a uniform relation in value. We must therefore avoid the error of supposing that a broken, irregular, mixed figure, such as the line in the annexed diagram, or such as we dash off with the hand at a scribble, is a locus. On the contrary, a locus is, in a certain important sense, *homogeneous.* That is, throughout its whole extent, it is so far alike as to be represented by one equation, and but one. No point in it can be found whose co-ordinates do not satisfy this one equation.

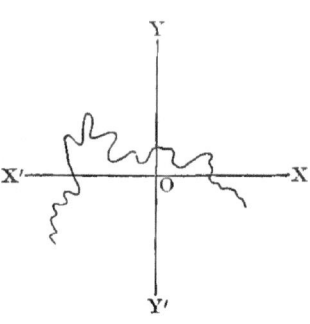

III. Variety of Meaning.—The idea of the locus should be conceived broadly enough to include, in addition to surfaces and lines, the various exceptional species enumerated in Art. 26. The attention has, perhaps, been sufficiently called to cases where it is a point, or a series of separate points, and where it is imaginary, or at infinity. But it is worth while to repeat that a locus is not necessarily a *single* figure. For example, the equation

$$x^2 - y^2 = 0$$

represents, as will be proved in the treatise which is to follow, *two* right lines, such as AA', BB', bisecting the supplemental angles between the axes. Again, as will also become evident upon a further acquaintance with the subject, the equation

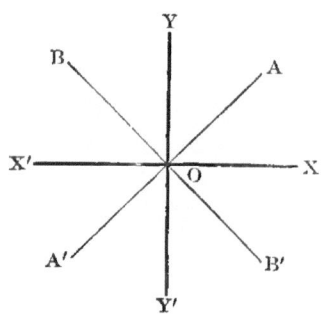

$$(x^2 + y^2 - r^2)(x^2 + y^2 - r'^2)(x^2 + y^2 - r''^2) = 0$$

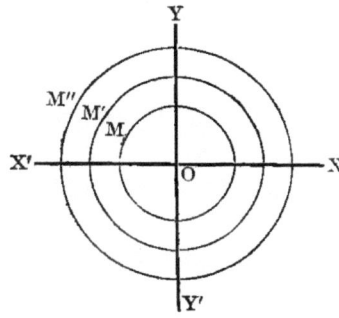

represents *three* circles, such as M, M', and M'', having a common center at the origin.

Examples of this kind might be greatly multiplied, but these are perhaps enough to render the principle clear, and to fix it in the memory.

II. THE METHOD OUTLINED; IN WHAT SENSE IT IS ANALYTIC.

29. The following is a very simple example of the method by which Analytic Geometry investigates the properties of figures. The beginner, of course, must accept upon authority the meaning of the equations employed.

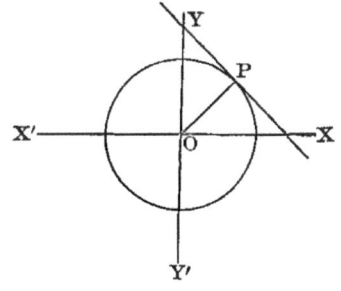

To prove that the tangent to a circle is perpendicular to the radius drawn to the point of contact.—Let the axes be rectangular, and the center of the circle at the origin. The equation to the circle is, in that case,

$$x^2 + y^2 = r^2.$$

The equation to the tangent at any point P, whose coordinates are x', y', is

$$x'x + y'y = r^2 \qquad (1).$$

The equation to the radius OP, referred to the same axes, is

$$x'y - y'x = 0 \qquad (2).$$

Now, it is known that when two equations of the first degree, referred to rectangular axes, interchange the co-efficients of x and y, at the same time changing the sign of one of them, they represent two right lines mutually perpendicular. Inspecting (1) and (2), we see that they answer to this condition. Hence, the lines which they represent are mutually perpendicular.

30. Generalizing from the foregoing illustration, we may sum up the method of our science as follows:

I. Any locus, the subject of investigation, is represented by its equation.

II. This equation is then subjected to such transformations, or such combinations with the equations to other loci, as the conditions of the problem may require.

III. The geometric meaning of these transformations and combinations, as derived from the convention of co-ordinates, is duly noted. In the same way, the form of the final result is interpreted. Thus the properties of the locus are deduced from the mere form of its equation.

31. We have now reached a point from which to obtain a clear view of the reasons why geometry, when treated by means of algebra, should be called analytic. We must, in the first place, warn the beginner that the reason most obviously suggested by the method just described, is not among them. It is true, certainly, that this method assumes the equation to any locus to be the *synthesis*, or expression in a single formula, of all its properties. It is true that these properties are drawn out from the equation by a process of real *analysis* — namely, by solving or transforming the equation, thus causing it to give out the several conditions which its original form unites in one symbol.

But this obvious fact, that we proceed from a complex unity to the elements that have vanished into it, by directly taking apart the unity itself, is not the distinctive reason for calling the Geometry of the Equation analytic; for it is a fact which does not distinguish it from the Geometry of the Diagram. In this last-named form of the science, the whole scheme of demonstration consists merely in developing what certain definitions and axioms imply: that is, the *basis* of the reasoning — all that gives it force and validity — is analytic. And, in fact, the same is true in every department of mathematics.

32. In short, the term *analytic* is applied to the Geometry of the Equation, not so much by way of contrast, as of emphasis. It should not be taken as implying that the Geometry of the Diagram is wholly *synthetic*, and the Geometry of the Equation wholly *analytic*, each to the exclusion of the other. Both are analytic, both synthetic; and in both, the vital principle of the proofs is analysis. But to the Geometry of the Equation, the character of analysis belongs in a special sense, and in a higher degree; just as that of synthesis belongs, in a special sense and higher degree, to the Geometry of the Diagram. It involves, moreover, certain phases of analysis, of a higher and more subtle kind than ordinary geometry attains.

33. Now, this special analytic character, it owes *to its use of the algebraic symbol.* The question therefore

naturally arises : How does the use of this symbol bring
with it this special character ? The answer is — In two
ways : first, by giving scope to the analytic tendency
inseparable from algebraic investigation; secondly, by
introducing the convention of co-ordinates, with its
added elements of analysis. We will illustrate both of
these ways somewhat in detail.

**34. Special Analytic Character of the Alge-
braic Calculus.**—This is due to two facts : the first,
that operations with symbols necessarily thrust into prom-
inence the analytic phase of the thinking they imply,
while they obscure the synthetic ; the second, that the
Theory of Equations — the essence of the science of
algebra, and the ground upon which all its investigations are based — is an application of analysis, peculiarly complex and subtle. That these are facts, will
best be seen by considering them separately.

I. To exhibit the first, let it be borne in mind that
every demonstration involves both analysis and synthesis — analysis, in thinking out the steps connecting
the premise with the conclusion; synthesis, in arrang-
ing those steps in their due order, and constructing the
conclusion as their unity. Now, if we use ordinary
language in making this array, clearness can not be
secured without stating these steps one by one ; thus we
seem to begin with *parts*, and to construct the conclusion
as the whole which they compose.

But if we employ algebraic language, our premise is
written down in a formula, and, at the outset, our attention is fixed upon it as a *whole.* The formula is the
permanent object of our thought; the operations, the
transformations it undergoes, seem transient and subordinate, and their results but dependent phases of its

original form. Derived from the formula by a particular series of transformations, while a number of others
are equally possible, the conclusion stands in the mind,
not as a whole but rather as a part. It appears to us
as but one of many elements involved in the original
formula—elements that may be made to show themselves, if we apply other transformations. Thus the
synthetic phase, though as real here as in using ordinary language, is lost to view, and we are only conscious of the analytic.

II. That the Theory of Equations involves a subtle
and peculiar form of analysis, is obvious, and need not
be enlarged upon. It is sufficient merely to recall its
topics and their accessories, such as the Doctrine of
Co-efficients, the Theory of Roots—their Number,
Form, Situation, and Limits, the Discussion of Series,
and the Binomial Theorem. But it is important to
mention, that, so controlling a part does this Theory
play in the whole science of algebra, and so emphat-
ically does it embody a method peculiarly analytic, the
science itself, from its earliest years, has been known by
the name of *Analysis.* And it was mainly in allusion to
the fact that the Geometry of the Equation brings the
discussion of Form within the scope of this Theory, that
the title *analytic* was originally applied to it.

**35. Elements of Analysis added by the Con-
vention of Co-ordinates.**—This convention has a
twofold analytic meaning:

I. First, it asserts that the conceptions of Position
and Form are merely *relative* ones, always implying
certain fixed limits to which they are referred — the
positions of points are their distances from these limits;
the forms of loci are the relative positions of their con-

stituent points. This assertion reaches the real essence
of Position and Form, and gives to the science based
upon it an element of analysis not attained in the Geometry of the Diagram.

II. Secondly, in referring the form of every locus to
fixed limits by means of the co-ordinates of *every* point,
the convention really determines that form by decom-
posing it into *infinitely small elements*. It thus brings
the form under the highest analytic conception known
to mathematics, and prepares for its discussion by the
various branches of the Infinitesimal Calculus.

36. To recapitulate : The Geometry of the Equation
is called Analytic, *first*, because its use of algebraic
processes puts forward the analytic, and retires the
synthetic phase of every demonstration, thus rendering
us conscious of investigation rather than of proof;
secondly, because its method consists in applying those
special modes of analysis which mark the Theory of
Equations; *thirdly*, because its convention of co-ordinates penetrates to the real nature of Position and
Form, resolving them into their essential constituents —
Fixed Limits and Distance; *finally and most significantly*, because it resolves all Forms into elements
infinitely small, and thus brings the discussion of loci
within the sphere of the Infinitesimal Calculus — the
highest expression of mathematical analysis.

37. These facts constitute a sufficient reason for pre-
ferring to call the science Analytic Geometry, rather
than Algebraic. The former title, more forcibly than
the latter, calls up the characteristics of its method, as
they have been detailed above. Moreover, the term
algebraic is ambiguous; for it is generally used, in contrast to *transcendental*, to characterize operations which

involve only addition, subtraction, multiplication, division, or involution and evolution with constant indices. But Analytic Geometry considers all loci whatsoever, not only *Algebraic* but *Transcendental*, whether the latter be *Exponential*, *Logarithmic*, or *Trigonometric*.

38. The superiority of the Geometry of the Equation to that of the Diagram consists partly in its greater brevity and elegance, but mainly in its greater power of generalization. For since the equation to any locus is the complete synthesis of all its properties, our power of investigating and discovering these is limited only by our ability to transform the equation and to determine the form, limits, number, and situation of its roots. And since *every* equation denotes some locus, the equations of the several degrees may be discussed in their most general forms. Loci may thus be grouped into Orders, according to the degree of their equations, and properties common to an entire Order may be discovered by absolute deduction — properties which, if they could be established at all by ordinary geometry, would involve the most tedious processes of comparison and induction.

III. THE SUBDIVISIONS OF THE SCIENCE.

39. The subdivisions of Indeterminate Geometry refer to the nature of the loci discussed in each; and these are classified, primarily, according to the form of their equations; secondarily, according to their situation in a plane, or in space.

40. Hence, the first division of Indeterminate Geometry is into TRANSCENDENTAL and ALGEBRAIC.

Transcendental Geometry discusses those loci whose equations involve transcendental functions; that is, functions which depend on either a *variable exponent*, a *logarithm*, or one of the expressions *sin*, *cos*, *tan*, etc.

Algebraic Geometry, those whose equations involve none but algebraic functions; that is, functions which imply only the operations of addition, subtraction, multiplication, division, or involution and evolution with constant indices.

41. Algebraic loci are classed into Orders, according to the degree of their equations referred to rectilinear axes. Thus, the locus whose equation is of the first degree, is called the locus of the First order; those whose equations are of the second degree are called loci of the Second order; and so on.

The loci of the First and Second orders, on account of their simplicity, symmetry, and limited number, are considered to form a class by themselves; and those of all higher orders are grouped together as a second class.

42. Accordingly, the second division of our subject is that of Algebraic Geometry into ELEMENTARY and HIGHER.

Elementary Geometry is the doctrine of loci of the First and Second orders.

Higher Geometry is the doctrine of loci of higher orders than the Second.

43. Each of these divisions based upon the form of the equations considered, falls into the province of Plane or of Solid Geometry, according as the equations are between plane co-ordinates, or co-ordinates in space; that is, according as the system of reference is *two* intersecting *lines*, giving rise to *two* co-ordinates for every point; or *three* intersecting *planes*, giving rise to *three* co-ordinates.

Hence, we have GEOMETRY OF TWO DIMENSIONS, and
GEOMETRY OF THREE DIMENSIONS.

44. The relations which the various divisions of Analytic Geometry sustain to each other, will be best
understood from the following

SYNOPTICAL TABLE OF DIVISIONS.

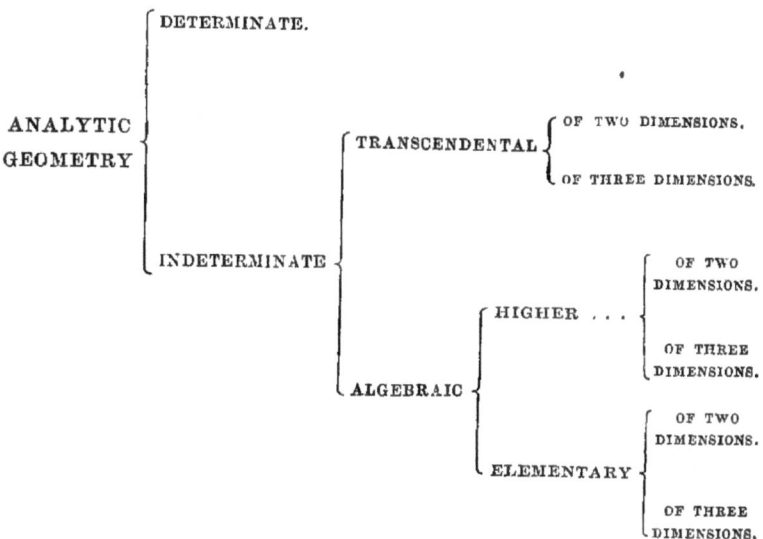

45. In the present treatise, we do not purpose more
than an introduction into the wide domain which the
foregoing scheme presents. We shall confine ourselves
to Elementary Geometry, not entering upon the other
departments any further than may prove necessary in
order to present the subject in its true bearings. And,
restricting ourselves to the discussion of loci of the First
and Second orders, we can not within that compass
give more than a sketch of the doctrine, methods,
and resources of the science, in its present advanced
condition. We shall, however, consider the loci of the

first two orders, both in a plane and in space. Accordingly, our treatise falls naturally into two Books: the first, upon Plane Co-ordinates; the second, upon Co-ordinates in Space.

NOTE.

As Greek characters are extensively used in all analytic investigations, and as we shall very frequently employ them in the following pages, we subjoin a list for the benefit of readers unacquainted with Greek.

A	α	*alpha.*	I	ι	*iota.*	P ρ	*rho.*
B	β	*beta.*	K	κ	*kappa.*	Σ σ	*sigma.*
Γ	γ	*gamma.*	Λ	λ	*lambda.*	T τ	*tau.*
Δ	δ	*delta.*	M	μ	*mu.*	Υ υ	*upsilon.*
E	ε	*epsilon.*	N	ν	*nu.*	Φ φ	*phi.*
Z	ζ	*zeta.*	Ξ	ξ	*xi.*	X χ	*chi.*
Η	η	*eta.*	O	ο	*omicron.*	Ψ ψ	*psi.*
Θ	θ	*theta.*	Π	π	*pi.*	Ω ω	*omega.*

BOOK FIRST:
PLANE CO-ORDINATES.

PLANE CO-ORDINATES.

PART I.

THE REPRESENTATION OF FORM BY ANALYTIC SYMBOLS.

46. In applying to plane curves of the First and Second orders the method sketched in the foregoing pages, our work will naturally divide itself into two portions: we shall first have to determine the equations which represent the several lines to be discussed; and then deduce from these equations the various properties of the corresponding lines. Accordingly, our First Book falls into two parts:

PART I.—ON THE REPRESENTATION OF FORM BY ANALYTIC SYMBOLS.

PART II.—ON THE PROPERTIES OF CONICS.

47. We say *Properties of Conics*, because, as will be shown hereafter, all the lines of the First and Second orders may be formed by passing a plane through a right cone on a circular base. By varying the position of the cutting plane, its sections with the conic surface

(47)

will assume the forms of the several lines; and these
may therefore be conveniently grouped under the general name of Conic Sections, or Conics.

It must be added, however, that this use of the term *Conics* is
wider than ordinary. For, speaking strictly, we mean by the
Conics the curves of the Second order alone; namely, the Ellipse,
the Hyperbola, and the Parabola; and ordinarily the line of the
First order is not included in the term.

48. We shall therefore proceed to develop the modes
of representing in algebraic language the Right Line,
the Circle, the Ellipse, the Hyperbola, and the Parabola.
And as these modes of representation are all derived
from the conventions adopted for representing a point,
we shall begin by explaining in full the principal forms
of those conventions, which were merely sketched in
the Introduction.

We shall obtain, first, the formulæ in ordinary use, or
those which may be said to constitute the Older Geometry; and, afterward, those belonging to the Modern
Geometry, based on what is called the Abridged Notation.

CHAPTER FIRST.

THE OLDER GEOMETRY · BILINEAR AND POLAR CO-ORDINATES.

SECTION I.—THE POINT.

BILINEAR OR CARTESIAN SYSTEM OF CO-ORDINATES.

49. Resuming the topic and diagram of Art. 19,
let us examine the Cartesian* system of co-ordinates

---

* *Cartesian,* from *Cartesius,* the latinized form of Descartes' name.

more minutely, and develop its elements in complete detail.

P being any point on a given plane, two right lines $X'X$ and $Y'Y$ are drawn in that plane, intersecting each other in *O*. From *P*, a line *PM* is drawn parallel to $Y'Y$.

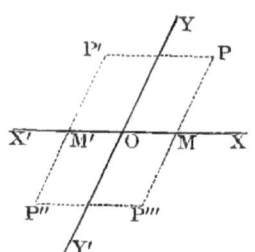

The distances *OM, MP* being known, the position of *P* is determined. These distances are called the *bilinear co-ordinates* of the point *P*. They are also termed *rectilinear*, and sometimes *parallel*, co-ordinates. They are frequently cited as the *Cartesian* co-ordinates of the point.

The co-ordinate *MP*, drawn parallel to $Y'Y$, is called the *ordinate* of the point *P*. The co-ordinate *OM*, which the former *cuts off* from $X'X$, is called the *abscissa* of the point. The abscissa of a point is represented by the symbol *x;* its ordinate, by the symbol *y.*

The two lines $X'X$ and $Y'Y$ are called the *axes of reference,* or simply the *axes.* $X'X$, on which the abscissas are measured, is called the *axis of abscissas;* or, more briefly, the *axis of* x. $Y'Y$, parallel to which the ordinates are drawn, is called the *axis of ordinates;* or, for brevity, the *axis of* y.

The point *O*, in which the axis of *x* cuts the axis of *y*, is called the *origin.*

The angle *YOX* is called the *inclination of the axes.* It is designated by the Greek letter ω, and may have any value from 0 to 180°. If $\omega = 90°$, the axes and co-ordinates are said to be *rectangular;* if ω has any other value, they are said to be *oblique.*

The two axes, being of infinite length, divide the

whole planar space about O into four angles. These are numbered to the *left*, beginning at the line OX. XOY is the *first* angle; YOX' is the *second;* $X'OY'$, the *third;* and $Y'OX$, the *fourth.*

The signs $+$ and $-$ are used in connection with the co-ordinates, and are taken to signify *measurement in opposite directions. Positive* abscissas are measured to the *right* from O, as OM; *negative* ones, to the *left;* as OM'. *Positive* ordinates are measured *upward* from $X'X$, as MP; *negative* ones, *downward;* as MP'''.

By attributing proper values to the co-ordinates x and y, and taking account of their signs, we may represent any point in either of the four angles. Thus,

$$\left.\begin{array}{l} x = + a \\ y = + b \end{array}\right\} \text{denotes a point in the } \textit{first} \text{ angle.}$$

$$\left.\begin{array}{l} x = - a \\ y = + b \end{array}\right\} \qquad `` \qquad `` \qquad `` \qquad \textit{second} \quad ``$$

$$\left.\begin{array}{l} x = - a \\ y = - b \end{array}\right\} \qquad `` \qquad `` \qquad `` \qquad \textit{third} \quad ``$$

$$\left.\begin{array}{l} x = + a \\ y = - b \end{array}\right\} \qquad `` \qquad `` \qquad `` \qquad \textit{fourth} \quad ``$$

Corollary 1.—For any point on the axis of x, we shall evidently have

$$y = 0,$$

while x, being susceptible of any value whatever, is indeterminate. Hence, the equation just written is the *equation to the axis of* x.

Corollary 2.—For any point on the axis of y, we shall have

$$x = 0,$$

while y is indeterminate. Hence, the equation last written is the *equation to the axis of* y.

Corollary 3.—For the origin, we shall obviously have

$$\dot{x} = 0,$$
$$y = 0;$$

and these expressions are therefore *the symbol of the origin.*

Remark 1.—For the sake of brevity, any point designated by Cartesian co-ordinates is written and cited as the point $x\,y$, the point $a\,b$, the point (3, 5), etc. These expressions are not to be confounded with algebraic products.

Remark 2.—The symbols x and y are used for coordinates of a *variable* point, and are therefore *general* in their signification. But it is often convenient to represent *particular*, or *fixed*, points by the variable symbols; especially when their positions, though fixed, are arbitrary. In such cases accents, or else inferiors, are used with the x and y. Thus $x'\,y'$, $x''\,y''$, $x_1\,y_1$, $x_2\,y_2$, all represent points which are to be considered as *fixed*, but fixed in positions chosen at pleasure. This distinction between the point $x\,y$ as *general*, and points such as $x'\,y'$, $x_2\,y_2$ as *particular*, should be carefully remembered.

Remark 3.—A point is said to be given by its co-ordinates, when their values and that of the angle ω are known. And when so given, the point may always be represented in position to the eye. For we have only to draw a pair of axes with the given inclination, and lay off by any scale of equal parts we please the given abscissa and ordinate. In practice, it is most convenient to lay off the co-ordinates on the *axes*, and draw through the points thus determined, lines parallel to the axes; the intersection of the latter will be the point required. The truth of this will be apparent on inspecting the diagram at the head of this article.

EXAMPLES.

1. Represent the point (− 5, 3) in rectangular co-ordinates.

2. Represent the point (− 3, − 7), axes oblique and $\omega = 60°$.

3. With the same axes as in Ex. 2, represent the points (1, 2), (− 3, 4), (− 5, − 6), (7, − 8).

4. Represent the points corresponding to the co-ordinates given in Ex. 3, axes being rectangular.

5. Given $\omega = 135°$, to represent the points (− 4, − 1), (8, 2) (2, 8).

6. Given $\omega = 90°$, represent the points (3, 4), (3, − 4), (− 3, 4), (− 3, − 4).

7. With same axes, represent (6, 8) and (8, 6); also (6, − 8) and (− 8, 6).

8. With ω still $= 90°$, represent the distance between (2, 3) and (4, 5).

9. With same axes, represent the distance between (4, 5) and (− 3, 2).

10. Axes rectangular, represent the distance between (0, 6) and (− 5, − 5);—the distance between (0, 0) and (6, 0). Does the latter distance depend on the value of ω, or not?

POLAR SYSTEM OF CO-ORDINATES.

50. A second method of representing the position of a point on a given plane, is founded on the fact that we naturally determine the position of any object by finding its direction and distance from our own.

Hence, if we are given a fixed point O and a fixed right line OX passing through it, we shall evidently know the position of any point P, so soon as we have determined the angle XOP and the distance OP. This method of representing a point is known as the method of Polar Co-ordinates.

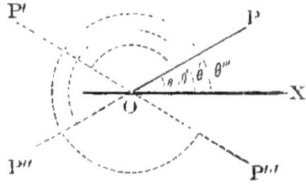

The fixed point O is called the *pole;* the fixed line OX, the *initial line.*

The distance OP is termed the *radius vector;* the angle XOP, the *vectorial angle.* It is customary to represent the former by the letter ρ, and the latter by θ. In this system, accordingly, a point is cited as the point $\rho\ \theta$, the point $\rho'\ \theta'$, etc.

By attributing proper values to ρ and θ, we may represent any point whatever in the plane PXO. Thus,

$$\left.\begin{array}{l} \rho\ =\ OP \\ \theta\ =XOP \end{array}\right\} \text{ denotes the point } P.$$

$$\left.\begin{array}{l} \rho'\ =\ OP' \\ \theta'\ =XOP' \end{array}\right\} \quad `` \quad `` \quad `` \quad P'.$$

$$\left.\begin{array}{l} \rho''\ =\ OP'' \\ \theta''\ =XOP'' \end{array}\right\} \quad `` \quad `` \quad `` \quad P''.$$

$$\left.\begin{array}{l} \rho'''=\ OP''' \\ \theta'''=XOP''' \end{array}\right\} \quad `` \quad `` \quad `` \quad P'''.$$

The student will observe that all the angles θ, θ', θ'', θ''' are estimated from XO toward the *left.*

Corollary 1.—For the *pole,* we evidently have

$$\rho = 0:$$

which may therefore be considered as the equation to that point.

Corollary 2.—For any point on the *initial line* to the *right* of the pole, we have

$$\theta = 2\,n\,\pi\,*.$$

For any point on the same line to the *left* of the pole, we have

$$\theta = (2\,n + 1)\,\pi.$$

---

* We shall frequently employ the symbol π = the semi-circumference to radius 1, to denote the angle 180°; on the principle that angles are measured by the arcs which subtend them.

In these expressions, n may have any integral value from 0 upward.

Note.—It is customary to measure the angle θ from XO toward the *left;* and the radius vector ρ, from O in such a direction *as to bound the angle.* When any distinctions of sign are admitted in polar co-ordinates, the directions just named are considered *positive;* while an angle measured from OX toward the *right,* and a radius vector measured from O in the direction *opposite* to that which bounds its angle, are considered *negative.* Thus, the point P is commonly denoted by the *positive* angle $\theta = XOP$ and the *positive* vector $\rho = OP$; but it may also be represented by the *negative* angle $XOP'' = -(\pi - \theta)$ and the then *negative* vector $\rho = OP$. And, again, the same point may be represented by the *positive* angle $\theta'' = \theta + \pi = XOP''$ and the *negative* vector $\rho = OP$.

The student will not fail to note that the signs $+$ and $-$, as applied to a line revolving about a fixed point, have a signification quite different from that in connection with bilinear co-ordinates. They discriminate between radii vectores, not necessarily as measured in *opposite* directions, but in directions *having opposite relations to the bounding of the vectorial angle.* We may therefore define the *positive* direction of a radius vector to be *that which extends from the pole along the front of the vectorial angle;* and the *negative,* to be *that extending opposite.*

In practice, negative values of ρ and θ are excluded from the ordinary formulæ; but the distinction of sign just explained has an important bearing on the principles of the Modern Geometry. For this reason, it should be mastered at the outset.

Remark.—To represent any point given in polar co-ordinates, we have only to draw the initial line, and lay off at any point taken for the pole, an angle equal to the given angle θ: then the distance ρ being measured from the pole, the required point is obtained.

EXAMPLES.

1. Represent in polar co-ordinates the point ($\rho = 8$; $\theta = \pi$).

2. Represent ($\rho = -8$; $\theta = 0$) and ($\rho = -8$; $\theta = \pi$).

3. Represent $\left(\rho = 15;\ \theta = \dfrac{\pi}{2}\right)$ and $\left(\rho = 5;\ \theta = \dfrac{3\pi}{2}\right)$.

4. Represent $\left(\rho = 6; \ \theta = \frac{\pi}{6}\right)$, $\left(\rho = -6; \ \theta = -\frac{5\pi}{6}\right)$, and $\left(\rho = -6; \ \theta = \frac{7\pi}{6}\right)$.

5. Represent the distance between $\left(\rho = 8; \ \theta = \frac{\pi}{4}\right)$ and $\left(\rho = 2; \ \theta = \frac{3\pi}{2}\right)$.

DISTANCE BETWEEN ANY TWO POINTS.

51. Any two points being given by their co-ordinates, the distance δ between them is given. For,

First: let the two points be $x'\ y'$ and $x''\ y''$. Taking P and P' to represent the points, we have

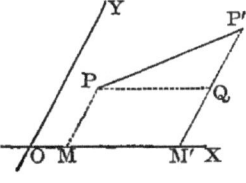

$$P'P = \delta,$$
$$OM = x', \ MP = y';$$
$$OM' = x'', \ M'P' = y''.$$

By Trig., 865,

$$P'P^2 = PQ^2 + P'Q^2 - 2\,PQ.\,P'Q \cos PQP'$$
$$= PQ^2 + P'Q^2 + 2\,PQ.\,P'Q \cos YOX. \quad \text{(Trig., 825)}.$$

That is,

$$\delta^2 = (x'' - x')^2 + (y'' - y')^2 + 2(x'' - x')(y'' - y') \cos \omega.$$

Corollary 1.—If $\omega = 90°$, then (Trig., 834) the last term in the foregoing expression vanishes, and we have, for the distance between two points in *rectangular* co-ordinates,

$$\delta^2 = (x'' - x')^2 + (y'' - y')^2.$$

Corollary 2.—For the distance of any point $x\ y$ from the origin, we have (Art. 49, Cor. 3)

$$\delta^2 = x^2 + y^2 + 2\,xy \cos \omega:$$

which, in *rectangular* co-ordinates, becomes

$$\delta^2 = x^2 + y^2.$$

An. Ge. 8.

Second: let the two points be $\rho'\,\theta'$ and $\rho''\,\theta''$. In the diagram, for this case,

$$OP = \rho',\ XOP = \theta';$$
$$OP' = \rho'',\ XOP' = \theta''.$$

Then, as before,

$$P'P^2 = OP^2 + P'O^2 - 2\ OP.\ P'O \cos P'OP;$$

that is,

$$\delta^2 = \rho'^2 + \rho''^2 - 2\,\rho'\rho'' \cos(\theta'' - \theta').$$

Corollary.—For the distance of any point $\rho\ \theta$ from the pole, we have (Art. 50, Cor. 1)

$$\delta^2 = \rho^2;\ \text{or}\ \delta = \rho :$$

which agrees with our definition of the radius vector.

Note.—In using the formulæ of this article, be careful to observe the *signs* of $r'\,y'$, $x''\,y''$

POINT DIVIDING IN A GIVEN RATIO THE DISTANCE
BETWEEN TWO GIVEN POINTS.

52. Let the given points be $x_1\,y_1$, $x_2\,y_2$; and the given ratio, $m : n$. Denote the co-ordinates of the required point by x and y.

By Geom., 313, we have in the diagram annexed, where $OM = x$, $MP = y$; $OM' = x_1$, $M'P' = y_1$; $OM'' = x_2$, $M''P'' = y_2$,

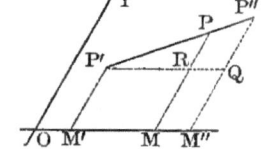

$$P'R : RQ :: P'P : PP'';$$

or, $\quad x - x_1 : x_2 - x :: m : n.$

$$\therefore\ x = \frac{mx_2 + nx_1}{m + n}.$$

By like reasoning, we find

$$y = \frac{my_2 + ny_1}{m + n}.$$

If the distance between two points $x_1 \ y_1, \ x_2 \ y_2$ were cut *externally* in the given ratio, we should have

$$x - x_1 : x - x_2 :: m : n.$$

$$\therefore \quad x = \frac{mx_2 - nx_1}{m - n};$$

and, similarly,

$$y = \frac{my_2 - ny_1}{m - n}.$$

And this we should expect: for, if the point P fell beyond P'', the segment PP'' would be measured in the direction opposite to $P'P$, and n would have the negative sign.

53. As the student may have surmised from what he has already noticed, formulæ referred to rectangular axes are generally simpler than those referred to oblique. For this reason, it is preferable to use rectangular axes whenever it is practicable. Hereafter, then, the attention should be fixed chiefly upon those formulæ which correspond to rectangular axes. In some cases, formulæ are true for any value of ω. Such are those deduced in the last article. In the examples given hereafter, the axes are *supposed to be rectangular, unless the contrary is mentioned.*

EXAMPLES.

1. Draw the triangle whose vertices are $(2, 5)$, $(-4, 1)$, $(-2, -6)$.

2. Find the lengths of the three sides of the same triangle.

3. Express algebraically the condition that $x \ y$ is equidistant from $(2, 3)$ and $(4, 5)$.   *Ans.* $x + y = 7$.

4. Find the distance between $(-3, 0)$ and $(2, -5)$.

5. Determine the co-ordinates of the point equidistant from the three points $(1, 2)$, $(0, 0)$, and $(-5, -6)$.

6. Solve Ex. 2, supposing ω successively $\dfrac{\pi}{3}$ and $\dfrac{\pi}{4}$.

7. Find the co-ordinates of the point bisecting the distance between $x_1\,y_1$ and $x_2\,y_2$.

8. The point $x\ y$ is midway between $(3,\ 4)$ and $(-5,\ -8)$: find its distance from the origin.

9. $x\ y$ divides externally the distance between $(2,\ -8)$ and $(-5,\ -3)$ in the ratio $6 : 7$. What is its distance from the point midway between $(3, 4)$ and $(6, 8)$?

10. Given the points $(\rho = 5;\ \theta = 30°)$ and $(\rho = 6;\ \theta = 225°)$ to find the distance between them, and the polar co-ordinates of its middle point.

TRANSFORMATION OF CO-ORDINATES.

51. A point being given by its co-ordinates, we can at pleasure change either the axes or the system to which they refer it. The process is called Transformation of Co-ordinates.

The position of the point is of course not affected by such a change. Its co-ordinates merely assume new values corresponding to the new axes or new system.

The transformation is effected *by substituting for the given co-ordinates their values in terms of the elements belonging to the new limits.* General formulæ for these substitutions are easily obtained. In investigating them, it is convenient to consider the subject in four cases; namely,

I. To change the Origin, the direction of the axes remaining the same.

II. To change the Inclination of the Axes, the origin remaining the same.

III. To change System—from Bilinears to Polars, and conversely.

IV. To change the Origin, at the same time transforming by II or III.

55. Case First: *To transform to parallel axes through a new origin.*

Let x and y be the co-ordinates of the point for the primitive axes; and X and Y its co-ordinates for the proposed parallel axes. Let m, n be the co-ordinates of the new origin. Then, if OY, 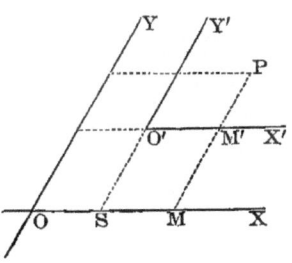 OX represent the primitive axes, and $O'Y'$, $O'X'$ the new: we shall have $x = OM$, $y = MP$; $X = O'M'$, $Y = M'P$; and $m = OS$, $n = SO'$. Now, from the diagram,

$$OM = OS + O'M' \quad \text{and} \quad MP = SO' + M'P;$$

that is, $$x = m + X,$$
$$y = n + Y:$$

which are the required formulæ of transformation.

56. Case Second: *To transform to new axes through the primitive origin, the inclination being changed.*

Let ω represent the inclination of the primitive axes. Let $a = $ the angle made by the new axis of x with the primitive; and $\beta = $ that made by the new axis of y.

Drawing the annexed diagram, we have $x = OM$, $y = MP$; $X = OM'$, $Y = M'P$; $a = X'OX = M'OR$, and $\beta = Y'OX = PM'S$. 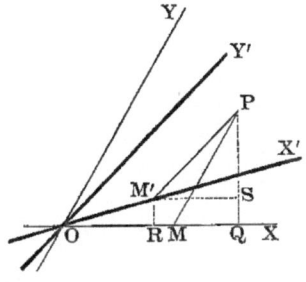 Then, letting fall PQ and $M'R$ perpendicular to OX, and $M'S$ perpendicular to PQ, we obtain (Trig., 858)

$$MP \sin PMQ = QP = RM' + SP.$$

But $RM' + SP = OM'$ sin $M'OR + M'P$ sin $PM'S$.

$\therefore MP$ sin $PMQ = OM'$ sin $M'OR + M'P$ sin $PM'S$;

or, y sin $\omega = X$ sin $\alpha + Y$ sin β.

By dropping perpendiculars from P and M' upon OY, and completing the diagram, we should obtain by the same principles

$$x \text{ sin } \omega = X \text{ sin } (\omega - \alpha) + Y \text{ sin } (\omega - \beta).$$

The details of the proof are left to the student.

We have, then, as the required formulæ of transformation,

$$x \text{ sin } \omega = X \sin(\omega - \alpha) + Y \sin (\omega - \beta),$$
$$y \text{ sin } \omega = X \sin \alpha + Y \sin \beta :$$

which include all cases of bilinear transformation. The particular transformations which may arise under this general case, are various; those of most importance are as follows:

Corollary 1.—*To transform from rectangular axes to oblique, the origin remaining the same.* Making $\omega = 90°$ in our general formulæ, we obtain (Trig., 834, 841)

$$x = X \cos \alpha + Y \cos \beta,$$
$$y = X \sin \alpha + Y \sin \beta.$$

Or we may obtain the formulæ geometrically, as follows: Supposing the angle YOX to be a right angle, OM will obviously coincide with OQ, and MP with QP, and we shall have

$$OM = OR + M'S = OM' \cos M'OR + M'P \cos PM'S$$
$$\therefore x = X \cos \alpha + Y \cos \beta;$$
$$MP = RM' + SP = OM' \sin M'OR + M'P \sin PM'S$$
$$\therefore y = X \sin \alpha + Y \sin \beta.$$

Corollary 2.—*To transform from oblique axes to rectangular, the origin and the axis of* x *remaining the same.* Here $a = 0$, and $\beta = 90°$: hence, (Trig., 829, 834,)

$$x \sin \omega = X \sin \omega - Y \cos \omega,$$
$$y \sin \omega = Y.$$

Geometrically as follows : OX, OY being the primitive axes, and OX, OY' the new : $x = OM$, $y = MP$; $X = OM'$ $Y = M'P$. By Trig., 859,

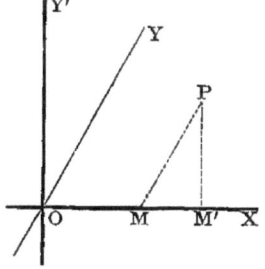

$$OM = OM' - M'M$$
$$\quad = OM' - M'P \cot PMM';$$
or, $\quad x = X - Y \cot \omega.$

$\therefore$ $x \sin \omega = X \sin \omega - Y \cos \omega.$

Again, by Trig., 858,

$$MP \sin PMM' = M'P;$$
or, $\quad y \sin \omega = Y.$

Corollary 3.—*To revolve the rectangular axes through any angle* θ. Here $a = \theta$, $\beta = 90° + \theta$, and $\omega = 90°$. Substituting in the general formulæ, we obtain

$$x = X \cos \theta + Y \cos (90° + \theta),$$
$$y = X \sin \theta + Y \sin (90° + \theta):$$

expressions which, by Trig., 843, become

$$x = X \cos \theta - Y \sin \theta;$$
$$y = X \sin \theta + Y \cos \theta.$$

Or we may deduce the formulæ independently, from the diagram annexed. Here $x = OM, y = MP;$ $X = OM'$, $Y = M'P;$ $\theta = M'OS$ $= ROM = M'PQ$. Drawing $M'S$ and $M'Q$ perpendicular respectively to OS and PQ, we shall have

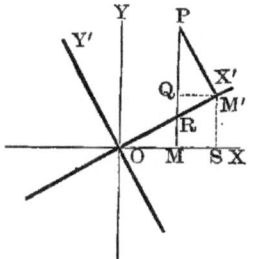

$$OM = OS - M'Q = OM' \cos M'OS - M'P \sin M'PQ,$$
$$MP = SM' + QP = OM' \sin M'OS + M'P \cos M'PQ.$$

That is,

$$x = X \cos \theta - Y \sin \theta,$$
$$y = X \sin \theta + Y \cos \theta.$$

57. Case Third: *To transform from a bilinear to a polar system of co-ordinates, or conversely.*

Let OX, OY be the primitive rectangular axes, and OX' the initial line of the proposed polar system. Let $a =$ the angle which the initial line makes with the axis of x: it will be *positive* or *negative* according as OX' lies *above* or *below* OX. It is obvious from the diagram that we shall have

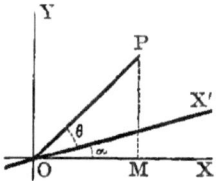

$$x = \rho \cos (\theta + a),$$
$$y = \rho \sin (\theta + a):$$

formulæ by which we can either find $x\ y$ in terms of $\rho\ \theta$, or $\rho\ \theta$ in terms of $x\ y$. In applying them, strict attention must be paid to the sign of a, according to the convention named above.

Note.—We have confined the discussion of this case to the change from *rectangular* axes, as this alone is of very frequent occurrence in practice. It may be well, however, to give the formulæ of the general case, in which ω is supposed to have any value whatever. Assuming, in the above diagram, the angle YOX to be oblique, we should have in the triangle OMP, (Trig., 867,)

$$\rho : x :: \sin \omega : \sin \{\omega - (\theta + a)\} \quad \therefore \quad x = \frac{\rho \sin \{\omega - (\theta + a)\}}{\sin \omega},$$

$$\rho : y :: \sin \omega : \sin (\theta + a) \quad \therefore \quad y = \frac{\rho \sin (\theta + a)}{\sin \omega}.$$

These formulæ evidently become those obtained above, when $\omega = 90°$.

Corollary.—If the axis of x is taken as the initial line, $a = 0$; and we have

$$x = \rho \cos \theta,$$
$$y = \rho \sin \theta:$$

a set of formulæ in very extensive use.

58. Case Fourth: *To combine a change of origin with any other transformation.*

To effect this, we first apply the formulæ of Art. 55, and thus pass to the new origin with a system of axes parallel to the primitive. That is, in effect, we remove the original system to the new position which the proposed origin requires. The formulæ for the special transformation in hand are then applied, and the whole change is accomplished.

From the nature of the formulæ in Art. 55, it is obvious that the present case is solved analytically by merely adding to the expressions for x and y, the co-ordinates m and n of the new origin. Thus, in general, by Art. 56,

$$x = m + \frac{X \sin (\omega - a) + Y \sin (\omega - \beta)}{\sin \; \omega},$$

$$y = n + \frac{X \sin a + Y \sin \beta}{\sin \omega},$$

To pass from bilinears to polars when the pole is a different point from the origin, we have

$$x = m + \rho \cos (\theta + a),$$
$$y = n + \rho \sin (\theta + a).'$$

59. It should be observed that in applying these various formulæ, great care is to be exercised in respect to the *signs* of the *constants* involved. And in the examples which follow, where given points are to be transformed, the same care should be taken with respect to all

An. Ge. 9.

the known co-ordinates. It is recommended that the student reduce every problem to drawing, at least in the earlier stages of his studies. In no other way will he readily acquire the habit of bringing every analytic process to the test of geometric interpretation.

EXAMPLES.

1. Given the point $(5, 6)$: what are its co-ordinates for parallel axes through the origin $(2, 3)$?

2. Transform $(-3, 0)$ to parallel axes through $(-4, -5)$;— to parallel axes through $(5, -3)$; — through $(-3, 5)$; — through $(-3, 0)$.

3. Given in rectangular co-ordinates the points $(1, 1)$, $(-1, 1)$, $(2, -1)$, and $(-3, -3)$: find their polar co-ordinates, the origin being the pole, and the axis of x the initial line.

4. Transform the points in Ex. 3, supposing the pole to be at $(-4, 5)$, and the initial line to make with the axis of x an angle $a = -30°$.

5. Solve Ex. 4, on the supposition that $a = 45°$.

6. Find the rectangular co-ordinates of $(\rho = 3; \theta = 60°)$ and $(\rho = -3; \theta = -60°)$, the origin and axis of x coinciding respectively with the pole and the initial line. Find the same, supposing the origin at $(-2, -1)$, and the angle $a = 30°$.

7. The co-ordinates of a point for a set of axes in which $\omega = 60°$, satisfy the equation $3x + 4y - 8 = 0$: what will the equation become when transformed to $\omega' = 45°$, $a = 15°$? What, when in addition the origin is moved to $(-3, 2)$?

8. Transform $x^2 + y^2 = r^2$ to parallel axes through $(-a, -b)$.

9. It is evident that when we change from one set of rectangular axes to another having the same origin, $x^2 + y^2$ must be equal to $x'^2 + y'^2$, since both express the square of the distance of the point from the origin. Verify this by squaring the expressions for x and y given in Art. 56, Cor. 3, and adding the results.

10. Transform to rectangular co-ordinates the following equations in polar; origin same as pole, and axis of x as initial line:

$$\rho^2 \sin 2\theta = 2 c^2; \quad \rho^{\frac{1}{2}} \cos \tfrac{1}{2}\theta = c^{\frac{1}{2}}; \quad \rho^2 = c^2 \cos 2\theta.$$

60. In an important sense, the whole science of Analytic Geometry may be said to consist in knowing how to translate algebraic symbols into geometric facts. Supposing that we have solved any geometric problem by analytic methods, our result must be some algebraic expression. Hence, in the end the question is, How shall we interpret that expression into the geometric property which we are seeking?

The principles governing such an interpretation are to be mastered in their fullness, only through an exhaustive study of the whole field of Analytic Geometry. But there are a few of them, which, lying at the root of all the others, are of universal application, and must be determined at the outset. In the following articles, we will state and establish them. The student may notice that the illustrations and, in some cases, the phraseology refer to Cartesian co-ordinates; but this does not affect the generality of the principles, since we can convert any geometric expression into one relating to the Cartesian system by transformation of co-ordinates.

61. *A single equation between plane co-ordinates represents a plane locus.*

This theorem follows directly from the corollary of Art. 27. It may be well, however, to add here some illustrations of the principle.

It is plain that if we have any equation between two variables, as

$$ax^3 + bx^2y + cxy^2 + dy^3 + f = 0,$$

we may assign to x any value we please, and obtain a corresponding value for y. The unknown quantities in such an equation are therefore not determinate. On the contrary, there is a *series* of values, infinite in number, any of which will satisfy the equation. Taking these values as denoting the co-ordinates of a point, the equation must represent an infinite number of points. But, though infinite in number, these points *can not be taken at random;* for the equation can not be satisfied by values arbitrary for *y* as well as *x*, but only by such values of *y* as its own conditions require in answer to the assigned values of *x*. Hence, the infinite series of points which the equation represents, conforms in all its members to the same law of position: a law expressed in the uniform relation which the equation establishes between the values of its variables. Such a series of points must constitute a *line.*

To illustrate by a diagram, we may suppose that in a given equation between two variables, *x* has the value *Om*. Corresponding to this there will be, let us say, three values of *y*, represented by *mp*, *mq*, *mr*. We thus determine three points, *p*, *q*, *r*. Again, supposing *x = Om'*, we determine three other points, *p'*, *q'*, *r'*. And again, making *x = Om''*, we obtain the points *p''*, *q''*, *r''*. We may continue this process as long as we please, and determine any number of points, by assigning successive values to *x*. By taking these sufficiently near each other, and drawing a line through the points thus found, we may determine the figure of the locus which the equation represents.

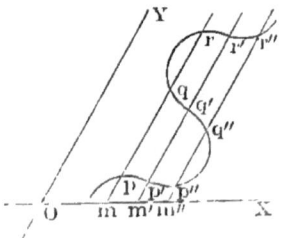

Remark.—We must here carefully recall the proposition, stated in the Introduction, that to render the principle of this article universally true we must take into account *imaginary* loci, loci *at infinity*, and cases where a locus degenerates into disconnected points or a single point. For example, the equation

$$x^2 + y^2 + 1 = 0$$

can not be satisfied by any real values of x and y: consequently, in order to bring it within the terms of our principle, we must say that it denotes an *imaginary* locus. It should be borne in mind that this amounts to saying that it has *no* geometric locus.

Again, the equation

$$0x + 0y + c = 0$$

can be satisfied by none but infinite values of x and y. All the points on its locus are therefore at an infinite distance from the origin, and it can be brought within the terms of our principle only by saying that it denotes a locus *at infinity*. This, too, is only another way of saying that it has *no* geometric locus.

Again, the equation

$$(x - a)^2 + y^2 = 0$$

obviously can not be satisfied unless we have, at the same time,

$$(x - a)^2 = 0 \text{ and } y^2 = 0;$$

that is, it admits of *no* values except

$$x = a \text{ and } y = 0.$$

Accordingly, if we would bring it within our principle, we must say that it denotes a locus *which has degenerated into a point* situated on the axis of x, at a distance a from the origin. We shall learn hereafter that this point is an infinitely small circle, having $(a, 0)$ for its center.

62. *Any two simultaneous equations between plane coordinates represent determinate points in a given plane.*

For, given *two* equations between two variables, we can determine the values of x and y by elimination.

Moreover, the points which such a pair of simultaneous equations determines, are the *points of intersection* common to the two lines which the equations respectively represent. For it is obvious that the values of x and y found by elimination, must satisfy *both* of the equations : hence, the points which these values represent must lie on *both* of the lines represented by them : that is, they are the *points common* to those lines.

Corollary 1.—Hence, *To find the points of intersection of two lines given by their equations, solve the equations for* x *and* y.

Remark.—The geometric meaning of simultaneity and elimination may be made clearer by the accompanying diagram. Let the curve A be represented by the equation

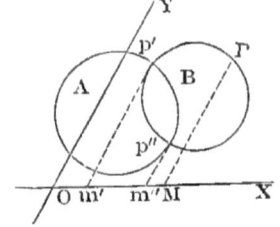

$$y = f(x) \qquad (1),$$

and the curve B by a second equation

$$y = \varphi(x)^* \qquad (2):$$

then, supposing A to intersect B in the points p' and p'', and in these points only, it is obvious that the x and y of equation (1) will become identical with the x and y of equation (2) when we substitute in both equations the co-ordinates of p' and p''; for we shall then have, in both equations,

$$x = Om' \text{ or } Om'',$$
$$y = m'p' \text{ or } m''p''.$$

And it is equally plain that the x and y of the two equations will *not* be the same, but different, so long as they represent the co-ordinates of any other point. Thus, if in (2) we make $x = OM$, we shall have $y = MP$: values which, it is manifest from the figure, the x and y of (1) can not have.

Since, then, the variables in the equations to different curves will in general have different values; and since, even in curves that intersect, the variables in their respective equations will become identical in value only at

---

* Equations (1) and (2) are read "$y =$ any function of x" and "$y =$ any *other* function of x."

the *points of intersection :* we learn the important principle, that to suppose two geometric equations simultaneous is to suppose that their loci intersect. In short, *simultaneity means intersection ;* and *elimination determines the intersecting points.*

Corollary 2.—*Two equations, of the* mth *and* nth *degree respectively, represent* mn *points.*

For (Alg., 246), elimination between them involves the solution of an equation of the mn^{th} degree; and such an equation (Alg., 396, 397) will have mn roots.

Two lines, therefore, of the mth and n^{th} order respectively, intersect in mn points. Two lines of the *first* order, for example, have but *one* point of intersection; two of the *second*, have *four ;* a line of the *first* order intersects one of the *second*, in *two* points; a line of the *second* order cuts one of the *third*, in *six ;* and so on.

It should be observed that any number of these mn points may become coincident: a fact which will be indicated, of course, by the existence of a corresponding number of *equal roots* in the equation obtained by elimination. Or, any number of them may become imaginary ; or, in certain cases, be situated at infinity : facts respectively indicated by the presence of *imaginary* and *infinite* roots.

63. *An equation lacking the absolute term represents a line passing through the origin.*

For every such equation will be satisfied by the values $x = 0$, $y = 0$; and these (Art. 49, Cor. 3) are the co-ordinates of the origin.

64. *Transformation of co-ordinates does not alter the degree of an equation, nor affect the form of the locus which it represents.*

For, supposing the equation to be originally of the nth degree, the term which tests that degree may be written $Mx^r y^s$, in which $r + s = n$. Now the most general case of transformation (compare Arts. 56, 58) will require us to substitute for this an expression of the form

$$M(aX + bY + c)^r (a'X + b'Y + c')^s :$$

which when expanded will certainly contain terms of the form $M'X^p Y^q$, where $p + q = r + s = n$, but can contain none in which the sum of the exponents of X and Y is greater than n. Hence the degree of the equation, and therefore the order of its locus, will remain unchanged through any number of transformations.

Nor will transformation affect the form of the locus at all. For, obviously, the figure of a curve does not depend upon limits to which we arbitrarily refer its points.

EXAMPLES.

1. What point is represented by the equations $3x + 5y = 13$ and $4x - y = 2$?

2. Given the two curves $x^2 + y^2 = 5$ and $xy = 2$, in how many points will they intersect? Find the points of intersection.

3. Find the points in which $x - y = 1$ intersects $x^2 + y^2 = 25$

4. Of what order is the curve $y^2 = 4px$? Show, by actual transformation, that it continues of the same order when passed from its original rectangular axes to oblique ones through $a, 2\sqrt{pa}$: the new axis of x being parallel to the old, and the inclination of the new axes being the angle whose $\tan^2 = p : a$.

5. Decide whether the following curves pass through the origin:

$$y = mx + b ; \; x^2 - y^2 = 0 ; \; x^2 - y^2 = 1 ; \; y^2 = 4px ;$$
$$3x^3 - 5xy + 7x^2 - 8y = 0.$$

SPECIAL INTERPRETATION OF PARTICULAR EQUATIONS.

65. The foregoing principles illustrate the doctrine that there is a *general* connection between an equation and the locus of a point. But every curve * has its own *particular* equation, and we may appropriately close our discussion of the Theory of Points by explaining briefly how to discover the form and situation of a curve from its equation.

66. The **Special Interpretation of an Equation** consists in tracing, by means of determined points, the curve which it represents.

67. In order to trace any curve from its equation, we solve the equation for either of its variables, say for y. We then assign to x various values at pleasure, and compute the corresponding values of y. Then, drawing the axes, we lay down the points corresponding to the co-ordinates thus found. A curve traced through these points will approximately represent the locus of the equation. Could we take the points infinitely near each other, we should obtain the exact curve.

68. Attention to certain characteristics of the given equation and of the values of the variable for which it is solved, will enable us to decide certain questions concerning the peculiar form of the corresponding curve. These algebraic characteristics, and their geometric meaning, we will now specify.

69. If the given equation is of a degree higher than the first, for every value assigned to x there will arise

---

* It is customary to call *any* plane locus a *curve*, even though this involves the apparent harshness of saying that the right line or an isolated point is a curve.

two or more values of y. The several points corresponding to the common abscissa are
said to lie on different PORTIONS
of the curve. Thus, in the figure,
the points p, p', p'' lie on one
portion of the curve represented;
the points q, q', q'' on another;
and the points r, r', r'' on a third.
The LIMITS of a portion — that
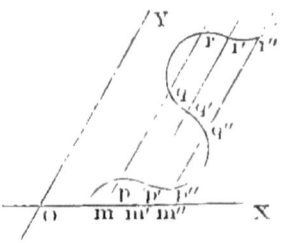
is, the points where it merges into another portion —
are the points whose abscissas cause two values of the
ordinate to become equal.

Corollary.—Hence, *To test whether a curve consists of
several portions, note whether its equation is of a degree
higher than the first.* To find the *limits* of the portions,
observe what values of x give rise to *equal roots* for y.

70. If all the values assigned to x within the limits
separating two portions of a curve, make the y's of the
two portions *numerically equal* but of *opposite sign*, the
corresponding points of these portions will be equally
distant from the axis of x. Similar conditions with
respect to the axis of y will determine points equally
distant from that axis.

Two portions of a curve, whose points are thus situated
with reference to either axis, are said to be SYMMETRICAL
to that axis. A *curve* is symmetrical, when all its portions taken two and two are symmetrical.

Corollary.—Hence, *To test for symmetry, note whether the
values of either variable, corresponding to all values of the
other between the limits of two portions of a curve, appear
in pairs, numerically equal with contrary signs.*

71. If any value assigned to x gives rise to *imaginary* values for y, the corresponding point or points will be imaginary. That is, the curve is *interrupted* at such points. And if, between any *two* values of either variable, the corresponding values of the other are *all* imaginary, the curve does not exist between the corresponding limits. Thus, in the curve

$$\frac{x^2}{a^2} - \frac{y^2}{b^2} = 1,$$

by solving for y we obtain

$$y = \pm \frac{b}{a} \sqrt{x^2 - a^2} :$$

so that y is real for every value of x which lies *beyond* the limits $x = a$ and $x = -a$, but is imaginary for every value of x lying *between* them; and the curve is interrupted in the latter region.

When the extent of a curve is nowhere interrupted, and it suffers no abrupt changes in curvature,* it is said to be CONTINUOUS. A curve may be either continuous throughout or composed of continuous parts.

Corollary.—*To test for continuity in* extent, *note whether the equation to a curve gives rise to limiting values of either variable, beyond or between which the values of the other are imaginary.*

Remark.—To test for continuity in *curvature*, we employ the Differential Calculus.

72. The continuous *parts* of a curve are called its BRANCHES. A *branch* should be distinguished from a *portion* of a curve: a *branch* may consist of several *portions;* or a *portion,* of several *branches.*

---

* *Curvature* i. e. *the rate at which a curve deviates from a right line.*

A branch of a curve may degenerate into isolated points, or a single point: such points are called CONJU-GATE POINTS.

Corollary.—The *number* and *extent* of the branches belonging to a curve may often be determined by *examining the limits beyond or between which its equation gives rise to imaginary values of the variables.* Thus, if

$$y = \pm \frac{b}{a} \sqrt{a^2 - x^2},$$

y will be real for all values of x lying *between* the limits $x = -a$ and $x = a$, but imaginary for all lying *beyond*. The curve therefore consists of a *single* branch, *surrounding* a portion of the axis of x whose length $= 2a$. If

$$y = \pm \frac{b}{a} \sqrt{x^2 - a^2},$$

y is real for all values of x lying *beyond* the limits $x = -a$ and $x = a$, but imaginary for all lying *between*. Hence, the curve consists of *two* branches, *separated* by a portion of the axis of x whose length $= 2a$. If

$$y = 2 \sqrt{px},$$

y is imaginary for all *negative* values of x, but real for all *positive* values. Hence, the curve consists of a *single infinite* branch, extending from the origin toward the right.

Remark.—Conjugate points belong to a class, known as SINGULAR POINTS, whose existence can not in general be tested without the aid of the Differential Calculus. If, however, a given equation is obviously satisfied by none but *isolated* values between certain limits, its locus between those limits will consist of conjugate points.

73. We add a few examples, merely premising that it is often convenient first to find the situation of the axes to which the given equation is referred. This is done by making the x and y of the equation successively equal to zero: the resulting values of y and x (Art. 62, Cor. 1) are the intercepts made by the curve on the axis of y and of x respectively.

We have given, for the sake of widening a little the student's view of the subject, a few equations to Higher Plane Curves, both Algebraic and Transcendental. These curves, of course, are beyond the province of the present work; and the reader who desires full information in regard to them is referred to SALMON's *Higher Plane Curves* or to the writings of PLÜCKER, PONCELET, and CHASLES.

It will be most convenient, in tracing the curves of the following examples, to use paper ruled in small squares, whose constant side may be taken for the linear unit. Let the limits of the imaginary values, if such exist in any equation, be first found: then, within the sphere of real values, let the abscissas be taken near enough together to determine the figure of the curve.

The axes are rectangular: as we shall always suppose them in examples, unless the contrary is indicated.

EXAMPLES.

1. Represent the curve denoted by $y = 2x + 3$.

Making, successively, $x = 0$ and $y = 0$, we obtain

$$y = 3 \text{ and } x = -\frac{3}{2}.$$

The curve therefore cuts the axis of y at a distance 3 above the origin, and the axis of x at a distance $\frac{3}{2}$ to the left of the origin. Draw the axes and lay down the corresponding points.

The equation being of the first degree, the curve consists of but one portion. y is obviously real for all real values of x: the curve is therefore of infinite extent. Making x successively $-3, -2, -1, 1, 2, 3$, the corresponding values of y are $-3, -1, 1, 5, 7, 9$. Laying down the points

$$(-3, -3), (-2, -1), (-1, 1), (1, 5), (2, 7), (3, 9),$$

we find that they all come upon the right line drawn through $(0, 3)$ and $\left(-\dfrac{3}{2}, 0\right)$, which is therefore the curve represented by the given equation.

2. Interpret $\dfrac{x^2}{9} + \dfrac{y^2}{4} = 1.$

Making $x = 0$, we obtain

$$y = \pm 2:$$

or the curve cuts the axis of y in two points: one at the distance 2 above the origin, and the other at the same distance below it.

Making $y = 0$, we obtain

$$x = \pm 3:$$

whence the curve cuts the axis of x in two points equally distant from the origin, and on opposite sides of it.

Since the equation is of the second degree, the curve consists of two portions; and as the values of y coincide and $= 0$ when $x = \pm 3$, these portions are separated by the axis of x. Solving for y, we find

$$y = \pm \frac{2}{3} \sqrt{9 - x^2}:$$

hence, y will become imaginary when $x > 3$ or $x < -3$. The curve therefore has no point beyond its intersections with the axis of x. But for every value of x between the limits -3 and 3, y is real; that is, the curve consists of a single continuous branch.

Making, now, x successively equal to $-3, -2.5, -2, -1, 0, 1, 2, 2.5, 3$, the corresponding values of y are $0, \pm 1.2,$ $\pm 1.5, \pm 1.9, \pm 2, \pm 1.9, \pm 1.5, \pm 1.2, 0$. From these values, we see that the curve is symmetrical to both axes; and, laying down the sixteen points thus found, we determine the figure of the curve as annexed. It is an *ellipse.*

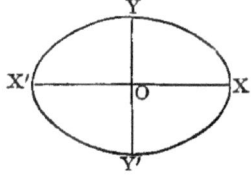

3. Interpret the equation $y = mx$.

4. Interpret $\dfrac{x^2}{9} - \dfrac{y^2}{4} = 1$. This curve is an *hyperbola.*

5. Interpret $y^2 = 4x$. This curve is a *parabola.*

6. Interpret $y = x^3$ and $y^2 = x^3$. The first of these curves is a *cubic parabola;* and the second, a *semi-cubic parabola.*

7. Interpret $x^3 - (a - x) y^2 = 0$. This is known as the *Cissoid of Diocles.*

8. Interpret $x^2 y^2 = (a^2 - y^2) (b + y)^2$. This is the *Conchoid of Nicomedes.*

9. Interpret $x = \text{versin}^{-1} y$ * $- \sqrt{2 ry - y^2}$. This is the *Common Cycloid.*

10. Interpret $y = \sin x$, the *Curve of Sines;* $y = \cos x$, the *Curve of Cosines;* and $y = \log x$, the *Logarithmic Curve.*

SECTION II.—THE RIGHT LINE.

THE RIGHT LINE UNDER GENERAL CONDITIONS.

74. In discussing the mode of representing the Right Line by analytic symbols, we shall in the first place have to determine the various forms of the equation which represents *any* right line: that is, our problem will be to represent the Right Line under *general* conditions only. This accomplished, we shall then pass to the more particular forms of the equation — those which represent the Right Line under such *special* conditions as *passing through two given points, passing in a given direction through one given point,* etc.

75. In showing that a certain curve is represented by a certain equation, we may proceed in either of two ways.

---

* This is the notation for an *inverse trigonometric function*, and is in this case read "the arc whose versed-sine is y." Similar expressions occur in terms of the other trigonometric functions: as, $x = \sin^{-1} y$; $x = \tan^{-1} y$; etc., read "$x =$ the arc whose sine is y," "$x =$ the arc whose tangent is y," and so on.

First, we may begin by assuming some fundamental property of the curve, define the curve by means of it, and, with the help of a diagram which brings it into relation with elementary geometric theorems, embody it in an equation between the co-ordinates of any point on the curve — an equation from which all other properties may be deduced by suitable transformations. Or, secondly, we may begin without any geometric assumptions except those on which the convention of co-ordinates is founded; may take an equation of any degree, in its most general form; and, by the purely analytic processes of algebraic, trigonometric, or co-ordinate transformation, reduce the equation to such simpler forms as will show us the species, figure, and properties of the corresponding curve. The latter method is the purely analytic one; the former mingles the processes of geometry and analysis.

76. In the present Book, both of these methods will be applied in succession. It will be natural to set out from the geometric point of view: for in this way we shall secure simplicity and clearness, by constantly bringing the analytic formulæ and operations to the test of interpretation by a diagram. After the characteristic forms of the equation to any locus have been obtained by the aid of geometry, and the beginner has become familiar with their geometric meaning, he may safely ascend to the higher analytic standpoint, and will be able to descend from it with some real appreciation of the scientific beauty which it brings to light.

We proceed to apply to the Right Line the two methods mentioned, and shall follow the order which has just been indicated.

I. GEOMETRIC POINT OF VIEW: — THE EQUATION TO THE
RIGHT LINE IS ALWAYS OF THE FIRST DEGREE.

77. There are three principal forms of the equation
to the Right Line, arising out of the three sets of data
by which the position of the line is supposed to be determined. Each of these will prove to be of the first
degree.

**78. Equation to the Right Line in terms of its
angle with the axis of x and its intercept on the
axis of y.**—It is obvious, on inspecting the diagram, that
the position of the line DT is given

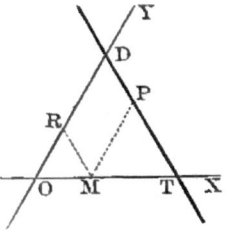

when the angle DTX* and the
intercept OD are given.

Let x and y denote the co-ordinates OM, MP of any point P on
the line. Let the angle $DTX = a$,
and let $OD = b$. Then, drawing
MR parallel to DT, we shall have
(Trig., 867)

$$OR : OM :: \sin OMR : \sin ORM.$$

That is, $b - y : x :: \sin a : \sin (a - \omega)$.

$$\therefore \quad y = \frac{\sin a}{\sin (\omega - a)} x + b;$$

or, putting $m = \dfrac{\sin a}{\sin (\omega - a)}$.

$$y = mx + b:$$

which is the equation in the terms required.

---

* The student will not fail to notice that the angle which a line makes
with the axis of x is always measured, *positively*, from that axis toward
the *left;* an angle measured from the axis toward the *right,* is *negative.*
This principle holds true of the angle between *any* two lines.

An. Ge. 10..

Corollary 1.—In the equation just obtained, the axes are supposed to have any inclination whatever. If the axes are rectangular, $\omega = 90°$ and $m = \tan a$. Hence, in

$$y = m.x + b,$$

when referred to *rectangular* axes, m denotes the *tangent of the angle* which the line makes with the axis of x; but when the equation is referred to *oblique* axes, m denotes the *ratio of the sines of the angles* which the line makes with the two axes respectively.

Remark.—In interpreting an equation of the form $y = m.x + b$, and tracing the line corresponding to the values m and b have in it, account must be taken of the *signs* of those constants. * The constant m will be positive or negative *according as the angle a is less or greater than ω.* For $m = \dfrac{\sin a}{\sin (\omega - a)}$: which is positive or negative upon the condition named, according to Trig., 829; *since a is supposed not to exceed* 180°. The constant b (Art. 49) will be positive or negative *according as the intercept on the axis of* y *falls above or below the origin.* Thus, in the case of the line in the diagram, m is negative, and b positive.

If the axes are *rectangular*, the sign of m will be + or — according as a is *acute* or *obtuse.* For, in that case, $m = \tan a$; and the variation of sign is determined by Trig., 825.

Corollary 2.—We can thus determine *the position of a right line with respect to the angles about the axes, by merely inspecting the signs of its equation.*

---

* The quantities m and b are constants, since they can have but one value for any particular right line. But the equation is true for *any* right line, because we can assign to m and b any values wo please. They are hence called *arbitrary* constants.

If m is negative, and b positive, the line crosses the axis of y *above* the origin, and makes with the axis of x an angle *greater* than ω: it therefore crosses the latter at some point to the *right* of the origin, and so lies across the *first* angle.

If m and b are both positive, the line lies across the *second* angle.

If m and b are both negative, the line lies across the *third* angle.

If m is positive, and b negative, the line lies across the *fourth* angle.

[The student may draw a diagram and verify the last three statements.]

Corollary 3.—If $m = 0$, we shall have $\sin a = 0$ $\therefore$ $a = 0$, and the line will be parallel to the axis of x.

Corollary 4.—If $m = \infty$, $\sin (\omega - a) = 0$ $\therefore$ $a = \omega$, and the line will be parallel to the axis of y.

If $m = \infty$ when the axes are rectangular, we shall have $\tan a = \infty$ $\therefore$ $a = 90°$, and the line will be perpendicular to the axis of x: which is essentially the same result as before.

Corollary 5.—If $b = 0$, the line must pass through the origin. But, in that case, the equation becomes

$$y = mx :$$

which is therefore the *equation to a right line passing through the origin.*

79. Equation to the Right Line in terms of its intercepts on the two axes.—The diagram shows that the position of DT is given when OT and OD are given. Let $OT = a$, and $OD = b$; represent by x and y, as before, the co-ordinates OM, MP of any point P on the line. By similar triangles, we have

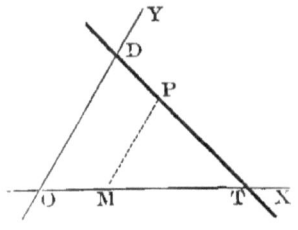

$$OT : OD :: MT : MP; \text{ that is, } a : b :: a - x : y.$$

$$\therefore \quad \frac{x}{a} + \frac{y}{b} = 1:$$

which is the symmetrical form of the required equation.

Remark 1.—When interpreting an equation of this form, the signs of the arbitrary constants a and b must be observed. By doing this, we can fix the position of the line with respect to the four angles, as in the preceding article.

When a and b are both positive, the line lies in the *first* angle, as in the diagram.

When a is negative, and b positive, the line lies in the *second* angle.

When a and b are both negative, the line lies in the *third* angle.

When a is positive, and b negative, the line lies in the *fourth* angle.

Remark 2.—This form of the equation to the Right Line is much used on account of its symmetry. It also deserves to be noticed on account of its resemblance to the analogous equations to the Conics, which we shall develop in due time. It is applicable, as is manifest from the investigation, to rectangular and oblique axes alike.

80. Equation to the Right Line in terms of its perpendicular from the origin and the angle made by the perpendicular with the axis of x.— By examining the diagram, it becomes evident that the position and direction of DT are given, if the length of OR perpendicular to DT, and the angle TOR which it makes with OX, are given. Let $OR = p$; and let the angle $TOR = a$, whence the

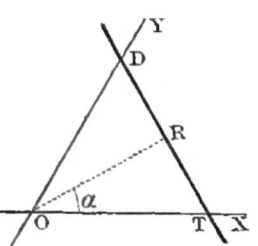

angle $DOR = \omega - a$. Then, a and b representing the intercepts OT and OD as before, we have (Trig., 359)

$$a = \frac{p}{\cos a} \; ; \; b = \frac{p}{\cos (\omega - a)}.$$

Substituting these values of a and b in the equation of Art. 79, we obtain

$$\frac{x \cos a}{p} + \frac{y \cos (\omega - a)}{p} = 1.$$

Clearing of fractions,

$$x \cos a + y \cos (\omega - a) = p;$$

which is the equation sought.

Remark.—The co-efficients of x and y in the foregoing equation are called the *direction-cosines* of the line which the equation represents. In using this form of the equation, it is most convenient to suppose that the angle a may have any value from 0 to $360°$, and that the perpendicular p is always *positive* — that is, (Art. 50, Note,) measured from O in such a direction as to bound the angle. This convention can not be too carefully remembered.

Corollary.—If the axes are *rectangular*, we shall have (Trig., 841)

$$x \cos a + y \sin a = p :$$

a form of the equation having the greatest importance, on account of its relations to the Abridged Notation.

81. The three forms of the equation to the Right Line are therefore as follows:

$$y = mx + b \qquad\qquad (1),$$

$$\frac{x}{a} + \frac{y}{b} = 1 \qquad\qquad (2),$$

$$x \cos a + y \cos (\omega - a) = p \qquad\qquad (3).$$

They are all of the *first degree*. Either of them may be derived from any other, by merely substituting for the constants in the latter their values in terms of those involved in the form sought. For (see diagram, Art. 80) we have

$$m = -\frac{b}{a}; \quad a = \frac{p}{\cos a}; \quad b = \frac{p}{\cos (\omega - a)}.$$

In the case of rectangular axes, we shall have $b = \dfrac{p}{\sin a}$.

82. Polar Equation to the Right Line.—Let ρ and θ be the co-ordinates OP, XOP of any point P on the line PT. Let OR, the perpendicular from the pole to the line, $= p$; and let XOR, the angle which the perpendicular makes with the initial line, $= a$. Then (Trig., 858) we have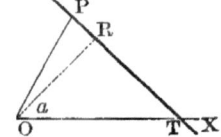

$$OP \cos ROP = OR;$$

that is,

$$\rho \cos (\theta - a) = p :$$

which is the equation required.

Corollary.—If the initial line were perpendicular to the right line, we should have $a = 0$, and the equation would become
$$\rho \cos \theta = p:$$
the *equation to a right line perpendicular to the initial line.*

Remark.—In applying the equation of the present article, it will be convenient to regard the sign of the angle a. This will be $+$ if the perpendicular *OR* falls *above* the initial line; but $-$, if it falls *below*.

83. To trace a Right Line.—The most direct method of solving this problem, consists in finding the intercepts made by the line on the axes, and laying them off according to any chosen scale of equal parts.

Hence, to trace a right line given by its equation: *Make the* y *and* x *of the equation successively* $= 0$: *the resulting values of* x *and* y *will be the intercepts on the axis of* x *and of* y *respectively. Lay off on the axes these intercepts, and the line drawn through their extremities will be the line required.*

Remark.—If the line is given by its polar equation, we find its intercept on the initial line by making $\theta = 0$. When this and the perpendicular from the pole are laid off, we draw the line through their extremities.

Note.—This method fails when the line passes through the origin or the pole, or is parallel to either axis. In the former case, in the Cartesian equation, make $x = 1$, construct the corresponding ordinate, and join its extremity to the origin; in the polar system, lay off the constant vectorial angle of the line. A parallel to either axis must be drawn as such, at the distance its equation requires.

84. We add a few miscellaneous exercises on the foregoing articles.

EXAMPLES.

1. Across which of the four angles does the line $y = 3x + 5$ lie?—the line $y = -6x + 2$?—the line $y = -2x - 4$?—the line $y = x - 1$?

2. What is the situation and direction of the line $y = x$?

3. Axes being oblique, what angle does the line $y = x$ make with the axis of x?—the line $y = \dfrac{x}{\sqrt{3}} + 2$?

4. What is the direction of the line $y = 4$?

5. What is the direction of the line $y = \dfrac{x}{0} + 6$? What are the intercepts made by the preceding lines on the axis of y?

6. Trace the line $y = 5x + 3$; — the line $y - x = 0$.

7. Trace the line $\dfrac{x}{2} + \dfrac{y}{3} = 1$; —the line $\dfrac{x}{3} - \dfrac{y}{2} = 1$.

8. Trace the line $\dfrac{1}{2} x \sqrt{3} + \dfrac{1}{2} y = 1$.

9. What is the value of the angle a in the line of the previous example?—of the angle ω?—of the perpendicular p? [Here $\cos a = \dfrac{1}{2} \sqrt{3}$; $\cos (\omega - a) = \dfrac{1}{2}$.]

10. $x + \dfrac{1}{2} y = 3$ is in the form $x \cos a + y \cos (\omega - a) = p$: determine the values of p, a, and ω, and trace the line.

11. $\dfrac{1}{2} x \sqrt{3} + \dfrac{1}{2} y = 5$ being referred to rectangular axes, find the values of a and p, and lay off the line by means of them.

12. In which of the angles lie the lines $\dfrac{x}{2} + \dfrac{y}{3} = 1$, $\dfrac{x}{3} - \dfrac{y}{2} = 1$, $\dfrac{x}{3} + \dfrac{y}{2} = -1$, and $\dfrac{y}{5} - x = 1$?

13. In which angle does the line $\dfrac{1}{2} x \sqrt{3} - \dfrac{1}{2} y = 2$ lie?

14. Trace the lines $\rho \cos (\theta - 45°) = 8$ and $\rho \sin \theta = -6$. What is the value of a in the second line, and on which side of the initial line is it measured?

15. Find the intercepts of the line $5x + 7y - 9 = 0$, and trace the line.

II. ANALYTIC POINT OF VIEW:—EVERY EQUATION OF THE FIRST DEGREE REPRESENTS A RIGHT LINE.

85. The most general form of the equation of the first degree in two variables, is

$$Ax + By + C = 0,$$

in which A, B and C are arbitrary constants, and may have any value whatever. We propose to prove that this equation, no matter what the values of A, B and C, always represents a right line.

Solving the equation for y, we obtain

$$y = -\frac{A}{B} x - \frac{C}{B}.$$

That is, y is always equal to x multiplied by an arbitrary constant, *plus* an arbitrary constant. In other words, the equation is always reducible to the form

$$y = mx + b,$$

and therefore (Art. 78) always represents a *right line*.

86. A second proof of the same proposition, by means of the *trigonometric function* which the equation implies, is as follows:

Write the equation in the form (to which we have just shown that it is always reducible)

$$y = mx + b:$$

in which m and b are merely abbreviations for the arbitrary constants $-\frac{A}{B}$ and $-\frac{C}{B}$. Now the equation, being true for *every* point of its locus, must be true for any three points $x'y'$, $x''y''$, $x'''y'''$. Hence,

$$y' = mx' + b \qquad (1),$$
$$y'' = mx'' + b \qquad (2),$$
$$y''' = mx''' + b \qquad (3).$$

An. Ge. 11.

Supposing, then, that the abscissas are taken in the order of magnitude, the equation $y = mx + b$ shows that the ordinates will also be in the order of magnitude; that is, if we take x'' greater than x', and x''' greater than x'', we shall either have y'' greater than y', and y''' greater than y'', or else y'' less than y', and y''' less than y''. Accordingly, we subtract (1) from (2), and (2) from (3), and by comparison obtain

$$\frac{y'' - y'}{x'' - x'} = \frac{y''' - y''}{x''' - x''} \qquad (4).$$

Since the form of the locus we are seeking, whatever it be, is (Art. 64) independent of the axes, let us for convenience refer the equation to rectangular axes, OX and OY. Draw the indefinite curve AB, to represent for the time being the unknown locus. Take P', P'', P''' as the three points $x'y'$, $x''y''$, $x'''y'''$; let fall the corresponding ordinates $P'M'$, $P''M''$, $P'''M'''$; draw the chords $P'P''$, $P''P'''$; and make $P'R$, $P''S$ parallel to OX. Then, from (4), we have

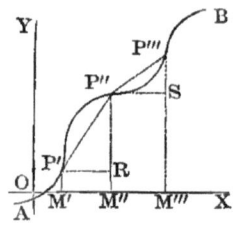

$$\frac{P''R}{P'R} = \frac{P'''S}{P''S};$$

that is, (Trig., 818,)

$$\tan P''P'R = \tan P'''P''S \ \therefore \ P''P'R = P'''P''S.$$

Hence, the three points P', P'', P''' lie on one right line.

But P', P'', P''' are *any* three points of the locus. P'' may therefore be *anywhere* on it between P' and P''', and is independent * of them. Hence, as we may take the points as near each other as we please, *all* the points of the locus lie on one right line; that is, the locus itself is a *right line.*

87. A third proof of the same proposition is furnished by *transformation of co-ordinates.*

$$A x + B y + C = 0$$

being given for geometric interpretation, is of course referred to *some* bilinear system of co-ordinates. Suppose the original axes to be rectangular, and let us transform the equation to a new rectangular system having the origin at the point $x'y'$.

To effect this, write (Art. 56, Cor. 3, cf. Art. 58) for x and y in the given equation $x' + x \cos \theta - y \sin \theta$ and $y' + x \sin \theta + y \cos \theta$. This gives us

$$(A \cos \theta + B \sin \theta) x - (A \sin \theta - B \cos \theta) y + A x' + B y' + C = 0$$

as the equation to the unknown locus, referred to the new axes.

Since x', y', and θ are arbitrary constants, we may subject them to any conditions we please. Let us then suppose that x' and y' satisfy the relation

$$A x' + B y' + C = 0,$$

and that the value of θ is such that

$$A \cos \theta + B \sin \theta = 0 \ i.\,e. \ \tan \theta = -\frac{A}{B}.$$

---

* This is essential to the argument. For three points of a *curve* may lie on one right line, if the third is *determined* by the other two. Thus, in the annexed diagram, $P . P'$, P'' are three points on the curve AB; yet they all lie on the right line $P P''$. The reason is, that P' is determined by joining P and P''.

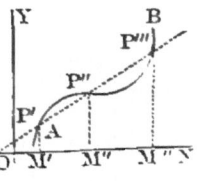

The first of these suppositions means that the new origin is taken somewhere on the unknown locus, since its co-ordinates satisfy the given equation; the second, that the new axis of x makes with the old, an angle whose tangent is $-\dfrac{A}{B}$.

Applying these suppositions to the transformed equation, we obtain, after reductions,

$$y = 0.$$

Hence (Art. 49, Cor. 1) *the locus coincides with the new axis of* x.

And, in general, since the equation $Ax + By + C = 0$ can, upon the suppositions above made, always be reduced to the form $y = 0$, we conclude that it represents a *right line*, which passes through the arbitrary point $x'y'$, and makes with the primitive axis of x an angle whose tangent is found by taking the negative of the ratio between the co-efficients of x and y. That is to say, since these co-efficients are also arbitrary, $Ax + By + C = 0$ is the *Equation to any Right Line.*

88. We have thus shown, by three independent demonstrations, that we can take the empty form of the General Equation of the First Degree, and, merely granting that it is to be interpreted according to the convention of co-ordinates, evoke from it the figure which it represents. It must not be supposed, however, *that we were ignorant of the figure of the Right Line* when we set out upon the foregoing transformations. On the contrary, each of the three demonstrations just given *presupposes* the figure of the Right Line, and certain of its properties. What we did *not* know is, that the equation $Ax + By + C = 0$ represents, and always represents, that figure.

It is important to call attention to this, because the significance of the result just obtained is sometimes over-estimated; and because the case of the Right Line is different in this respect from that of any higher locus. In the strictly analytic investigation of loci of higher orders than the First, not even the figure of the curves is presupposed, but is conceived as being learned for the first time from their equations. But the whole scheme of Analytic Geometry takes the figure and elementary properties of the Right Line for granted; as is obvious from the nature of the convention of co-ordinates and of the theorems for transformation.

89. Starting, then, from $Ax + By + C = 0$ as the equation to the Right Line in its most general form, our next step will naturally be to determine the meaning of the constants A, B, and C. This meaning will be found to vary according to the data by which we may suppose the position of a right line to be fixed. In discussing the Right Line from the geometric point of view, we found that its equation assumed three forms, depending upon the three sets into which the data for its position naturally fall. We shall now see that the general equation

$$Ax + By + C = 0$$

will assume one or another of those forms, according as the constants in it are interpreted by one or another of the sets of data.

90. The first step toward a correct interpretation of these, is to observe that the *arbitrary* constants in our equation *are but two:* — a proposition which we might infer from the fact that in an *equation* we are concerned only with the *mutual ratios* of the co-efficients. But its truth will be obvious, if we consider that an equation may be divided by any constant without affecting the relation between its variables, and therefore without affecting the locus which it represents. Accordingly, if we divide $Ax + By + C = 0$ by either of its constants, for example C, we obtain

$$\frac{A}{C}x + \frac{B}{C}y + 1 = 0 :$$

a form in which there are only two arbitrary constants.

Corollary.—Hence, *Two conditions determine a right line.* Conversely, *A right line may be made to satisfy any two conditions.* This agrees with the fact that the data upon which the three forms of the equation to the Right Line were developed geometrically, are taken by twos. (See Arts. 78, 79, 80.)

91. We now proceed to the analytic deduction of those forms. In this process, the meaning of the ratios among the constants A, B, and C will duly appear.

I. Let the data be *the angle which the line makes with the axis of* x, *and its intercept on the axis of* y. We use the symbols a, ω, m, b to denote the same quantities as in Art. 79. 7

If in $Ax + By + C = 0$ we make $y = 0$, we obtain

$$x = -\frac{C}{A} = (\text{Art. } 83)\ OT \qquad (1).$$

If we make $x = 0$, we obtain

$$y = -\frac{C}{B} = (\text{Art. } 83)\ OD = b \ (2).$$

Dividing (2) by (1), $\dfrac{OD}{OT} = \dfrac{A}{B}$.

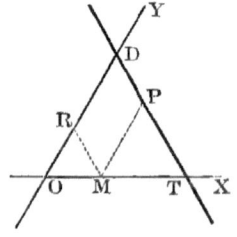

Hence, (see Trig., 867,) $\dfrac{A}{B} = \dfrac{\sin a}{\sin (a - \omega)} = -m$ (3).

From (2), we have $B = -\dfrac{C}{b}$; and from (3), $A = -mB = \dfrac{mC}{b}$. Substituting these values in the original equation, we obtain

$$\frac{mC}{b} x - \frac{C}{b} y + C = 0.$$

$$\therefore \quad y = mx + b.$$

Corollary 1.—By (3), $m = -\dfrac{A}{B}$; and by (2), $b = -\dfrac{C}{B}$. Hence, in the equation to a given right line the ratio between the co-efficients of x and y, taken with a contrary sign, denotes *the ratio between the sines of the angles which the line makes with the two axes:* or, when the axes are *rectangular*, it denotes *the tangent of the angle made with the axis of* x; and the ratio of the absolute term to the co-efficient of y, taken with a contrary sign, denotes *the intercept of the line on the axis of* y.

Corollary 2.—Hence, to reduce an equation in the form $Ax + By + C = 0$ to the form $y = mx + b$, we merely solve the equation for y.

II. Let the data be *the intercepts of the line on the two axes.* Here a and b have their usual signification.

Making $y = 0$ in $Ax + By + C = 0$, we find as before

$$x = -\frac{C}{A} = OT = a \quad (1).$$

Making $x = 0$, we obtain

$$y = -\frac{C}{B} = OD = b \quad (2).$$

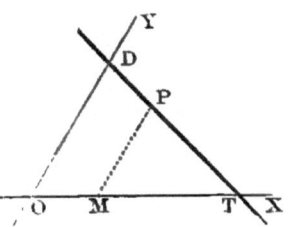

From (1), $A = -\dfrac{C}{a}$; and from (2), $B = -\dfrac{C}{b}$. Substituting for A and B in the equation, and reducing,

$$\frac{x}{a} + \frac{y}{b} = 1.$$

Corollary 1.—From (1) and (2), we see that the ratios of the absolute term to the co-efficients of x and y respectively, taken with contrary signs, denote *the intercepts of the line on the axis of* x *and the axis of* y.

Corollary 2.—To reduce $Ax + By + C = 0$ to the form $\dfrac{x}{a} + \dfrac{y}{b} = 1$, we divide it by its absolute term, and, if necessary, change its signs.

III. Let the data be *the perpendicular from the origin on the line, and the angle of that perpendicular with the axis of* x. We use p and a in the same sense as in Art. 80.

Making y and x successively equal to 0 in the general equation, we obtain, as before,

$$x = -\frac{C}{A} = OT = \text{(Trig., 859)} \ \frac{p}{\cos a} \qquad (1),$$

$$y = -\frac{C}{B} = OD = \text{(Trig., 859)} \ \frac{p}{\cos (\omega - a)} \qquad (2).$$

From (1), $A = -\dfrac{C \cos a}{p}$.

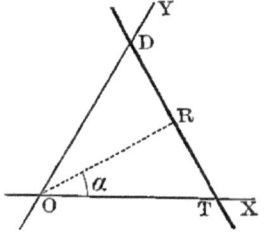

From (2), $B = -\dfrac{C \cos (\omega - a)}{p}$.

Substituting for A and B in the general equation, we have

$$-\frac{C \cos a}{p} x - \frac{C \cos (\omega - a)}{p} y + C = 0 :$$

$$\therefore \quad x \cos a + y \cos (\omega - a) = p.$$

Corollary.—From (1) and (2) we learn that

$$\frac{A}{B} = \frac{\cos a}{\cos (\omega - a)} \; ;$$

or, the ratio between the co-efficients of x and y, denotes *the ratio between the direction-cosines.*

Remark.—The reduction of $Ax + By + C = 0$ to the form $x \cos a + y \cos (\omega - a) = p$ is of such importance that we shall discuss its method in a separate article.

92. Reduction of $Ax + By + C = 0$ to the form $x \cos a + y \cos (\omega - a) = p$.—The problem may be more precisely stated: *To find the values of* $\cos a$, $\cos (\omega - a)$, *and* p, *in terms of* A, B, *and* C.

If, in the preceding article, we divide (1) by (2), we find $\dfrac{B}{A} = \dfrac{\cos (\omega - a)}{\cos a}$. That is, (Trig., 845, IV; 839,)

$$\frac{B}{A} = \cos \omega + \sin \omega \tan a.$$

$$\therefore \; \tan a = \frac{B - A \cos \omega}{A \sin \omega} \cdot$$

$$\therefore \; \cos a = \frac{1}{\sqrt{(1 + \tan^2 a)}} = \frac{A \sin \omega}{\sqrt{(A^2 + B^2 - 2 AB \cos \omega)}} \cdot$$

$$\therefore \; \cos (\omega - a) = \frac{B \cos a}{A} = \frac{B \sin \omega}{\sqrt{(A^2 + B^2 - 2 AB \cos \omega)}} \cdot$$

$$\therefore \; p = - \frac{C \cos a}{A} = - \frac{C \sin \omega}{\sqrt{(A^2 + B^2 - 2 AB \cos \omega)}} \cdot$$

Therefore, to make the required reduction, *Multiply the equation throughout by* $\dfrac{\sin \omega}{\sqrt{(A^2 + B^2 - 2 AB \cos \omega)}}$. *

---

* The following elegant solution of this problem is by SALMON: *Conic Sections,* p. 20:

"Suppose that the given equation when multiplied by a certain factor R is reduced to the required form, then $RA = \cos a$, $RB = \cos \beta$. But it can easily be proved that, if a and β be any two angles whose sum is ω, we shall have

Remark.—Since we have agreed always to consider p a *positive* quantity, it may be necessary to *change the signs* of the given equation, before multiplying, so that its absolute term when transposed to the second member may be positive.

Corollary 1.—If the axes are rectangular, we shall have

$$\cos a = \frac{A}{\sqrt{(A^2 + B^2)}}; \quad \sin a = \frac{B}{\sqrt{(A^2 + B^2)}};$$

$$p = -\frac{C}{\sqrt{(A^2 + B^2)}}.$$

Accordingly, $Ax + By + C = 0$ is reduced to the form $x \cos a + y \sin a = p$, *by dividing all its terms by* $\sqrt{A^2 + B^2}$, *after the necessary changes of sign.*

Great importance pertains to the transformation explained in this corollary. The general reduction to $x \cos a + y \cos (\omega - a) = p$ is of comparatively infrequent use.

Corollary 2.—From the values of p above obtained, we learn that the length of the perpendicular from the origin upon the line $Ax + By + C = 0$ is

---

$$\cos^2 a + \cos^2 \beta - 2 \cos a \cos \beta \cos \omega = \sin^2 \omega.$$

Hence, $R^2(A^2 + B^2 - 2AB \cos \omega) = \sin^2 \omega$; and the equation reduced to the required form is

$$\left. \frac{A \sin \omega}{\sqrt{(A^2 + B^2 - 2AB \cos \omega)}} x + \frac{B \sin \omega}{\sqrt{(A^2 + B^2 - 2AB \cos \omega)}} y \right\}$$
$$\left. + \frac{C \sin \omega}{\sqrt{(A^2 + B^2 - 2AB \cos \omega)}} \right\} = 0.$$

And we learn that

$$\frac{A \sin \omega}{\sqrt{(A^2 + B^2 - 2AB \cos \omega)}}, \quad \frac{B \sin \omega}{\sqrt{(A^2 + B^2 - 2AB \cos \omega)}}$$

are respectively the cosines of the angles that the perpendicular from the origin on the line $Ax + By + C = 0$ makes with the axes of x and y; and that $\dfrac{C \sin \omega}{\sqrt{(A^2 + B^2 - 2AB \cos \omega)}}$ is the length of that perpendicular."

$$-\frac{C\sin\omega}{\sqrt{(A^2 + B^2 - 2AB\cos\omega)}};$$

and that this length becomes

$$-\frac{C}{\sqrt{(A^2 + B^2)}},$$

when the axes are rectangular.

93. Polar Equation to the Right Line, deduced analytically.—The polar equation may be obtained from the Cartesian as follows : Let $Ax + By + C = 0$, referred to rectangular axes, be reduced to the form

$$x\cos a + y\sin a = p.$$

Transforming to polar co-ordinates, we have (Art. 57, Cor.) $x = \rho\cos\theta$, and $y = \rho\sin\theta$: and the equation becomes

$$\rho(\cos\theta\cos a + \sin\theta\sin a) = p;$$

that is, $\qquad \rho\cos(\theta - a) = p,$

the equation of Art. 82.

Corollary.—Transforming the original equation to polars, we have

$$\rho(A\cos\theta + B\sin\theta) + C = 0.$$

Hence, an equation in the form $\rho(A\cos\theta + B\sin\theta) = C$ may be reduced to the form $\rho\cos(\theta - a) = p$ by dividing each term by $\sqrt{A^2 + B^2}$.

94. Before advancing to the more particular forms of the equation to the Right Line, the student should make sure of having mastered the general ones which precede, and the principles which have been developed in the course of the discussions just closed. To this end, let the following exercises be performed.

EXAMPLES.

1. Transform $3x - 5y + 6 = 0$ to $y = 0$. What is the angle made by the new axis of x with the old?

2. What angle does the line $8x + 12y + 2 = 0$ make with the axis of x? What is the length of its intercept on the axis of y? In which of the four angles does it lie?

3. Find the intercepts of the line $x + 3y - 3 = 0$.

4. Find the ratio between the direction-cosines of the right line $2x + 3y + 4 = 0$, ω being $= 60°$.

5. What is the tangent of the angle made with the axis of x by the perpendicular from the origin on $3x - 2y - 6 = 0$?

6. Reduce all the equations in the previous examples to the form $x \cos a + y \sin a = p$, and determine a and p for each line.

7. Reduce $3x + 4y = 12$ to the form $x \cos a + y \cos (\omega - a) = p$. What are the values of a and $(\omega - a)$, supposing ω successively equal to $30°$, $45°$, $60°$, and $\sin^{-1} \frac{3}{4}$?

8. Find the length of the perpendicular from the origin on $3x + 4y + 12 = 0$, under the several values of ω last supposed.

9. Find the length of the perpendicular from the origin on $3x - 4y - 12 = 0$, axes being rectangular.

10. Reduce $2\rho \cos \theta - 3\rho \sin \theta = 5$ to the form $\rho \cos (\theta - a) = p$. What are the values of a and p?

THE RIGHT LINE UNDER SPECIAL CONDITIONS.

95. Equation to the right line passing through Two Fixed Points.—Let the two points be $x'y'$, $x''y''$. Since they are points on a right line, their co-ordinates must satisfy the equation

$$Ax + By + C = 0 \qquad (1),$$

and we therefore have

$$Ax' + By' + C = 0 \qquad (2),$$
$$Ax'' + By'' + C = 0 \qquad (3).$$

Subtracting (2) from (1) and (3) successively, we obtain

$$A(x - x') + B(y - y') = 0,$$
$$A(x'' - x') + B(y'' - y') = 0,$$

and thence

$$\frac{y - y'}{x - x'} = \frac{y'' - y'}{x'' - x'} \qquad (4):$$

which is the equation required. For it is the equation to *some* right line, since it is of the first degree; and it is the equation to the line passing through the two given points, because it vanishes when either x' and y' or x'' and y'' are substituted in it for x and y.

Corollary 1.—Equation (4) may evidently be written

$$(y' - y'')x - (x' - x'')y + x'y'' - y'x'' = 0:$$

a form often useful, though the form (4) is more easily remembered.

Corollary 2.—If in the last equation we suppose $x'' = 0$ and $y'' = 0$, we obtain

$$y'x - x'y = 0:$$

the *equation to the right line passing through a fixed point and the origin.*

Remark.—The same equation, (4), might have been obtained geometrically. For, since the triangles PRP', $P'SP''$ are similar, we have

$$\frac{PR}{RP'} = \frac{P'S}{SP''};$$

that is, after changing the signs of the equation,

$$\frac{y - y'}{x - x'} = \frac{y'' - y'}{x'' - x'}.$$

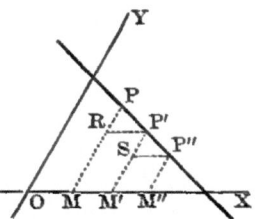

It is worth while to place the analytic and geometric proofs thus side by side, in order to make sure that the geometric meaning of all the symbols in the equations developed, shall be clearly understood.

96. Angle between two right lines given by their equations.—All formulæ for angles are greatly simplified by the use of rectangular axes. We therefore present the subject of this article first in the form which those axes determine.

Let the two lines be $y = mx + b$ and $y = m'x + b'$; and let $\varphi =$ the angle between them. From the diagram, $\varphi = a' - a$; and we have (Trig., 845, VI)

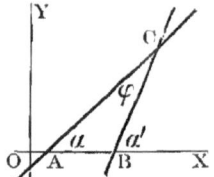

$$\tan \varphi = \frac{\tan a' - \tan a}{1 + \tan a \tan a'}$$

$$= \frac{m' - m}{1 + mm'} \quad \text{(Art. 78, Cor. 1)}.$$

Corollary 1.—If the equations were given in the form $Ax + By + C = 0$, $A'x + B'y + C' = 0$, we should have (Art. 91, I, Cor. 1) $m = -\dfrac{A}{B}$, and $m' = -\dfrac{A'}{B'}$. Hence, in that case,

$$\tan \varphi = \frac{AB' - A'B}{AA' + BB'}.$$

Corollary 2.—If $\varphi = 0$ or π, $\tan \varphi = 0$; and we have

$$m' - m = 0 \quad \text{or} \quad AB' - A'B = 0:$$

the *condition that two right lines shall be parallel.*

Corollary 3.—If $\varphi = 90°$, $\tan \varphi = \infty$; and we have

$$1 + mm' = 0 \quad \text{or} \quad AA' + BB' = 0:$$

the *condition that two right lines shall be mutually perpendicular.*

97. Angle between two lines, axes being oblique.—From Art. 92 we shall have $\cos a = \dfrac{A \sin \omega}{\sqrt{(A^2 + B^2 - 2AB \cos \omega)}}$, and $\cos a' = \dfrac{A' \sin \omega}{\sqrt{(A'^2 + B'^2 - 2A'B' \cos \omega)}}$. Hence (Trig., 838)

$$\sin a' = \frac{B' - A' \cos \omega}{\sqrt{(A'^2 + B'^2 - 2A'B' \cos \omega)}},$$

$$\sin a = \frac{B - A \cos \omega}{\sqrt{(A^2 + B^2 - 2AB \cos \omega)}}.$$

Therefore (Trig., 845, III and IV)

$$\sin (a' - a) = \frac{(AB' - A'B) \sin \omega}{\sqrt{(A^2+B^2-2AB\cos\omega)}\,\sqrt{(A'^2+B'^2-2A'B'\cos\omega)}},$$

$$\cos (a' - a) = \frac{AA' + BB' - (AB' + A'B) \cos \omega}{\sqrt{(A^2+B^2-2AB\cos\omega)}\,\sqrt{(A'^2+B'^2-2A'B'\cos\omega)}}.$$

Whence (Trig., 839)

$$\tan \varphi = \frac{(AB' - A'B) \sin \omega}{AA' + BB' - (AB' + A'B) \cos \omega}$$

a formula which evidently becomes that of Art. 96, Cor. 1, if $\omega = 90°$.

Corollary 1.—The condition that the two lines shall be parallel, is

$$AB' - A'B = 0:$$

from the identity of which with the condition of Art. 96, Cor. 2, we learn that *the condition of parallelism is independent of the value of ω*. The same follows from the fact that the condition itself is not a function of ω.

Corollary 2.—The condition that the lines shall be perpendicular to each other, is

$$AA' + BB' - (AB' + A'B) \cos \omega = 0.$$

98. Equation to a right line parallel to a given one.—Let the given line be $y = mx + b$. The required equation will be of the form

$$y = m'x + b'.$$

But in this, the condition of parallelism (Art. 96, Cor. 2; Art. 97, Cor. 1) gives us $m' = m$. The required equation is therefore

$$y = mx + b'.$$

Corollary.—If the given line were $Ax + By + C = 0$, the equation would be of the form $A'x + B'y + C' = 0$. In that case, $m = -\dfrac{A}{B}$, and $b' = -\dfrac{C'}{B'}$. The required equation would therefore be, after writing C'' for $\dfrac{BC'}{B'}$,

$$Ax + By + C'' = 0.$$

From this we infer that *the equations to parallel right lines differ only in their constant terms.*

99. Rectangular equation to a right line perpendicular to a given one.—The equation will be of the form $y = m'x + b'$, in which m' is determined by the condition (Art. 96, Cor. 3)

$$1 + mm' = 0.$$

Hence, $m' = -\dfrac{1}{m}$; and the equation is

$$y = -\frac{1}{m}x + b'.$$

Corollary.—When the given line is $Ax + By + C = 0$, the equation to its perpendicular is (Art. 91, I, Cor. 1)

$$Ay - Bx + C_1 = 0.$$

Hence, *if two right lines are perpendicular to each other, their rectangular equations interchange the co-efficients of* x *and* y, *and change the sign of one of them.*

100. Equation to a right line perpendicular to a given one, axes being oblique.—From the condition of perpendicularity, (Art. 97, Cor. 2.)

$$A'(A - B\cos\omega) + B'(B - A\cos\omega) = 0$$

Hence,
$$-\frac{A'}{B'} = \frac{B - A\cos\omega}{A - B\cos\omega};$$

and the required equation (Art. 91, I, Cor. 1) is

$$(A - B\cos\omega)y - (B - A\cos\omega)x + C_2 = 0.$$

101. Equation to a right line parallel to a given one, passing through a Fixed Point.—By Art. 98 we have the *form* of the equation,

$$y = mx + b' \qquad (1).$$

Calling the fixed point $x'y'$, we therefore obtain

$$y' = mx' + b':$$
$$\therefore \quad b' = y' - mx'.$$

Substituting for b' in (1), the equation sought is

$$y - y' = m(x - x').$$

Corollary 1.—If in this equation we suppose m indeterminate, the direction of the line is indeterminate; and we have the *equation to any right line passing through a fixed point.*

Corollary 2.—Let $\sin\alpha : \sin\omega = k$, and $\sin(\omega - \alpha) : \sin\omega = h$. Then $k : h + \sin\alpha : \sin(\omega - \alpha) = m$; and the equation may be written.

$$\frac{y - y'}{k} = \frac{x - x'}{h};$$

or, if we denote either of these equal ratios by l,

$$\frac{y - y'}{k} = \frac{x - x'}{h} = l:$$

An. Ge. 12.

a formula often convenient, and known as the *Symmetrical Equation to the Right Line*.

Corollary 3.—If the axes are *rectangular*, we have $k : h = \sin a : \cos a$; and the equation may be more conveniently written

$$\frac{y - y'}{s} = \frac{x - x'}{c} = l:$$

where s and c are abbreviations for the sine and cosine of the angle which the line makes with the axis of x.

102. Geometric meaning of the ratio l.—If, in the annexed diagram, P' denote the fixed point $x'y'$, and P any point of a right line passing through it, $P'R$ being parallel to OX,

$$\frac{y - y'}{k} = \frac{PR \sin \omega}{\sin PP'R} ,$$

$$\frac{x - x'}{h} = \frac{P'R \sin \omega}{\sin P'PR} .$$

But (Trig., 871)

$$\frac{PR \sin \omega}{\sin PP'R} = P'P = \frac{P'R \sin \omega}{\sin P'PR} .$$

Hence,

$$\frac{y - y'}{k} = \frac{x - x'}{h} = l$$

denotes *the distance from a fixed point* x'y' *to any point* xy *of a right line passing through it.*

Remark 1.—So long as the point xy is variable, l is of course indeterminate. But the formula enables us to find the distance from $x'y'$ to any *given* point on the line, by merely substituting for x or y the abscissa or ordinate of such given point.

Thus, supposing the given point of the line to be $x''y''$, we should have (assuming the axes for convenience to be rectangular)

$$\frac{y'' - y'}{l} = s,$$

$$\frac{x'' - x'}{l} = c.$$

Squaring both sides of these equations, adding, and remembering (Trig., 838) that $s^2 + c^2 = 1$, we obtain

$$l^2 = (x'' - x')^2 + (y'' - y')^2:$$

which agrees with the formula (Art. 51. I, Cor. 1) for *the distance between two given points.*

Remark 2.—The signs $+$ and $-$ in connection with l denote distances measured along the line in opposite directions from $x'y'$. Thus, if $+ l$ were measured in the direction $P'D$, $- l$ would be measured in the direction $P'T$: and *vice versâ.* Speaking with entire generality, $+ l$ must be laid off from $x'y'$ in the positive direction of the *line* (Art. 50, Note), and $- l$ in the negative.

103. Rectangular equation to a right line passing through a Fixed Point, and cutting a given line at a Given Angle.—Let the given line be $y = mx + b$, and let $\theta =$ the given angle. The equation sought (Art. 101. Cor. 1) is of the form

$$y - y' = m' (x - x'),$$

in which m' is to be determined from the conditions of the problem.

The line in question obviously makes with the axis of x an angle $\alpha' = \alpha + \theta$. Hence, (Trig., 845. v.)

$$m' = \frac{m + \tan \theta}{1 - m \tan \theta} .$$

Substituting this value for m', the required equation is

$$y - y' = \frac{m + \tan \theta}{1 - m \tan \theta} (x - x').$$

Corollary 1.—Dividing both terms of m' by $\tan \theta$, this equation may be written

$$y - y' = \frac{m \cot \theta + 1}{\cot \theta - m} (x - x').$$

Accordingly, if $\theta = 90°$, it becomes

$$y - y' = -\frac{1}{m} (x - x') :$$

the *equation to a right line passing through a fixed point, and perpendicular to a given line.*

Corollary 2.—By substituting for $-m$ its value $A : B$, we obtain

$$A (y - y') - B (x - x') = 0$$

as the equation to the perpendicular of $Ax + By + C = 0$, passing through $x'y'$: the co-efficients of which evidently satisfy the criterion of the corollary to Art. 99.

104. *Equation to a right line passing through a Fixed Point and cutting a given line at a Given Angle, axes being oblique.*—Let the given line be $Ax + By + C = 0$. Since $a' = a + \theta$, $\theta = a' - a$.

By putting for m' its value, the required equation assumes the form

$$y - y' = -\frac{A'}{B'} (x - x').$$

From Art. 97, we have

$$\tan \theta = \frac{(AB' - A'B) \sin \omega}{AA' + BB' - (AB' + A'B) \cos \omega} ;$$

therefore

$$\frac{A'}{B'} = \frac{A \sin \omega - (B - A \cos \omega) \tan \theta}{B \sin \omega + (A - B \cos \omega) \tan \theta} ,$$

and the required equation is

$$
\left. \begin{array}{l}
\{A \sin \omega - (B - A \cos \omega) \tan \theta\} \, (x - x') \\
+ \{B \sin \omega + (A - B \cos \omega) \tan \theta\} \, (y - y')
\end{array} \right\} = 0.
$$

Corollary 1.—When the given line is $y = mx + b$, the equation assumes the form

$$
y - y' = \frac{m \sin \omega + (1 + m \cos \omega) \tan \theta}{\sin \omega - (m + \cos \omega) \tan \theta} (x - x').
$$

This evidently becomes the equation of Art. 103 when $\omega = 90°$.

Corollary 2.—Dividing the equation by $\tan \theta$, and supposing $\theta = 90°$, we obtain

$$
(B - A \cos \omega)(x - x') - (A - B \cos \omega)(y - y') = 0 :
$$

the *general equation to the perpendicular of a given line, passing through a fixed point.*

This might have been obtained directly from the equation of Art. 100.

105. Length of the perpendicular from any point to a given right line.—Let the given point be xy, and the given line $x \cos a + y \cos (\omega - a) - p = 0$.

Represent the given point at P, and the given line by DT. Let PQ be the required perpendicular. Draw PM parallel to OD, OS parallel to PQ, and PS, MN parallel to DT. Then,

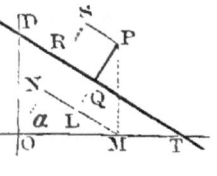

$$
ON = OM \cos MON = x \cos a,
$$
$$
NS = PM \cos MPL = y \cos (\omega - a) :
$$

$$
PQ = ON + NS - OR = x \cos a + y \cos (\omega - a) - p.
$$

In the foregoing discussion, the point P and the origin were assumed to be on *opposite* sides of the given line; had they been supposed on the *same* side, OR would have been greater than OS, and we should have had

$$
PQ = OR - (ON + NS) = p - x \cos a - y \cos (\omega - a).
$$

Hence, if a perpendicular be let fall from any point xy to the line

$$x \cos a + y \cos (\omega - a) - p = 0,$$

its *length* will be

$$\pm (x \cos a + y \cos (\omega - a) - p)$$

according as the point and the origin lie on *opposite* sides of the line, or on the *same* side.

Remark.—The student will observe that we here use the same symbols xy to denote the co-ordinates of the point from which the perpendicular is dropped, and those of any point on the given line. But it must not be supposed that the xy of the point and the xy of the line have necessarily the same values. Generally, of course, they have not; for the point from which the perpendicular falls, is generally supposed *not* to be a point on the line. It *may* be; and when it is, the length of the perpendicular vanishes. Hence, supposing the quantity $x \cos a + y \cos (\omega - a) - p$ to denote the length of the perpendicular from xy on a given line, *the equation*

$$x \cos a + y \cos (\omega - a) - p = 0$$

signifies that the point lies somewhere on the line in question.

The double use of xy may cause some confusion at first, but its advantages more than compensate for the attention required to overcome this.

Corollary 1.—If the axes are rectangular, the length of the perpendicular is

$$\pm (x \cos a + y \sin a - p)$$

according as the point and the origin lie on opposite sides of the line, or on the same side.

Corollary 2.—The length of the perpendicular from xy on the line $Ax + By + C = 0$ (Art. 92 cf. Cor. 1) is

$$\pm \frac{(Ax + By + C) \sin \omega}{\sqrt{(A^2 + B^2 - 2AB \cos \omega)}} \text{ or } \pm \frac{Ax + By + C}{\sqrt{(A^2 + B^2)}}$$

according as the axes are *oblique* or *rectangular*.

Corollary 3.—The perpendicular from a point on the same side of a line as the origin must have the same sign as p. But we have agreed (Art. 80, Rem.) that p shall always be *positive;* and we have seen that the perpendicular changes sign in passing from one side of the line to the other. Hence, *Perpendiculars falling on the side of a line next the origin are positive; and those falling on the side remote, are negative.*

106. To find the point of intersection of two right lines given by their equations.—Eliminate between the equations to the two lines: the resulting values of x and y (Art. 62) are the co-ordinates of the required point.

Thus, in general, the lines being $Ax + By + C = 0$ and $A'x + B'y + C' = 0$, we have

$$x = \frac{BC' - B'C}{AB' - A'B}, \quad y = \frac{CA' - C'A}{AB' - A'B}$$

as the co-ordinates of the common point.

107. Equation to a right line passing through the intersection of two given ones.—If we multiply the equations to two given lines each by an arbitrary constant, and add the results, thus:

$$l\,(Ax + By + C) + m\,(A'x + B'y + C') = 0,$$

the new equation will represent a right line passing through the intersection of the lines $Ax + By + C = 0$ and $A'x + B'y + C' = 0$.

For it manifestly denotes *some* right line, since it is of the first degree. Moreover, it is satisfied by any values of x and y that satisfy $Ax + By + C = 0$ and $A'x + B'y + C' = 0$ simultaneously: for its left member must vanish whenever the quantities $Ax + By + C$ and

$A'x + B'y + C'$ are both equal to zero. That is, it passes through a point whose co-ordinates satisfy the equations of both the given lines. But such a point (Arts. 62, 106) is the intersection of the two lines.

Corollary.—By varying the values of the constants l and m, we can cause the above equation to represent *as many different lines as we please*, all passing through the intersection of the two given ones.

Remark.—The truth of the above equation is independent of ω, and of the form of the equations to the given lines. This is manifest from the method by which it was obtained. It may therefore, when convenience requires, be written

$$\left. \begin{array}{l} l \left\{ x \cos a + y \cos (\omega - a) - p \right\} \\ + m \left\{ x \cos \beta + y \cos (\omega - \beta) - p' \right\} \end{array} \right\} = 0$$

or

$$l (x \cos a + y \sin a - p) + m (x \cos \beta + y \sin \beta - p') = 0.$$

108. Meaning of the equation $L + kL' = 0$.—If we put $k = m : l$, and represent by L and L' the quantities which are equated to zero in the equations of the two given lines, the equation of the preceding article becomes

$$L + kL' = 0.$$

We shall now prove that this is the equation to *any* right line passing through the intersection of two given ones. *

---

* The beginner may suppose that this has been done already, in the preceding article. But we merely proved there, that the equation may represent an *infinite number* of lines answering to the given condition. Now, that an *infinite number* of lines is not the same as *all* the lines passing through the intersection of two others, is evident. For between two intersecting right lines there are *two angles*, supplemental to each other, in *each* of which there may be an infinite number of lines passing through the common point.

Let $\theta =$ the angle made with the axis of x by the line which the equation represents. Then (Art. 91, I, Cor. 1)

$$\frac{\sin \theta}{\sin (\omega - \theta)} = -\frac{A + kA'}{B + kB'} .$$

Therefore

$$\tan \theta = \frac{(A + kA') \sin \omega}{(A \cos \omega - B) + k (A' \cos \omega - B')} .$$

Hence, as k may have any value from 0 to ∞, and be either positive or negative, $\tan \theta$ may have any value, either positive or negative, from 0 to ∞. That is, the equation is consistent with any value of θ whatever.

Therefore, if $L = 0$ and $L' = 0$ are the equations to two right lines,

$$L + kL' = 0$$

is the equation to *any* right line passing through the intersection of the two.

Corollary 1.—The equation to a *particular* line intersecting two others in their common point, is formed from the above by assigning to k such a value, in terms of the conditions which the line must satisfy. as the relation $L + kL' = 0$ implies.

Thus, if the condition were that the intersecting line make with the axis of x a given angle $= \theta$, we should have, from the value of $\tan \theta$ above,

$$k = -\frac{A \sin \omega - (A \cos \omega - B) \tan \theta}{A' \sin \omega - (A' \cos \omega - B') \tan \theta} :$$

or, in case the axes were rectangular,

$$k = -\frac{A - B \tan \theta}{A' - B' \tan \theta} .$$

An. Ge. 13.

If the condition were that the intersecting line pass through a fixed point $x'y'$, we should have

$$(Ax' + By' + C) + k(A'x' + B'y' + C') = 0:$$

$$\therefore \quad k = -\frac{A\,x' + B\,y' + C}{A'x' + B'y' + C'}.$$

Corollary 2.—The *sign* of k has a most important geometric meaning. Obviously, in order to change its sign, a quantity must pass through the value 0 or ∞. Now if $k = 0$, we have by Cor. 1

$$A + B \tan \theta = 0 \quad \therefore \quad \tan \theta = -\frac{A}{B}.$$

And if $k = \infty$,

$$A' + B' \tan \theta = 0 \quad \therefore \quad \tan \theta = -\frac{A'}{B'}.$$

That is, (Art. 91, I, Cor. 1,) *at the instant when* k *changes sign, the line which passes through the intersection of two given ones coincides with one of them.* Hence, of the lines $L + kL' = 0$ and $L - kL' = 0$, one lies in the angle between $L = 0$ and $L' = 0$ *supplemental* to that in which the other lies.

It now remains to determine which of these supplemental angles corresponds to $+ k$, and which to $- k$. If the two lines are $x \cos a + y \cos (\omega - a) - p = 0$ and $x \cos \beta + y \cos (\omega - \beta) - p' = 0$, we shall have

$$k = -\frac{x \cos a + y \cos (\omega - a) - p}{x \cos \beta + y \cos (\omega - \beta) - p'};$$

that is, (Art. 105,) one of the geometric meanings of k is, *the negative of the ratio between the perpendiculars let fall on two given lines from any point of a line passing through their intersection.* Hence, when k is *positive*, those perpendiculars have *unlike* signs; and when k is *negative*, their signs are *like*. That is, (Art. 105, Cor. 3,) the perpendiculars corresponding to $+ k$ fall *one* on the side of one line *next* the origin, and the *other* on the *remote* side of the other line; while those corresponding to — *k* fall *both* on the side of the lines *next* the origin or *both* on the side *remote* from it. In other words, the *line* corresponding to + *k* lies in the

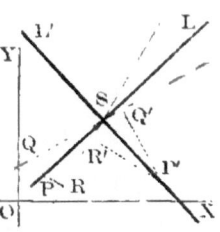

angle *remote* from the origin, e. g. *P'L'*: and the line corresponding to —*k* lies in the *same* angle as the origin, e. g. *PL*.

For convenience, we shall call the angle *R'SQ'*, remote from the origin, the *external* angle between the given lines; and the angle *RSQ*, in which the origin lies, the *internal* angle.

Hence, if $L = 0$ and $L' = 0$ are the equations to any two right lines,

$$L + kL' = 0$$

denotes a line passing through their intersection and lying in the *external* angle between them; and

$$L - kL' = 0$$

denotes one lying in the *internal* angle.

109. Equation to a right line bisecting the angle between two given ones.—Any point on the bisector being equally distant from the two given lines, we have (Art. 105 cf. Cor. 2)

$$\frac{Ax + By - C}{\sqrt{(A^2 + B^2 - 2AB\cos\omega)}} = \pm \frac{A'x + B'y + C'}{\sqrt{(A'^2 - B'^2 - 2A'B'\cos\omega)}} \quad (1)$$

or

$$x\cos a - y\cos(\omega - a) - p = \pm \{x\cos\beta + y\cos(\omega - \beta) - p'\}* \quad (2).$$

---

* The student may at first think that *both* members of (1) and (2) should have the double sign. But since an equation always implies the possibility of changing its signs, it is evident that we should write the expressions as above.

according as the given lines are $Ax + By + C = 0$ and $A'x + B'y + C' = 0$ or $x \cos a + y \cos (\omega - a) - p = 0$ and $x \cos \beta + y \cos (\omega - \beta) - p' = 0$. Expressions (1) and (2) are the principal forms of the required equation.

Corollary 1.—When the axes are rectangular, the equation becomes

$$\frac{Ax + By + C}{\sqrt{(A^2 + B^2)}} = \pm \frac{A'x + B'y + C'}{\sqrt{(A'^2 + B'^2)}} \tag{3}$$

or

$$x \cos a + y \sin a - p = \pm (x \cos \beta + y \sin \beta - p') \quad (4).$$

Corollary 2.—Expressions (1), (2), (3), (4) are evidently in the form $L \pm kL' = 0$. Hence (Art. 108, Cor. 2) there are *two* bisectors: one lying in the *external* angle of the given lines, and the other in the *internal*. For the sake of brevity, we shall call the former the *external* bisector; and the latter, the *internal* bisector.

Corollary 3.—If we put, as we conveniently may,

$$a = x \cos a + y \cos (\omega - a) - p,$$
$$\beta = x \cos \beta + y \cos (\omega - \beta) - p',$$

expressions (2) and (4) will be included in the brief and striking form

$$a \pm \beta = 0.$$

Hence, if $a = 0$, $\beta = 0$ are the equations to any two right lines, the line

$$a + \beta = 0$$

bisects the *external* angle between them; and the line

$$a - \beta = 0$$

bisects their *internal* angle.

Caution.—From the proposition just reached, the student is apt to rush to the conclusion that

$$L \pm L' = 0$$

is the equation to the bisector of the angle between the lines $L = 0$ and $L' = 0$, without regard to the values of L and L'. *This is a grave error.* It assumes that the value of k, in the case of a bisector, is *always* ± 1. When the equations to two lines are in terms of the perpendicular from the origin and its angle with the axis of x; that is, when $L = x \cos a + y \cos (\omega - a) - p$ and $L' = x \cos \beta + y \cos (\omega - \beta) - p'$, $k = \pm 1$. But from equation (1) we have

$$k = \pm \frac{1}{1} \frac{(A^2 + B^2 - 2\,A\,B \cos \omega)}{(A'^2 + B'^2 - 2\,A'B' \cos \omega)} :$$

which is obviously not in general equal to ± 1. The condition that it shall have that value is

$$A^2 + B^2 - 2\,AB \cos \omega = A'^2 + B'^2 - 2\,A'B' \cos \omega$$

or, when the axes are rectangular,

$$A^2 + B^2 = A'^2 + B'^2.$$

Corollary 4.—If we denote by r the particular value which k assumes in the case of a bisector, then

$$L + rL' = 0$$

represents the *external*, and

$$L - rL' = 0$$

the *internal* bisector of the angle between the lines $L = 0$ and $L' = 0$.

In these expressions, r is to be determined from the relation

$$r = \pm \frac{\sqrt{(A^2 + B^2 - 2AB\cos\omega)}}{1\ (A'^2 + B'^2 - 2A'B'\cos\omega)} :$$

which, in case the axes are rectangular, becomes

$$r = \pm \frac{\sqrt{(A^2 + B^2)}}{1\ (A'^2 + B'^2)} .$$

110. Equation to a right line situated at infinity.—To assume that a right line is at an infinite distance from the origin is to assume that its intercepts on the axes are infinite. Hence, we have

$$-\frac{C}{A} = \infty \quad \text{and} \quad -\frac{C}{B} = \infty.$$

That is, supposing C to be *finite*,

$$A = 0 \text{ and } B = 0.$$

The required equation is therefore

$$0x + 0y + C = 0 :$$

in which C is *finite*. We shall cite it in the somewhat inaccurate but very convenient form

$$C = 0.$$

Remark.—The student will of course remember that a line *at infinity* is not a *geometric* conception at all—in fact does not *exist*, in any sense known to pure geometry. As an *analytic* conception, however, it has important bearings; and the equation just obtained is useful in some of the higher investigations of curves.

111. Equations of Condition.—When elements of position and form sustain certain geometric relations to each other, the constants which enter their analytic equivalents must sustain corresponding relations. In other words, if the geometric relations exist, the constants satisfy certain equations. Such equations are called Equations of Condition.

Thus we saw (Art. 97, Cor. 1) that if two right lines $Ax + By + C = 0$ and $A'x + B'y + C' = 0$ are parallel, the constants A, B, A', B' must satisfy the equation

$$AB' - A'B = 0 ;$$

and that if they are mutually perpendicular, the constants must satisfy the equation

$$AA' + BB' - (AB' + A'B) \cos \omega = 0.$$

112. Condition that Three Points shall lie on One Right Line.—If $x_1 y_1$, $x_2 y_2$, $x_3 y_3$ lie on one right line, $x_3 y_3$ must satisfy the equation to the line which passes through $x_1 y_1$ and $x_2 y_2$. Hence, (Art. 95,) the equation of condition is

$$(y_1 - y_2) x_3 - (x_1 - x_2) y_3 + x_1 y_2 - y_1 x_2 = 0 :$$

which may for the sake of symmetry be written

$$y_1 (x_2 - x_3) + y_2 (x_3 - x_1) + y_3 (x_1 - x_2) = 0.$$

Remark.—It is worth while to notice the *order* of the elements which enter into the latter form of this equation. In writing symmetrical forms, the *analogous symbols must be taken in a fixed order*, which will be best understood by conceiving of the successive symbols as forming a *circuit*, about which we move according to the annexed diagram. Thus, as in the last equation, we pass from x_1 to x_2, from x_2 to x_3, and from x_3 to x_1; and so round again: *always going back to the* FIRST *element when the list has been completed*, and then proceeding as before *in numerical succession*. The advantage of symmetrical forms is very decided, especially when we have to compare or combine analogous equations. But unless this order is observed, the methods of reasoning based upon it will of course fail; and, in some cases, false conclusions may be drawn by combining equations according to rules which presuppose it.

113. Condition that Three Right Lines shall meet in One Point.—If three lines $Ax + By + C = 0$, $A'x + B'y + C' = 0$, $A''x + B''y + C'' = 0$ pass through the same point, the co-ordinates of intersection for the first two must satisfy the equation to the third. Hence, (Art. 106,) the required condition is

$$A'' (BC' - B'C) + B'' (CA' - C'A) + C'' (AB' - A'B) = 0.$$

114. A condition often more convenient in practice, is derived from the principle of Arts. 107, 108. For if three right lines meet in one point, the equation to the third must be in the form of the equation to a right line passing through the intersection of the first and second. Therefore, supposing the three lines to be $L = 0$, $L' = 0$, $L'' = 0$, we can always find some three constants, l, m, and $-n$, such that

$$-nL'' = lL + mL',$$

where $-n$ may be either a positive or a negative quantity. Hence, the condition is

$$lL + mL' + nL'' = 0.$$

That is, three right lines meet in one point *when their equations*, upon being multiplied respectively by any three constants and added, *vanish identically*.

Remark.—For brevity, we shall often refer to three right lines passing through one point by the name of *convergents*.

115. Condition that a Movable Right Line shall pass through a Fixed Point.—Comparing Art. 101, Cor. 1 with Art. 91, I, Cor. 1, it is evident that the equation to a right line passing through a fixed point whose co-ordinates are $l : n$ and $m : n$, may be written

$$A (nx - l) + B (ny - m) = 0.$$

And comparing this with the general equation to a right line, we have

$$- (lA + mB) = nC.$$

The required condition is therefore

$$lA + mB + nC = 0.$$

Hence, *a movable right line passes through a fixed point, so long as the co-efficients in its equation suffer no change inconsistent with their vanishing when multiplied each by a fixed constant and added together.*

Corollary.—The criterion of this article may be otherwise taken as the *condition that any number of lines shall pass through one point.* For every line whose co-efficients satisfy the equation $lA + mB + nC = 0$, must pass through the point $l : n$, $m : n$.

116. In this connection also. Art. 108 furnishes us with a second condition. For, since we may always regard a fixed point as the intersection of two given right lines, the most general expression for a right line passing through a fixed point is

$$L + kL' = 0.$$

By writing L and L' in full, and collecting the terms, this becomes

$$(A + kA') x + (B + kB') y + (C + kC') = 0.$$

Now the condition that the line shall be movable is that k be indeterminate. Hence, *a movable right line passes through a fixed point whenever its equation involves an indeterminate quantity in the first degree.*

Corollary.—The co-ordinates of the fixed point may be found by throwing the given equation into the form $L + kL' = 0$. and solving $L = 0$ and $L' = 0$ for x and y.

117. A third condition may be obtained as follows: Suppose that we have an equation of the form

$$\left.\begin{aligned}(Ax' + By' + C)\, x + (A'\, x' + B'\, y' + C')\, y \\ + (A''x' + B''y' + C'')\end{aligned}\right\} = 0,$$

in which x', y' are the co-ordinates of any point on the line

$$Mx' + Ny' + P = 0.$$

By means of the latter relation, we can eliminate y' from the given equation, which will .then contain the indeterminate x' in the first degree. Therefore, *a movable right line passes through a fixed point whenever its equation involves in the* first degree *the co-ordinates of a point which moves along a given right line.*

118. If in any equation $Ax + By + C = 0$, $C : A$ or $C : B$ is constant, the corresponding line (Art. 91, II, Cor. 1) makes a constant intercept on one of the axes.

Hence, as a fourth condition, *a movable right line passes through a fixed point whenever its equation satisfies the relation* $C : A ==$ constant *or* $C : B =$ constant.

Corollary.—A particular case of this is, that the line passes through a fixed point when $C = 0$. And, in fact, (Art. 63) every such line does pass through the *origin.*

Remark.—It has seemed worth while to present the condition of this article separately, as it is often convenient in practice. It is obviously, however, only a particular case of Art. 116.

EXAMPLES ON THE RIGHT LINE.

I. NOTATION AND CONDITIONS.

119. In some of the exercises which follow, the student must use his judgment as to the selection of the axes. The labor of solving will be much lessened by a judicious choice. A few hints have been given where they seemed necessary.

1. Form the equations to the sides of a triangle, the co-ordinates of whose vertices are $(2, 1)$, $(3, -2)$, $(-4, -1)$.

2. The equations to the sides of a triangle are $x + y = 2$, $x - 3y = 4$ and $3x + 5y + 7 = 0$. find the co-ordinates of its vertices.

3. Form the equations to the lines joining the vertices of the triangle in Ex. 1 to the middle points of the opposite sides.

4. Form the equation to the line joining $x'y'$ to the point midway between $x''y''$ and $x'''y'''$: or, show that, in general, the equations to the lines from the vertices of a triangle to the middle points of the opposite sides are

$$(y'' +y'''-2y')x-(x'' +x'''-2x')y+(x'' y' -y'' x')+(x'''y' -y'''x')=0,$$

$$(y'''-y' -2y'')x-(x'''-x' -2x')y+(x'''y''-y'''x'')+(x' y'' -y' x'')=0,$$

$$(y' +y'' -2y''')x-(x' +x'' -2x''')y+(x' y'''-y' x''')+(x''y'''-y'y''')=0.$$

5. In the triangle of Ex. 1, form the equations to the perpendiculars from each vertex to the opposite side. What inference as to the shape of the triangle?

6. Prove that, in general, the equations to such perpendiculars are

$$(x'' -x''')x + (y'' -y''')y + (x' x'''+y' y''')-(x' x'' +y' y'')= 0.$$

$$(x'''-x')x + (y'''-y')y + (x'' x' +y'' y')-(x'' x'''+y'' y''')= 0,$$

$$(x' -x'')x + (y' -y'')y + (x'''x'' +y'''y'')-(x'''x' +y'''y')= 0.$$

7. Prove that the general equations to the perpendiculars through the middle points of the sides are

$$(x'' -x''') x + (y'' -y''')y = \tfrac{1}{2}(x''^2-x'''^2) + \tfrac{1}{2}(y''^2-y'''^2),$$

$$(x'''-x') x + (y'''-y')y = \tfrac{1}{2}(x'''^2-x'^2) + \tfrac{1}{2}(y'''^2-y'^2),$$

$$(x' -x'') x + (y' -y'')y = \tfrac{1}{2}(x'^2-x''^2) + \tfrac{1}{2}(y'^2-y''^2).$$

8. Find the angle between the lines $x + y = 1$ and $y - x = 2$. and determine their point of intersection.

9. Write the equation to any parallel of $x \cos a + y \sin a - p = 0$. Decide whether $x \sin a - y \cos a = p'$ or $x \sin a + y \cos a = p''$ may be parallel to it; and, if so, on what condition.

10. Taking for axes the sides a and b of any triangle, form the equation to the line which cuts off the mth part of each, and show that it is parallel to the base. What condition follows from this?

11. Prove that $y = constant$ is the equation to any parallel of the axis of x, and $x = constant$ the equation to any parallel of the axis of y, whether the axes are rectangular or oblique.

12. Two lines AB, CD intersect in O; the lines AC, BD join their extremities and meet in E; the lines AD, BC join their extremities and meet in F: required the condition that EF may be parallel to AB.

13. Form the equation to the line which passes through $(2, 3)$ and makes with the line $y = 3x$ an angle $\theta = 60°$.

14. Form the equation to the line which passes through $(2, -3)$ and makes with the line $3x - 4y = 0$ an angle $\theta = -45°$.

15. We have shown that, in rectangular axes, $-A : B = \tan a$; but that, in oblique axes, $-A : B = \sin a : \sin (\omega - a)$. Prove that, in *all* cases, $\tan a = A \sin \omega : (A \cos \omega - B)$.

16. Axes being oblique, show that two lines will make with the axis of x angles equal but estimated in opposite directions (one above, the other below) upon the condition

$$\frac{B}{A} + \frac{B'}{A'} = 2 \cos \omega.$$

17. Find the length of the perpendicular from $(3, -4)$ on the line $4x + 2y - 7 = 0$, when $\omega = 60°$. On which side of the line is the given point?

18. Find the length of the perpendicular from the origin on the line $a (x - a) + b (y - b) = 0$.

19. Given the equations to two parallel lines: to find the distance between them.

20. What points on the axis of x are at the distance a from the line $\dfrac{x}{a} + \dfrac{y}{b} = 1$?

21. Form the equation to the bisectors of the angles between $3x + 4y - 9 = 0$ and $12x + 5y - 3 = 0$; —— the equation to *any* right line passing through their intersection.

22. Prove that whether the axes be rectangular or oblique the lines $x + y = 0$, $x - y = 0$ are at right angles to each other, and bisect the supplemental angles between the axes. Show analytically that *all* bisectors of supplemental angles are mutually perpendicular.

23. Find the equation to the line passing through the intersection of $3x - 5y + 6 = 0$ and $2x + y - 8 = 0$, and striking the point $(5, 6)$.

24. Find the equation to the line joining the origin to the intersection of $Ax + By + C = 0$ and $A'x + B'y + C' = 0$.

25. Show that the equation to the line passing through the intersection of $Ax + By + C = 0$ and $A'x + B'y + C = 0$, and parallel to the axis of x, is $(AB' - A'B)y + (AC' - A'C) = 0$. Does this agree with the theorem of Ex. 11?

26. Find the equation to the line passing through the intersection of

$$x \cos a + y \sin a = p, \ x \cos \beta + y \sin \beta = p'$$

and cutting at right angles the line

$$x \cos \gamma + y \sin \gamma = p''.$$

27. Given any three parallel right lines of different lengths; join the adjacent extremities of the first and second, and produce the two lines thus formed until they meet; do the same with respect to the second and third, and the third and first: the three points of intersection lie on one right line.

28. Given the frustum of a triangle, with parallel bases: the intersection of its diagonals, the middle points of the bases, and the vertex of the triangle are on one right line. [Take vertex of triangle as origin, and sides for axes.]

29. On the sides of a right triangle squares are constructed; from the acute angles diagonals are drawn, crossing the triangle to the vertices of these squares; and from the right angle a perpendicular is let fall upon the hypotenuse: to prove that the diagonals and the perpendicular meet in one point. [Let the lengths of the sides be a and b, and take them for axes.]

30. Prove that the three lines which join the vertices of a triangle to the middle points of the opposite sides are convergents, taking for axes any two sides. [See also their equations above, Ex. 4.]

31. The three perpendiculars from the vertices on the sides, and the three that rise from the middle points of the sides, are each convergents. [See their equations, Exs. 6 and 7.]

32. The three bisectors of the angles in any triangle are convergents: for their equations are

$$(x \cos a + y \sin a - p\ ) - (x \cos \beta + y \sin \beta - p'\ ) = 0,$$
$$(x \cos \beta + y \sin \beta - p'\ ) - (x \cos \gamma + y \sin \gamma - p'') = 0,$$
$$(x \cos \gamma + y \sin \gamma - p'') - (x \cos a + y \sin a - p\ ) = 0,$$

if we suppose the origin to be *within the triangle.*

33. Through what point do all the lines $y = mx$, $Ax + By = 0$, $2x = 3y$, $x \cos a + y \sin a = 0$, $\rho \cos \theta = 0$ pass?

34. Decide whether the lines $2x - 3y + 6 = 0$, $4x + 3y - 6 = 0$, $5x - 5y + 10 = 0$, $7x + 2y - 4 = 0$, $x - y + 2 = 0$ pass through a fixed point.

35. Given the line $3x - 5y + 6 = 0$: form the equations to five lines passing through a fixed point, and determine the point.

36. Given three constants 2, 3, 5: form the equations to five lines passing through a fixed point, and determine the point.

37. Given the vertical angle of a triangle, and the sum or difference of the reciprocals of its sides: the base will move about a fixed point.

38. If a line be such that the sum of the perpendiculars, each multiplied by a constant, let fall upon it from n fixed points, is $= 0$, it will pass through a fixed point known as the *Center of Mean Position* to the given points.

The conditions of the problem (Art. 105, Cor. 1) give us

$$\left.\begin{array}{r} m' (x' \cos a + y' \sin a - p) + m'' (x'' \cos a + y'' \sin a - p) \\ + m''' (x''' \cos a + y''' \sin a - p) + \&c. \end{array}\right\} = 0;$$

or, putting $\Sigma(mx')$ as an arbitrary abbreviation for the sum of the mx's, $\Sigma(my')$ for the sum of the my's, and $\Sigma(m)$ for the sum of the m's,

$$\Sigma(mx') \cos a + \Sigma(my') \sin a - \Sigma(m) p = 0.$$

Solving for p, and substituting its value in $x \cos a + y \sin a - p = 0$, the equation to the movable line becomes

$$\Sigma(m)x - \Sigma(mx') + \tan a \left\{\Sigma(m)y - \Sigma(my')\right\} = 0:$$

which (Art. 116) proves the proposition. [Solution by Salmon.]

39. If the three vertices of a triangle move each on one of three convergents, and two of its sides pass through fixed points $x'y'$, $x''y''$. the third will also pass through a fixed point. [Take the two exterior convergents for axes]

40. If the vertex in which the two sides mentioned in Ex. 39 meet, does *not* move on a line convergent with the two on which the other vertices move, to find the condition that the third side may still pass through a fixed point.

II. RECTILINEAR LOCI.

120. The following examples illustrate the method of solving problems in which the path of a moving point is sought. When a point moves under given conditions. *we have only to discover what relation between its co-ordinates is implied in those conditions: then by writing this relation in algebraic symbols we at once obtain the equation to its locus.*

As the investigation of loci is one of the principal uses of Analytic Geometry, it is important that the student should early acquire skill in thus writing down any geometric condition. The relation between the co-ordinates is sometimes so patent as to require no investigation: for instance, if we were required to determine the locus of a point the sum of the squares on whose co-ordinates is constant, we should at once write the equation

$$x^2 + y^2 = r^2$$

and discover (Art. 25. II) that the locus is a circle. But as a general thing the relation must be developed from the conditions by means of the geometric or trigonometric properties which they imply. The process may be much simplified by a proper choice of axes. As a rule. the equations are rendered much shorter and easier to interpret by taking for axes *two prominent lines of the figure to which the conditions give rise.* Still, by taking axes distinct from the figure, we sometimes obtain equations whose symmetry more than compensates for their loss of simplicity. For instance, in the equations of Exs. 4, 6, and 7, when the first is obtained, the second and third can be written out at once by analogy.

1. Given the base of a triangle and the difference between the squares on its sides: to find the locus of its vertex.

Let us take for axes the base and a perpendicular at its middle point. Let the base $= 2m$, and the difference between the squares on the sides $= n^2$. Then, the co-ordinates of the vertex being x and y, we shall have $RP^2 = y^2 + (m + x)^2$ and $PQ^2 = y^2 + (m - x)^2$. That is, $y^2 + (m + x)^2 - \{ y^2 + (m - x)^2 \} = n^2$; or the equation to the required locus is

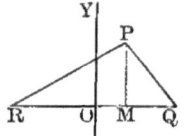

$$4\,mx = n^2.$$

Hence (Art. 25, I) the vertex moves upon *a right line perpendicular to the base.*

2. Find the locus of the vertex of a triangle, given the base and the sum of the cotangents of the base angles.

Using the same axes as in Ex. 1, and putting $\cot R + \cot Q = n$, we have from the diagram $\cot R = \dfrac{m + x}{y}$ and $\cot Q = \dfrac{m - x}{y}$. Hence, the equation to the required locus is

$$ny = 2m;$$

and (Art. 25, I) the vertex moves along *a line parallel to the base at a distance from it* $= 2m : n$.

3. Given the base and that $r \cot R \pm s \cot Q = p \pm q$, find the locus of the vertex.

4. Given the base and the sum of the sides, let the perpendicular to the base through the vertex be produced upward until its length equals that of one side: to find the locus of its extremity.

5. Given two fixed lines OM, ON; if any line MN parallel to a third fixed line OL intersect them: to find the locus of P where MN is cut in a given ratio, so that $MP = nMN$.

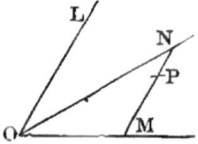

Here it will be most convenient to take for axes the two fixed lines OM, OL. Then, since N is a point on the line ON, we shall have $MN = mOM$; and, substituting in the given equation, we find the equation to the required locus, namely,

$$y = max.$$

Hence (Art. 78, Cor. 5) the point of proportional section moves on *a right line passing through the intersection of the two given lines.*

6. Find the locus of a point P, the sum of whose distances from OM, OL is constant.

7. A series of triangles whose bases are given in magnitude and position, and whose areas have a constant sum, have a common vertex : to find its locus.

8. Given the vertical angle of a triangle, and the sum of its sides: to find the locus of the point P where the base is cut in a given ratio, so that $mPX = nPY$.

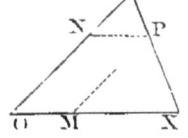

9. Determine the locus of P in the annexed diagram under the following successive conditions: PQ being perpendicular to OQ, and PR to $OR:$

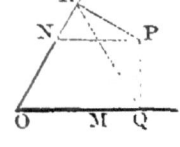

   I. $OQ + OR =$ constant.

   II. QR parallel to $y = mx$.

   III. QR cut in a given ratio by $y = mx + b$.

121. Hitherto we have found the equations to required loci by expressing the given conditions directly in terms of the variable co-ordinates. But it is often more convenient *to express them at first in terms of other lines, and then find some relation between these auxiliaries by which to eliminate the latter:* the result of eliminating will be an equation between the co-ordinates of the point whose locus is sought.

This process of forming an equation *by eliminating an indeterminate auxiliary* is extensively used, and is of especial advantage when investigating the intersections of movable lines.

An. Ge. 14.

10. Given the fixed point A on the axis of x, and the fixed point B on the axis of y; on the axis of x take any point A', and on the axis of y any point B', such that $OA' + OB' = OA + OB$: to find the locus of the intersection of AB' and $A'B$.

If $OA = a$, and $OB = b$ $\therefore$ $OA' = a + (k$ indeterminate$)$, and $OB' = b - k$. Hence, (Art. 79,) by clearing of fractions and collecting the terms, the equations to AB' and $A'B$ may be written

$$bx + ay - ab + k(a - x) = 0,$$
$$bx + ay - ab + k(y - b) = 0.$$

Subtracting, we eliminate the indeterminate k; and the equation to the required locus is

$$x + y = a + b.\,*$$

11. In a given triangle, to find the locus of the middle point of the inscribed rectangle.

12. In a given parallelogram, whose adjacent sides are a and b, to find the locus of the intersection of AB' and $A'B$: the lines AA', BB' being any two parallels to the respective sides.

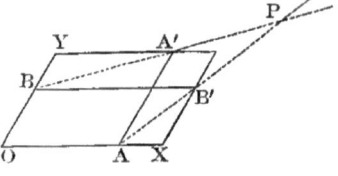

[The statement of this problem will apparently involve *two* indeterminate quantities; but both can be eliminated at one operation.]

13. A line is drawn parallel to the base of a triangle, and the points where it meets the two sides are joined transversely to the extremities of the base: to find the locus of the intersection of the joining lines.

14. Through any point in the base of a triangle is drawn a line of given length in a given direction: supposing it to be cut by the base in a given ratio, find the locus of the intersection of the lines joining its extremities to those of the base.

15. Given a point and two *fixed* right lines; through the point draw *any* two right lines, and join transversely the points where they meet the fixed ones: to find the locus of the intersection of the joining lines.

---

* See Salmon's *Conic Sections*, p. 46.

122. When we have to determine the locus of the extremity of a line drawn *through a fixed point* under given conditions, it is generally convenient to employ *polar co-ordinates.* We make the fixed point the *pole*, and then the distance from it to the extremity of the moving line becomes the *radius vector.*

16. Through a fixed point O is drawn a line OP, perpendicular to a line QR which passes through the fixed point Q : to find the locus of its extremity P, if $OP.OR =$ constant.

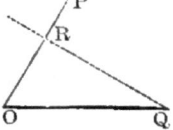

Let the distance $OQ = a$, and let $OP.OR = m^2$. From the diagram, $OR = OQ \cos ROQ$. Hence, the equation to the locus of P is

$$\rho\, a \cos \theta = m^2;$$

and (Art. 82, Cor.) P moves on *a right line perpendicular to* OQ.

17. One vertex of an equilateral triangle is fixed, and the second moves along a fixed right line : to find the locus of the third.

18. In a right triangle whose two sides are in a constant ratio, one acute angle has a fixed vertex, and the vertex of the right angle moves on a given right line: to find the locus of the remaining vertex.

19. Given the angles of *any* triangle: if one vertex is fixed, and the second moves on a given right line, to find the locus of the third.

[The student will readily perceive that Exs. 17 and 18 are particular cases of Ex. 19. He should trace this relation through the equations to the corresponding loci.]

20. Given the base of a triangle, and the sum of the sides; through either extremity of the base, a perpendicular to the adjacent side is drawn : to find the locus of its intersection with the bisector of the vertical angle.

Section III.—Pairs of Right Lines.

123. Since the equation of the first degree *always* represents a right line, there is but *one* locus whose equation is of the *first* degree. But there are *several* loci whose equations are of the *second* degree; and, in accordance with the principle of Art. 76, we shall consider these separately before discussing their relations to the locus of the Second order in general. In the case of each, we shall first obtain its equation, and then find the condition on which the *general* equation of the second degree will represent it. When this has been done for all of them, we can pass to the purely analytic ground, and show that the equation of the second degree *always* represents one of these lines.

We begin with the cases in which it represents *a pair of right lines.* These have a special interest, as furnishing the title by which the Right Line takes its place in the order of Conics.

I. GEOMETRIC POINT OF VIEW:—THE EQUATION TO A PAIR OF RIGHT LINES IS OF THE SECOND DEGREE.

124. Any equation in the type of $LMN\ldots = 0$ will obviously be satisfied by supposing either of the factors L, M, N, etc., equal to zero. If, then, $L = 0$, $M = 0$, $N = 0$, etc., are the equations to n different lines, their product will be satisfied by any values of x and y that satisfy *either* of them. That is, $LMN\ldots = 0$ will be satisfied by the co-ordinates of any point on *either* of the n lines: its locus is therefore the *group* of lines severally denoted by the separate factors. Hence, *we form the equation to a* group *of lines by multiplying together the equations to its constituents.*

125. Equation to a Pair of Right Lines.—By the principle just established, the required equation is

$$LL' = 0;$$

or, by writing the abbreviations in full, expanding, and collecting the terms,

$$A'A''x^2+A'B'' \begin{vmatrix} xy+B'B''y^2+A'C'' \\ +A''B' \end{vmatrix} \begin{vmatrix} x+B'C'' \\ +A''C'' \end{vmatrix} \begin{vmatrix} y+C'C'' \\ +B''C' \end{vmatrix} \Big\} = 0:$$

which is manifestly a particular case of the general equation of the second degree in two variables,

$$Ax^2 + 2Hxy + By^2 + 2Gx + 2Fy + C = 0, *$$

in which A, B, C, F, G, H are any six constants.

126. Equation to a Pair of Right Lines passing through a Fixed Point.—The equations to two right lines passing through the point $x'y'$ (Art. 101, Cor. 1) are

$$A'(x - x')+B' (y - y') = 0, \quad A''(x - x')+B''(y - y') = 0.$$

Hence, (Art. 124,) the required equation is

$$A'A''(x-x')^2+(A'B''+A''B')(x-x')(y-y')+B'B''(y-y')^2=0;$$

or, since $A'A''$, $A'B''+A''B'$, $B'B''$ are independent of each other,

$$A(x - x')^2 + 2H(x - x')(y - y') + B(y - y')^2 = 0.$$

Corollary 1.—The equation to a pair of right lines passing through the *origin* (Art. 49, Cor. 3) will be

$$Ax^2 + 2Hxy + By^2 = 0.$$

Corollary 2.—Of the two equations last obtained, the former is evidently homogeneous with respect to $(x - x')$ and $(y - y')$: and the latter, with respect to x and y. Hence, *Every homogeneous quadratic in* $(\mathbf{x} - \mathbf{x'})$ *and*

---

* The equation of the second degree in two variables is usually written

$$Ax^2 + Bxy + Cy^2 + Dx + Ey + F = 0.$$

I have followed Salmon in departing from this familiar expression. The sequel will show, however, that the new form imparts to the equations derived from it a simplicity and symmetry which far outweigh any inconvenience that may arise from its unfamiliar appearance.

$(y - y')$ *represents two right lines passing through the fixed point* x'y' ; and, *Every homogeneous quadratic in* x *and* y *represents two right lines passing through the origin.*

127. The equation $Ax^2 + 2Hxy + By^2 = 0$ deserves a fuller interpretation, as it leads to conditions which have a most important bearing on the relations of the Right Line to the Conics.

If we divide it by x^2, it assumes the form of a complete quadratic in $y : x$,

$$B\left(\frac{y}{x}\right)^2 + 2H\left(\frac{y}{x}\right) + A = 0 \qquad (1).$$

But (Arts. 124 ; 78, Cor. 5) it may also be written

$$\left(\frac{y}{x} - m\right)\left(\frac{y}{x} - m'\right) = 0 \qquad (2).$$

Hence, (Alg., 234, Prop. 2d,) its roots are *the tangents of the angles made with the axis of* x *by the two lines which it represents.* Now if we solve (1) for $y : x$, we obtain

$$\frac{y}{x} = -\frac{H \pm \sqrt{(H^2 - AB)}}{B} :$$

that is, the roots are real and *unequal* when $H^2 > AB$; real and *equal* when $H^2 = AB$; and *imaginary* when $H^2 < AB$. Therefore, if $H^2 - AB > 0$, the equation denotes two *real* right lines passing through the origin; if $H^2 - AB = 0$, two *coincident* ones; but in case $H^2 - AB < 0$, two *imaginary* ones.

Corollary.—The reasoning here employed is obviously applicable to the equation of Art. 126. The meaning of *any* homogeneous quadratic may therefore be determined according to the following table of corresponding analytic and geometric conditions :

$H^2 - AB > 0$ ∴. Two *real* right lines passing through a fixed point.

$H^2 - AB = 0$ ∴. Two *coincident* right lines passing through a fixed point.

$H^2 - AB < 0$ ∴. Two *imaginary* right lines passing through a fixed point.

Note.—By thus admitting the conception of coincident and imaginary lines as well as of real ones, we are enabled to assert that *every* quadratic which satisfies certain conditions represents *two* right lines. In fact, the result just obtained permits us to say that every equation between plane co-ordinates denotes a *line* (or *lines*), and to include in this statement such apparently exceptional equations as

$$(x - a)^2 + y^2 = 0.$$

Of this equation, we saw (Art. 61, Rem.) that the only *geometric* locus is the fixed point $(a, 0)$. But it is evidently homogeneous in $(x - a)$, $(y - 0)$ and fulfills the condition $H^2 - AB < 0$. We may therefore with greater *analytic* accuracy say that it denotes *two imaginary right lines passing through the point* $(a, 0)$; or, as we shall see hereafter, *two imaginary right lines whose intersection is the center* $(a, 0)$ *of an infinitely small circle.*

Such statements may seem to be mere fictions of terminology; but the farther we advance into our subject, the greater will appear the advantage of thus making the language of geometry correspond exactly to that of algebra. If we neglect to do so, we shall overlook many remarkable analogies among the various loci which we investigate.

128. Angle of a Pair of Right Lines.—From Art. 96 we have, as the expression for determining this,

$$\tan \varphi = \frac{m' - m}{1 + mm'}.$$

But we saw (Art. 127) that m and m' are the roots of the equation $Ax^2 + 2Hxy + By^2 = 0$. Hence,

$$m' - m = \frac{2\sqrt{(H^2 - AB)}}{B} \ ; \quad mm' = \frac{A}{B}.$$

Therefore $\tan \varphi = \dfrac{2\sqrt{(H^2 - AB)}}{A + B}$.

Corollary.—The *condition that a pair of right lines shall cut each other at right angles* is

$$A + B = 0.$$

Remark.—The student can readily convince himself that when the axes are oblique

$$\tan \varphi = \frac{2 \sin \omega \sqrt{(H^2 - AB)}}{A + B - 2H \cos \omega} .$$

129. Equation to the Bisectors of the angles between the Pair $Ax^2 + 2Hxy + By^2 = 0$.—The equation (compare Arts. 124; 109, Cor 1) will be

$$(A'x + B'y)^2 (A''^2 + B''^2) - (A''x + B''y)^2 (A'^2 + B'^2) = 0.$$

Expanding, collecting the terms, and dividing through by $A'B'' - A''B'$, this becomes

$$(A'B'' + A''B') x^2 - 2 (A'A'' - B'B'') xy - (A'B'' + A''B') y^2 = 0.$$

Comparing the co-efficients with those in the original form of the equation to the two given lines (Art. 126) we obtain

$$Hx^2 - (A - B) xy - Hy^2 = 0.$$

Corollary 1.—The co-efficients of this equation satisfy the condition (Art. 128, Cor.) $A + B = 0$, and show that the two bisectors are at right angles : which agrees with the result of Ex. 22, p. 123.

Corollary 2.—The equation to the bisectors is evidently a quadratic in $y : x$ of the first or second form : its roots are therefore *always* real, whether those of the equation to the given pair are real or not.   Hence we have the singular result, that *a pair of imaginary lines may have a real pair bisecting the angles between them.*   And it is a noticeable inference from the discussion in Art. 127, that *two imaginary lines may have a real point of intersection.*

Remark.—These two equations, $Ax^2 + 2Hxy + By^2 = 0$ and $Hx^2 - (A - B)xy - Hy^2 = 0$, merit the student's special attention. They will re-appear, in a somewhat unexpected quarter.

II. ANALYTIC POINT OF VIEW:—THE EQUATION OF THE SECOND DEGREE ON A DETERMINATE CONDITION REPRESENTS TWO RIGHT LINES.

130. This theorem is an immediate consequence of the method by which we form the equation to a pair of right lines. For, as that equation always originates in the expression

$$LL' = 0,$$

so, conversely, it must always be reducible to this form. Hence, *An equation of the second degree will represent two right lines whenever it can be resolved into two factors of the first degree.*

Corollary.—The same reasoning manifestly applies to $LMN... = 0$. Hence, an equation of *any* degree, which can be separated into factors of lower degrees, represents the assemblage of lines separately denoted by the several factors. In particular, *if an equation of the* n[th] *degree is separable into* n *factors of the first degree, it represents* n *right lines.*

131. We may express the condition just determined, in the form of a constant relation among the co-efficients in the general equation of the second degree.

The equation to any pair of right lines may be written

$$\{y - (m'x + b')\}\{y - (m''x + b'')\} = 0.$$

Hence, (Alg., 234. Prop. 2d.) in order that the equation of the second degree may represent two right lines, its roots must assume the form $y = mx + b$. Now if we solve $Ax^2 + 2Hxy + By^2 + 2Gx + 2Fy + C = 0$ as a complete quadratic in y, we obtain

An. Ge. 15.

$By = -(Hx+F) \pm \sqrt{\{(H^2-AB)x^2 + 2(HF-BG)x + (F^2-BC)\}}$;

and in order that this may assume the form $y = mx + b$, the expression under the radical must be a perfect square. But the condition for this (or, in other words, the condition that the general equation of the second degree may represent two right lines) is

$$(H^2 - AB)(F^2 - BC) = (HF - BG)^2 \qquad (1).$$

Expanding and reducing, we may write this in the striking symmetrical form

$$ABC + 2FGH - AF^2 - BG^2 - CH^2 = 0 \qquad (2).$$

Corollary.—It is evident from (1), that $H^2 - AB$ and $F^2 - BC$ will be *positive* together or *negative* together, but will not necessarily *vanish* together.

132. In the light of these results, it will be interesting to test the conditions $H^2 - AB > 0$, $H^2 - AB = 0$, $H^2 - AB < 0$ in the *general* equation to a pair of right lines.

From (1), it follows that the roots of this equation may be written

$$y = -\frac{(Hx+F) \pm \sqrt{\{(H^2-AB)x^2 + 2x\sqrt{(H^2-AB)(F^2-BC)} + (F^2-BC)\}}}{B};$$

hence (taking account of the preceding corollary) they will be *real and unequal*, when $H^2 > AB$; will *differ by a constant*, when $H^2 = AB$; and will be *imaginary*, when $H^2 < AB$. But these roots are *the ordinates of the two lines represented by the equation of Art.* 125: hence, in *any* equation of the second degree whose co-efficients satisfy the condition (2), we shall have the following criteria :

$H^2 - AB > 0 \therefore$ Two *real, intersecting* right lines.

$H^2 - AB = 0 \therefore$ Two *parallel* right lines.

$H^2 - AB < 0 \therefore$ Two *imaginary, intersecting* right lines.

Corollary.—If $H^2 - AB = 0$, and $F^2 - BC = 0$ at the same time, the two lines are *coincident.* Hence, *A right line formed by the coincidence of two others is the limit of two parallels.*

EXAMPLES.

1. Form the equation to the two lines passing through $(2, 3)$, $(4, 5)$ and $(1, 6)$, $(2, 5)$.

2. What locus is represented by $xy = 0$? By $x^2 - y^2 = 0$? By $x^2 - 5xy + 6y^2 = 0$? By $x^2 - 2xy \tan\theta - y^2 = 0$?

3. What lines are represented by $x^3 - 6x^2y + 11xy^2 - 6y^3 = 0$?

4. Show what loci are represented by the equations $x^2 + y^2 = 0$, $x^2 + xy = 0$, $x^2 + y^2 + a^2 = 0$, $xy - ax = 0$.

5. Interpret $(x - a)(y - b) = 0$, $(x - a)^2 + (y - b)^2 = 0$, and $(x - y + a)^2 + (x + y - a)^2 = 0$.

6. Show that $(y - 3x + 3)(3y + x - 9) = 0$ represents two right lines cutting each other at right angles.

7. Find the angles between the lines in Ex. 2.

8. What is the angle between the lines $x^2 + xy - 6y^2 = 0$?

9. Write the equation to the bisectors of the angles between the pair $x^2 - 5xy + 6y^2 = 0$.

10. Write the equations to the bisectors of the angles between the pairs $x^2 - y^2 = 0$ and $x^2 - xy + y^2 = 0$.

11. Show that the pair $6x^2 + 5xy - 6y^2 = 0$ intersect at right angles.

12. Show that $6x^2 + 5xy - 6y^2 = 0$ bisect the angles between the pair $2x^2 + 12xy + 7y^2 = 0$.

13. Verify that $x^2 - 5xy + 4y^2 + x + 2y - 2 = 0$ represents two right lines, and find the lines.

14. Show that $9x^2 - 12xy + 4y^2 - 2x + y - 3 = 0$ does *not* represent right lines, and find what value must be assigned to the coefficient of x in order that it may.

15. Show that $4x^2 - 12xy + 9y^2 - 4x + 6y - 12 = 0$ represents two parallel right lines, and find the lines.

16. Show that $4x^2 - 12xy + 9y^2 - 4x + 6y + 1 = 0$ denotes two coincident right lines, and find the line which is their limit.

17. Show that $5x^2 - 12xy + 9y^2 - 2x + 6y + 10 = 0$ represents two imaginary right lines, and find them.

18. Show that if A, B, C, L, N are any five constants, the equation to any pair of right lines may be written

$$(Ax + By + C)^2 = (Lx + N)^2;$$

that the lines are real, when $(Lx + N)^2$ is *positive;* imaginary, when $(Lx + N)^2$ is *negative;* parallel, when $L = 0$; and coincident, when L and N are both $= 0$.

19. Form an equation in the type of $Ax^2 + 2Hxy + By^2 + 2Gx + 2Fy + C = 0$ with numerical co-efficients, which shall represent two real right lines.

20. Form a similar equation representing two parallel lines, and one representing two coincident ones; also, one representing two imaginary lines.

Section IV.—The Circle.

I. GEOMETRIC POINT OF VIEW:—THE EQUATION TO THE CIRCLE IS OF THE SECOND DEGREE.

133. The Circle is distinguished by the following remarkable property : *The variable point of the curve is at a constant distance from a fixed point, called the center.*

134. Rectangular equation to the Circle.—If xy represent any point on the curve, and r its constant distance from the center gf, the required equation (Art. 51, I, Cor. 1) will be

$$(x - g)^2 + (y - f)^2 = r^2.$$

After expansion and reduction, this assumes the form

$$x^2 + y^2 - 2gx - 2fy + (g^2 + f^2 - r^2) = 0 :$$

which is evidently a particular case of the general equation

$$Ax^2 + 2Hxy + By^2 + 2Gx + 2Fy + C = 0.$$

Remark.—The annexed diagram will render clearer the geometric meaning of the equation just obtained. In this, we have $OM = x$, $MP = y$; $OM' = g$, $M'C = f$; and $CP = r$. Now, drawing PQ parallel to OX, we obtain by the Pythagorean theorem $PQ^2 + QC^2 = CP^2$. That is,

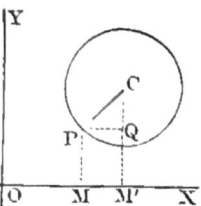

$$(x - g)^2 + (y - f)^2 = r^2.$$

135. Equation to the Circle, axes being oblique.—To obtain this, we have simply to express, in terms proper for oblique axes, the fact that the distance from xy to gf is constant. Hence, (Art. 51, I,) the equation is

$$(x - g)^2 + (y - f)^2 + 2(x - g)(y - f) \cos \omega = r^2;$$

or, in the expanded form,

$$x^2 + 2xy \cos \omega + y^2 - 2(g + f \cos \omega)x - 2(f + g\cos\omega)y + (g^2 - 2gf \cos\omega + f^2 - r^2) = 0.$$

136. Equation to the Circle, referred to rectangular axes with the Center as Origin.—Since the equation of Art. 134 is true for *any* origin, we have only to suppose in it $g = 0$ and $f = 0$, and the equation now sought is

$$x^2 + y^2 = r^2.$$

Remark.—The student will recognize this as the equation of Art. 25, II. Its great simplicity commends it to constant use. It may also be written

$$\frac{x^2}{r^2} + \frac{y^2}{r^2} = 1:$$

a form analogous to that of the equation to a right line,

$$\frac{x}{a} + \frac{y}{b} = 1.$$

137. Equation to the Circle, referred to any Diameter and the Tangent at its Vertex.—Since the diameter of a circle is perpendicular to its vertical tangent, in order to obtain the present equation we have merely to transform the last one to parallel axes through $(-r, 0)$. Writing, then, $x - r$ for x in $x^2 + y^2 = r^2$, and reducing, we find

$$x^2 + y^2 - 2rx = 0 \; ;$$

or,

$$y^2 = 2rx - x^2.$$

Remark.—In the equation just obtained, we suppose the origin to be at the *left-hand* vertex of the diameter. It is customary to adopt this convention. If the origin were at the *right-hand* vertex, we should have to replace the x of $x^2 + y^2 = r^2$ by $x + r$, and the equation would be

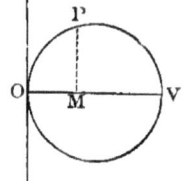

$$x^2 + y^2 + 2rx = 0; \text{ or, } y^2 = - (2rx + x^2).$$

The equation of this article is verified by the diagram. For (Geom., 325) $MP^2 = OM.MV$; that is,

$$y^2 = x(2r - x) = 2rx - x^2.$$

The form of this equation (Art. 63) shows that the origin is on the curve : which agrees with our hypothesis.

138. Polar Equation to the Circle.—The property that the distance from the center (d, a) to the variable point of the curve is constant, when expressed in polars, (Art. 51, II) gives us

$$\rho^2 + d^2 - 2\rho d \cos(\theta - a) = r^2.$$

Hence, the equation now sought is

$$\rho^2 - 2\rho d \cos(\theta - a) = r^2 - d^2.$$

Corollary 1.—Making $a = 0$ in the foregoing expression, we obtain

$$\rho^2 - 2\rho d \cos\theta = r^2 - d^2 :$$

the *polar equation to a circle whose center is on the initial line.*

Corollary 2.—Making $d = 0$, we obtain $\rho^2 = r^2$; or,

$$\rho = \text{constant}:$$

the *polar equation to a circle whose center is the pole.*

Remark.—We may verify these equations geometrically as follows: Let OX be the initial line, and C the center of the circle. Then $OP = \rho$, $XOP = \theta$; $OC = d$, $XOC = a$; and $CP = r$. But by Trig., 865, $OP^2 + OC^2 - 2OP \cdot OC \cos COP = CP^2$ Hence,

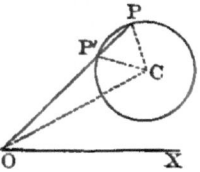

$$\rho^2 - 2\rho d \cos(\theta - a) = r^2 - d^2.$$

If the point C fell on OX, COP would become XOP; and we should have

$$\rho^2 - 2\rho d \cos\theta = r^2 - d^2.$$

If C coincided with O, OP would become CP; and we should have

$$\rho^2 = r^2; \text{ or. } \rho = \text{constant.}$$

These equations all imply that for every value of θ there will be *two* values of ρ: which is as it should be, since it is obvious from the diagram that the radius vector OP, corresponding to any angle XOP, cuts the curve in *two* points, P and P'.

EXAMPLES.

1. Form the equation to the circle whose center is $(3, 4)$, and whose radius $= 2$.

2. Lay down the center of the circle $(x - 2)^2 + (y - 6)^2 = 25$, and determine the length of its radius.

3. Do the same in the case of the circle $(x + 2)^2 + (y - 5)^2 = 1$; — in the case of the circle $x^2 - y^2 = 3$.

4. Form the equation to the circle whose center is $(5, -3)$ and whose radius $= \sqrt{7}$, when $\omega = 60°$.

5. Form the equation to the circle whose radius $= 3$, and whose center is $(3, 0)$. Transform the equation to the opposite vertex of the diameter.

6. Form the equation to the circle whose radius $= 6$, and whose center lies, at a distance from the pole $= 5$, on a line which makes with the initial line an angle $= 60°$.

7. Form the equation to the circle whose center is on the initial line at a distance of 16 inches from the pole, and whose radius $=$ 1 foot.

8. Transform the equation of Ex. 6 to rectangular axes, the origin being coincident with the pole.

9. Form the equation to the circle whose center is at the pole, and whose radius $= 3$. Transform it to rectangular axes, origin same as pole.

10. Form the equation to an infinitely small circle whose center is $(a, 0)$. [Compare the result with the Remark, Art. 61, and the Note, Art. 127.]

II. ANALYTIC POINT OF VIEW: — THE EQUATION OF THE SECOND DEGREE ON A DETERMINATE CONDITION REPRESENTS A CIRCLE.

139. The theorem is implied in the fact established in Arts. 134, 135, that the equation to the Circle is a particular case of the general equation of the second degree. To determine the condition on which the general equation will represent a circle, we have therefore merely to compare its co-efficients with those of the equation to the Circle written in its most general form.

140. We saw (Art. 134) that the rectangular equation to the Circle is

$$x^2 + y^2 - 2gx - 2fy + (g^2 + f^2 - r^2) = 0.$$

Since g and f may be either positive or negative, and r is not a function of either f or g, this may be written in the still more general form

$$A(x^2 + y^2) + 2Gx + 2Fy + C = 0.$$

If, then, the equation

$$A x^2 + 2Hxy + By^2 + 2Gx + 2Fy + C = 0$$

represent a circle, it must assume the form just obtained. But in order to this, we must have

$$H = 0 \text{ with } A = B:$$

which therefore constitutes the *condition that the general equation of the second degree shall represent a circle.*

Corollary.—It is obvious from the condition just determined, that when the equation of the second degree represents a circle, it also fulfills the condition

$$H^2 - AB < 0.$$

141. From the result of Art. 135, it follows that

$$H = A \cos \omega \text{ with } A = B$$

is the *condition in* oblique *axes that the equation of the second degree shall represent a circle.* And it is evident that in this case, too, we have the condition

$$H^2 - AB < 0.$$

142. If we are given an equation in the general form

$$A (x^2 + y^2) + 2Gx + 2Fy + C = 0,$$

we can at once *determine the position and magnitude of the corresponding circle.*

For, by transposing C, adding $\dfrac{G^2 + F^2}{A}$ to both members, and dividing through by A, the given equation may be thrown into the form

$$\left(x + \frac{G}{A} \right)^2 + \left(y + \frac{F}{A} \right)^2 = \frac{G^2 + F^2 - AC}{A^2}.$$

Comparing this with the equation of Art. 134, namely,

$$(x - g)^2 + (y - f)^2 = r^2,$$

we obtain, for determining the *co-ordinates of the center*,

$$g = -\frac{G}{A}, \quad f = -\frac{F}{A};$$

and for determining the *radius*,

$$r = \frac{1 \, '(G^2 + F^2 - AC)}{A}.$$

Corollary 1.—From the expressions for g and f, which are independent of C, we learn the important principle, that *the equations to concentric circles differ only in their constant terms.*

Corollary 2.—From the expression for r, we derive the following conclusions :

I. $G^2 + F^2 - AC > 0$ ∴ the circle is *real.*

II. $G^2 + F^2 - AC = 0$ ∴ the circle is *infinitely small.*

III. $G^2 + F^2 - AC < 0$ ∴ the circle is *imaginary.*

143. We may also determine the position and magnitude of the circle by finding its intercepts on the two axes.

Thus, if we make the y and x of the given equation successively $= 0$, we obtain the two equations

$$Ax^2 + 2Gx + C = 0, \quad Ay^2 + 2Fy + C = 0.$$

Since each of these is a quadratic, the circle *cuts each axis in two points;* and as a circle is completely determined by *three* points, its center and radius are *a fortiori* fixed by the *four* points thus found. We have therefore only to find the co-ordinates of the point equidistant from either *three* of the four, and we obtain the center : the distance between this and any *one* of the four, is the radius. Or we may proceed with greater brevity as follows :

Let the intercepts on the axis of x be x', x''; and those on the axis of y be y', y'': then (Alg., 234, **Props.** 3d and 4th) $2G : A = -(x' + x'')$, $2F : A = -(y' + y'')$, and $C : A = x'x'' = y'y''$. Hence, (Art. 142,) for determining the center,

$$g = -\frac{x' + x''}{2}, \quad f = \frac{y' + y''}{2};$$

and for determining the radius,

$$r = \tfrac{1}{2} \sqrt{(x'^2 + x''^2) + (y'^2 + y''^2)}.$$

Corollary.—Hence, the equation to any circle whose intercepts on the axis of x are given, is

$$x^2 + y^2 - (x' + x'') x - 2fy + x'x'' = 0;$$

the equation to any circle whose intercepts on the axis of y are given, is

$$x^2 + y^2 - 2gx - (y' + y'') y + y'y'' = 0:$$

and the equation to the circle whose intercepts on *both* axes are given, is

$$2(x^2 + y^2) - 2(x' + x'') x - 2(y' + y'') y + (x'x'' + y'y'') = 0.$$

144. Of the two equations

$$Ax^2 + 2Gx + C = 0, \quad Ay^2 + 2Fy + C = 0,$$

the first (Alg., 237, 1) will have equal roots when $G^2 = AC$; and the second, when $F^2 = AC$. Hence,

$$G^2 - AC = 0$$

is the *condition that a circle shall touch the axis of* x;

$$F^2 - AC = 0$$

is the *condition that it shall touch the axis of* y; and

$$G^2 = F^2 = AC$$

is the *condition that it shall touch* both *axes*.

EXAMPLES.

I. NOTATION AND CONDITIONS.

1. Decide whether the following equations represent circles:

$$3x^2 + 5xy - 7y^2 + 2x - 4y + 8 = 0;$$
$$5x^2 - 5y^2 + 3x - 2y + 7 = 0;$$
$$5x^2 + 5y^2 - 10x - 30y + 15 = 0.$$

2. Determine the center and radius of each of the circles

$$x^2 + y^2 + 4y - 4x - 1 = 0,$$
$$x^2 + y^2 + 6x - 3y - 1 = 0.$$

3. Form the equation to the circle which passes through the origin, and makes on the two axes respectively the intercepts $+ h$ and $+ k$.

4. Find the points in which the circle $x^2 + y^2 = 9$ intersects the lines $x + y + 1 = 0$, $x + y - 1 = 0$, and $2x + y \sqrt{5} = 9$.

5. What must be the inclination of the axes in order that each of the equations

$$x^2 - xy + y^2 - hx - hy = 0,$$
$$x^2 + xy + y^2 - hx - hy = 0,$$

may represent a circle? Determine the magnitude and position of each circle.

6. Write the equations to any three circles concentric with

$$2(x^2 + y^2) + 6x - 4y - 12 = 0.$$

7. Form the equation to the circle which makes on the axis of x the intercepts $(5, -12)$, and on the axis of y the intercepts $(4, -15)$. Determine the center and radius of the same.

8. What is the equation to the circle which touches the axes at distances from the origin, each $= a$? — at distances $= 5$ and 6 respectively?

9. ABC is an equilateral triangle: taking A as pole, and AB as initial line, form the polar equation to the circumscribed circle. Transform it to rectangular axes, origin same as pole, and axis of x as initial line.

10. If $S = 0$ and $S' = 0$ are the equations to any two circles, what does the equation $S - k^2S' = 0$ represent, k being arbitrary?

II. CIRCULAR LOCI.

1. Given the base of a triangle and the sum of the squares on its sides: to find the locus of its vertex.

Taking the base and a perpendicular through its middle point for axes, putting $2s^2 =$ the given sum of squares, and in other respects using the notation of Ex. 1, p. 126, we have

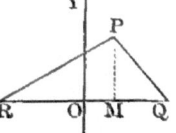

$$PR^2 = y^2 + (m+x)^2, \quad PQ^2 = y^2 + (m - x)^2.$$

Hence, the equation to the required locus is

$$x^2 + y^2 = s^2 - m^2;$$

and the vertex moves upon *a circle whose center is the middle point of the base, and whose radius* $= \sqrt{s^2 - m^2}$.

2. Given the base and the vertical angle of a triangle: to find the locus of the vertex.

3. Given the base and the vertical angle of a triangle: to find the locus of the center of the inscribed circle.

4. Find the locus of the middle points of chords drawn from the vertex of any diameter in a circle.

5. Given the base and the ratio of the sides of a triangle: to find the locus of the vertex.

6. Given the base and vertical angle: to find the locus of the intersection of the perpendiculars from the extremities of the base to the opposite sides.

7. When will the locus of a point be a circle, if the square of its distance from the base of a triangle bears a constant ratio to the product of its distances from the sides?

8. When will the locus of a point be a circle, if the sum of the squares of its distances from the three sides of a triangle is constant?

9. ABC is an equilateral triangle, and P is a point such that

$$PA = PB + PC:$$

find the locus of P.

10. ACB is the segment of a circle, and any chord AC is produced to a point P such that $AC = nCP$: to find the locus of P.

11. To find the locus of the middle point of any chord of a circle, when the chord passes through *any* fixed point.

12. On any circular radius vector OQ, OP is taken in a constant ratio to OQ : find the locus of P.

13. Find the locus of P, the square of whose distance from a fixed point O is proportional to its distance from a given right line.

14. O is a fixed point, and AB a fixed right line; a line is drawn from O to meet AB in Q, and on OQ a point P is taken so that $OQ.OP = k^2$: to find the locus of P.

15. A right line is drawn from a fixed point O to meet a fixed circle in Q, and on OQ the point P is so taken that $OQ.OP = k^2$: to find the locus of P.

Section V. — The Ellipse.

I. GEOMETRIC POINT OF VIEW :—THE EQUATION TO THE ELLIPSE IS OF THE SECOND DEGREE.

145. The Ellipse may be defined by the following property : *The sum of the distances from the variable point of the curve to two fixed points is constant.*

We may therefore trace the curve and discover its figure by the following process : — Take any two points F' and F, and fasten in them the extremities of a thread whose length is greater than $F'F$. Place the point of a pencil P against the thread, and slide it so as to keep the thread constantly stretched: P 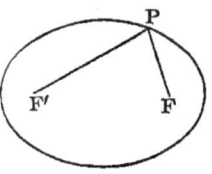 in its motion will describe an ellipse. For, in every position of P, we shall have

$$F'P + FP = \text{constant},$$

as the sum of these distances will always be equal to the fixed length of the thread.

146. The two fixed points, F' and F, are called the *foci;* and the distances with a constant sum, $F'P$ and FP, the *focal radii* of the curve.

The right line drawn through the foci to meet the curve in A' and A, is called the *transverse axis.* The point O, taken midway between F' and F, we may for the present call the *focal center.*

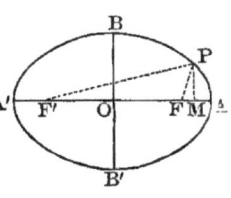

The line $B'B$, drawn through O at right angles to $A'A$, and terminated by the curve, is called the *conjugate axis.*

147. Equation to the Ellipse, referred to its Axes.—Let $2c =$ the constant distance between the foci, and $2a =$ the constant sum of the focal radii.

Then, from the diagram above, $F'P^2 = (x + c)^2 + y^2$, and $FP^2 = (x - c)^2 + y^2$. Hence, the fundamental property of the Ellipse, expressed in algebraic symbols, will be

$$\sqrt{\{(x + c)^2 + y^2\}} + \sqrt{\{(x - c)^2 + y^2\}} = 2a.$$

Freeing this expression from radicals, we obtain the required equation,

$$(a^2 - c^2)\, x^2 + a^2 y^2 = a^2\, (a^2 - c^2) \qquad (1).$$

To abbreviate, put $a^2 - c^2 = b^2$, and this becomes

$$b^2 x^2 + a^2 y^2 = a^2 b^2 \qquad (2):$$

which may be more symmetrically written

$$\frac{x^2}{a^2} + \frac{y^2}{b^2} = 1.$$

Remark.—The student will observe the analogy between the last form and the equation to the Right Line in terms of its intercepts.

148. It is important that we should get a clear conception of the *general* form of the equation just obtained. Let us for a moment return to the form (1),

$$(a^2 - c^2) x^2 + a^2y^2 = a^2 (a^2 - c^2).$$

In this, the definition of the Ellipse requires that $a^2 - c^2$ shall have the same sign as a^2; for the sum of the distances of a point from two fixed points can not be less than the distance between them: that is, a can not be numerically less than c, and consequently a^2 not less than c^2. Therefore, in the equation to the Ellipse, the co-efficients of x^2 and y^2 must have *like signs*.

Advancing now to the form (2),

$$b^2x^2 + a^2y^2 = a^2b^2,$$

it is obvious that the constant term a^2b^2 will have but one sign, whether the co-efficients of x^2 and y^2 be both *positive* or both *negative*. Supposing, then, that a^2 and b^2 are both *positive*, the equation after transposition would be

$$b^2x^2 + a^2y^2 - a^2b^2 = 0 \qquad (3).$$

Supposing them both *negative*, it would become, after transposition and the changing of its signs,

$$b^2x^2 + a^2y^2 + a^2b^2 = 0 \qquad (4).$$

Now, what is the meaning of the supposition that a^2 (and thence b^2) is negative? Plainly (since in that case we shall have, instead of a, $a \sqrt{-1}$) it signifies that the corresponding ellipse is *imaginary*.* Hence, admitting into our conception of this curve the imaginary locus of (4), we learn that the equation to the Ellipse is of the general form

$$A'x^2 + B'y^2 + C' = 0 :$$

---

* This interpretation is put beyond question by the form of equation (4) itself, which denotes an impossible relation.

in which A' and B' are *positive*, and C' is either *positive* or *negative* according as the curve is imaginary or real.

149. We may at this point derive from the equation to the Ellipse a single property of the curve, as we shall need it in discussing the general equation of the second degree.

Definition.—A **Center of a Curve** is a point which bisects every right line drawn through it to meet the curve.

Theorem.—*In any ellipse, the focal center is the center of the curve.* The equation to any right line drawn through the focal center (Art. 78, Cor. 5) is

$$y = mx.$$

Comparing this with the equation to any ellipse, namely,

$$\frac{x^2}{a^2} + \frac{y^2}{b^2} = 1,$$

we see that if x', y' satisfy both equations, $-x'$, $-y'$ will also satisfy both. In other words, the points in which the ellipse is cut by any right line drawn through the focal center may be represented by x', y' and $-x'$, $-y'$. But these symbols (Art. 51, I. Cor. 2) necessarily denote two points *equidistant* from the focal center: which proves the proposition.

150. Polar Equation to the Ellipse, the Center being the Pole.—Replacing (Art. 57, Cor.) the x and y of $a^2 y^2 + b^2 x^2 = a^2 b^2$ by $\rho \cos \theta$ and $\rho \sin \theta$, we find

$$\rho^2 = \frac{a^2 b^2}{a^2 \sin^2 \theta + b^2 \cos^2 \theta} :$$

that is, (Trig., 838.)

$$\rho^2 = \frac{a^2 b^2}{a^2 - (a^2 - b^2) \cos^2 \theta} .$$

An. Ge. 16.

Divide both terms of the second member of this expression by a^2, and, to abbreviate, put

$$\frac{a^2 - b^2}{a^2} = e^2 :$$

the result will be the form in which the required equation is usually written, namely,

$$\rho^2 = \frac{b^2}{1 - e^2 \cos^2\theta} .$$

Remark.—The equation indicates that for any value of θ there will be *two* radii vectores, numerically equal with contrary signs. The accompanying diagram verifies this result; for the two points of the ellipse, P and P', evidently correspond to the same angle θ, and the point P' has the radius vector $OP' = -OP$.

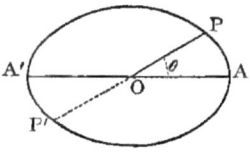

151. Special attention should be given to the two abbreviations used above,

$$a^2 - c^2 = b^2 \quad \text{and} \quad \frac{a^2 - b^2}{a^2} = e^2.$$

We shall find hereafter that they represent elements of great significance in the Conics. For the present, however, they are to be regarded as abbreviations *merely*. It is evident that by combining them we obtain the relation

$$c = ae.$$

Corollary.—Hence, the central polar equation to the Ellipse may be written

$$\rho^2 = \frac{a^2 (1 - e^2)}{1 - e^2 \cos^2\theta} :$$

a form which will frequently prove more convenient than that of Art. 150.

152. Polar Equation to the Ellipse, the Focus being the Pole.—From the annexed diagram, we have $F'P = \rho$, and $FP = \sqrt{(\rho^2 + 4c^2 - 4\rho c \cos \theta)}$. Hence, expressing the fundamental property of the Ellipse,

$$\rho + \sqrt{(\rho^2 + 4c^2 - 4\rho c \cos \theta)} = 2a :$$

$$\therefore \quad \rho = \frac{a^2 - c^2}{a - c \cos \theta}.$$

Replacing c by its equal ae, we obtain the usual form of the equation,

$$\rho = \frac{a(1 - e^2)}{1 - e \cos \theta}.$$

Remark.—The student should carefully discriminate between this equation and that of the corollary to Art. 151. From their striking similarity, the two are liable to be confounded.

In using this equation, it is to be remembered that in it the *left-hand* focus is taken for the pole. In practice, this assumption is generally found to be the more convenient. The student may show, however, that when the *right-hand* focus is the pole the equation to the Ellipse is

$$\rho = \frac{a(1 - e^2)}{1 + e \cos \theta}.$$

EXAMPLES.

1. Given the two points $(-3, 0)$ and $(3, 0)$; the extremities of a thread whose length $= 10$ are fastened in them: form the equation to the ellipse generated by pushing a pencil along the thread so as to keep it stretched.

2. In a given ellipse, half the sum of the focal radii $= 3$, and half the distance between the foci $= 2$: write its equation.

3. Form the equation to the ellipse whose focus is 3 inches from its center, and whose focal radii have lengths whose constant sum $= 1$ foot.

4. In a given ellipse, the sum of the focal radii $= 8$, and the difference between the squares of half that sum and half the distance between the foci $= 9$: write its equation.

5. Show that in each of the ellipses hitherto given, the focal center bisects the lines $y = 2x$, $y = 3x$, $y = 5x$, and any others whose equations are in the form $y = mx$.

6. Write the central polar equation to the ellipse in which the difference between the squares of half the sum of the focal radii and half the distance between the foci $= 9$, and the ratio of the distance between the foci to the sum of the focal radii $= 1 \,.\, 2$.

7. Write the central polar equation to the ellipse of Ex. 2.

8. Find, in the same ellipse, the ratio of the sum of the focal radii to the distance between the foci, and write the equation in the form corresponding to that in the corollary to Art. 151.

9. In a given ellipse, the sum of the focal radii $= 12$, and the ratio between that sum and the distance from the left-hand focus to the right-hand one $= 3$: write its polar equation, the focus being the pole.

10. The focus being the pole, form the polar equation to the ellipse of Ex. 3. What would the equation be, if the *right-hand* focus were the pole?

II. ANALYTIC POINT OF VIEW: — THE EQUATION OF THE SECOND DEGREE ON A DETERMINATE CONDITION REPRESENTS AN ELLIPSE.

153. We have seen (Art. 148) that the equation to the Ellipse may always be written in the form

$$A'x^2 + B'y^2 + C' = 0,$$

in which A' and B' are positive, and C' is indeterminate in sign. If, then, we can show that the general equation of the second degree is reducible to this form, and can find real conditions upon which the reduction may always be effected, we shall have established the theorem at the head of this article.

154. In order that the general equation of the second degree,

$$Ax^2 + 2Hxy + By^2 + 2Gx + 2Fy + C = 0,$$

may assume the form

$$A'x^2 + B'y^2 + C' = 0,$$

the co-efficients of its x, y, and xy must all vanish. The question therefore is : *Can we transform the equation to axes such as will cause these co-efficients to disappear?*

155. As we have seen, the equation $A'x^2 + B'y^2 + C' = 0$ is referred to the *center* of the Ellipse, and to its *axes*, which by definition cut each other at right angles. Assuming, then, as we may, that the general equation of the second degree as above written is referred to *rectangular* axes, our first step will naturally be *to determine, if possible, the center of the locus which it represents, and to reduce it to that center as origin.*

Let x', y' be the co-ordinates of the center sought, and let us transform

$$Ax^2 + 2Hxy + By^2 + 2Gx + 2Fy + C = 0 \qquad (1)$$

to parallel axes passing through $x'y'$. Replacing (Art. 55) the x and y of (1) by $(x' + x)$ and $(y' + y)$, we obtain, after reductions,

$$\left. \begin{array}{l} Ax^2 + 2Hxy + By^2 \\ + 2(Ax' + Hy' + G)x + 2(By' + Hx' + F)y \\ + Ax'^2 + 2Hx'y' + By'^2 + 2Gx' + 2Fy' + C \end{array} \right\} = 0.$$

Now, since this equation is referred to the center as origin, it must (Art. 149) be satisfied equally by x, y and $-x$, $-y$. But in order to this, we must have

$$Ax' + Hy' + G = 0, \quad By' + Hx' + F = 0.$$

Eliminating between these simultaneous equations, we find the *co-ordinates of the center,*

$$x' = \frac{BG - HF}{H^2 - AB}, \quad y' = \frac{AF - HG}{H^2 - AB}.$$

It is obvious that these values of x' and y' will be finite so long, and only so long, as $H^2 - AB$ is not equal to zero. Hence we conclude that the locus of (1) *has* a center, which is situated at a *finite* distance from the origin or at *infinity*, according as (1) does *not* or *does* fulfill the condition $H^2 - AB = 0$.

If, then, in the result of our first transformation above, we substitute these values of x' and y', we shall obtain an equation to the locus of (1), referred to its center. Now the only elements of that result which depend on x' and y', are the co-efficients of x and y, and the absolute term. Of these, the first two vanish, when the *finite* co-ordinates of the center are substituted in them; the third may be thrown into the form

$$(Ax' + Hy' + G)\, x' + (By' + Hx' + F)\, y' + (Gx' + Fy' + C):$$

and if in *this* we substitute the finite co-ordinates of the center, it becomes

$$\frac{AF^2 - 2FGH + BG^2}{H^2 - AB} + C \qquad (a);$$

or,

$$-\frac{ABC + 2FGH - AF^2 - BG^2 - CH^2}{H^2 - AB} \qquad (b).$$

Hence, putting C' to represent either (a) or (b), the *equation of the second degree, reduced to the* geometric *center of its locus,* is

$$Ax^2 + 2Hxy + By^2 + C' = 0 \qquad (2).$$

156. The transformation from (1) to (2) has destroyed the co-efficients of x and y, but the co-efficient of xy still remains. Reduction to the center as origin is therefore not sufficient to bring (1) into the form

$$A'x^2 + B'y^2 + C' = 0.$$

And, in fact, we might have anticipated as much; for the equation to the Ellipse, of which the required form is the type, is referred not to the center merely, but to the *axes of the curve.* To destroy the co-efficient of xy, then, we must resort to additional transformation: and our next step will naturally be *to determine, if possible, the axes of the locus represented by* (2), *and to revolve the reference-axes until they coincide with them.*

Let $\theta =$ the angle made with the reference-axes of (2) by the possible axes of the locus. Replacing (Art. 56, Cor. 3) the x and y of (2) by $x \cos \theta - y \sin \theta$ and $x \sin \theta + y \cos \theta$, we obtain, after reductions,

$$\left.\begin{array}{c} (A \cos^2 \theta + 2H \sin \theta \cos \theta + B \sin^2 \theta)\, x^2 \\ - 2\{(A-B)\sin \theta \cos \theta - H(\cos^2 \theta - \sin^2 \theta)\}\, xy \\ + (A \sin^2 \theta - 2H \sin \theta \cos \theta + B \cos^2 \theta)\, y^2 \end{array}\right\} + C' = 0:$$

that is, (Trig., 847: I, II, IV.)

$$\left.\begin{array}{c} \tfrac{1}{2}\{(A+B) - (A-B)\cos 2\theta - 2H \sin 2\theta\}x^2 \\ -\{(A-B)\sin 2\theta - 2H \cos 2\theta\}xy \\ + \tfrac{1}{2}\{(A+B) - (A-B)\cos 2\theta - 2H \sin 2\theta\}y^2 \end{array}\right\} + C' = 0.$$

If, in this expression, we equate to zero the co-efficient of xy, we shall have

$$(A - B)\sin 2\theta - 2H \cos 2\theta = 0:$$

$$\therefore \tan 2\theta = \frac{2H}{A-B} \qquad (c).$$

That is, since a tangent may have any value positive or negative from 0 to ∞, 2θ (and therefore θ) is a real angle; in other words, there do exist two real lines, at right angles to each other, which in virtue of their destroying the co-efficient of xy we may call *axes of the curve* to the locus of (2). Accordingly, if in the equation last obtained we substitute for the functions of 2θ their values as implied in (c),* the resulting equation will represent the locus, referred to these so-called axes.

From (c), $\sin 2\theta = \dfrac{2H}{\sqrt{\{(A-B)^2 + (2H)^2\}}}$;

$$\cos 2\theta = \dfrac{A-B}{\sqrt{\{(A-B)^2 + (2H)^2\}}} .$$

Substituting these values in our last equation, we find

$$\left. \begin{array}{l} \tfrac{1}{2}\{(A+B)+\sqrt{(A-B)^2+(2H)^2}\}\,x^2 \\ + \tfrac{1}{2}\{(A+B)-\sqrt{(A-B)^2+(2H)^2}\}\,y^2 \end{array} \right\} + C' = 0 :$$

whence, by writing

$$A' = \tfrac{1}{2}\{(A+B)+\sqrt{(A-B)^2+(2H)^2}\} \qquad (d),$$

$$B' = \tfrac{1}{2}\{(A+B)-\sqrt{(A-B)^2+(2H)^2}\} \qquad (e),$$

the *equation of the second degree, reduced to the axes of its locus,* is

$$A'x^2 + B'y^2 + C' = 0 \qquad\qquad (3).$$

---

* By Trig., 836, $\sin A = \dfrac{a}{c}$; $\cos A = \dfrac{b}{c}$. But since c represents the hypotenuse, and a and b the sides, of a right triangle: $c = \sqrt{a^2 + b^2}$. Hence, when the *tangent* is given, e. g. $\tan A = \dfrac{a}{b}$, we at once derive the *sine* and *cosine* by writing

$$\sin A = \dfrac{a}{\sqrt{(a^2+b^2)}} ; \quad \cos A = \dfrac{b}{\sqrt{(a^2+b^2)}} .$$

157. From (3) it appears, that, setting aside the question of *signs*, the general equation of the second degree *can* be reduced to the required form; provided it is not subject to the condition $H^2 - AB = 0$.

158. It remains, then, only to inquire what condition the general equation must fulfill in order that its reduced form (3) may have that combination of *signs* which (Art. 148) is characteristic of the Ellipse.

If A' and B' are both *positive*, we shall have

$$A'B' = \text{positive} ;$$

or, by substituting for A' and B' from (d) and (e) above, and reducing,

$$AB - H^2 = \text{positive} ;$$

that is, changing the signs of both sides of the expression,

$$H^2 - AB = \textit{negative.}$$

Hence, *The equation of the second degree represents an ellipse whenever its co-efficients fulfill the condition*

$$H^2 - AB < 0.$$

159. At the close of Art. 148 we saw that the sign of C' is *plus* or *minus* according as the ellipse is imaginary or real. Let us then seek the conditions which the general equation must fulfill in order to distinguish between these two states of the curve.

Applying the condition $H^2 - AB < 0$ to the value of C' as given in (b) of Art. 155, we see that, for C' to be *negative*, we must have

$$ABC - 2FGH - AF^2 - BG^2 - CH^2 < 0 ;$$

and, for C' to be *positive*,

$$ABC + 2FGH - AF^2 - BG^2 - CH^2 > 0.$$

An Ge. 17.

In other words, we find that the same quantity which (Art. 131) by *vanishing* indicates a pair of right lines as the locus of the general equation, by *changing sign* indicates the transition from the real to the imaginary ellipse. This quantity is called, in modern algebra, the *Discriminant* of the general equation; and we may appropriately represent it by the Greek letter $\varDelta$. Adopting this notation, we have

$$H^2 - AB < 0 \text{ with } \varDelta < 0$$

as the *condition that the equation of the second degree shall represent a* real *ellipse;* and

$$H^2 - AB < 0 \text{ with } \varDelta > 0$$

as the *condition that it shall represent an* imaginary one.*

160. We can now see, at least in part, the real bearing of the conditions in terms of $H^2 - AB$ which we some time ago developed respecting Pairs of Right Lines.

Comparing the results of Arts. 131, 132, we infer that

$$H^2 - AB < 0 \text{ with } \varDelta = 0 \qquad (l)$$

is the condition that the equation of the second degree shall represent two *imaginary intersecting* lines. But this condition evidently lies between the two criteria

$$H^2 - AB < 0 \text{ with } \varDelta < 0 \qquad (k),$$
$$H^2 - AB < 0 \text{ with } \varDelta > 0 \qquad (m);$$

so that we can not pass from (k) to (m) without passing through (l). We thus learn that *two imaginary right lines intersecting each other, form the limit between the real and the imaginary ellipse.*

If we now revert to the equation (Art. 132) denoting

---

* In testing any given equation by these criteria, we must see that its signs are so arranged that A (the co-efficient of x^2) may be *positive*. The conditions with respect to $\varDelta$, are derived on this assumption.

two right lines, and take its two roots separately, we see that the two lines are

$$(H + \imath\ \overline{H^2 - AB})\, x + By + (F + \imath\ \overline{F^2 - BC}) = 0,$$

$$(H - \imath\ \overline{H^2 - AB})\, x + By + (F - \imath\ \overline{F^2 - BC}) = 0.$$

Eliminating between these equations, and recollecting [Art. 131, (1)] that

$$F^2 - BC = \frac{(HF - BG)^2}{H^2 - AB},$$

we find, as the *co-ordinates of intersection* for the two lines,

$$x = \frac{BG - HF}{H^2 - AB}, \quad y = \frac{AF - HG}{H^2 - AB}.$$

That is, (Art. 155,) the lines intersect in *the center of the locus of the general equation.* But we have seen that this center is *real,* irrespective of the state of $H^2 - AB$: and is *finite,* so long as $H^2 - AB$ is not zero. Hence, whenever the equation of the second degree represents two intersecting lines, *their intersection is a finite real point, whether they be real or imaginary.*

Uniting the two conclusions thus reached, we obtain the following important theorem : *The Point, as the intersection of two imaginary right lines, is the limiting case of the Ellipse.*

Remark.—This result is corroborated by the equation (Art. 148) to the Ellipse itself. For if, in the expression

$$A'x^2 + B'y^2 + C' = 0,$$

we suppose $\Delta = 0$, then $C' = 0$, and the equation becomes

$$A'x^2 + B'y^2 = 0 :$$

which (Art. 126, Cor. 2 cf. Art. 127) denotes two imaginary lines passing through the origin; that is, in this case, through the center.

161. The Point and the Pair of Imaginary Intersecting Lines have thus been brought within the order of Conics. We shall now show that the Circle likewise belongs there.

The condition that the equation of the second degree shall represent a circle (Art. 140) is

$$H = 0 \text{ with } A = B.$$

But, as we noticed in the corollary to Art. 140, this is merely a special form of the condition

$$H^2 - AB < 0.$$

Hence, *the Circle is a particular case of the Ellipse.*

Resuming, then, the equation to the Ellipse, namely,

$$(a^2 - c^2)\, x^2 + a^2 y^2 = a^2 (a^2 - c^2),$$

we notice that it already fulfills the condition $H = 0$. Adding the condition $A = B$, necessary to make it represent a circle, we obtain, as characteristic of the Circle,

$$a^2 - c^2 = a^2 \; \therefore \; c = 0.$$

We hence learn that *the Circle is an ellipse whose two foci have become coincident at the center.*

Moreover, the Circle is real, vanishes, or is imaginary, on the same conditions as the Ellipse. For we saw (Art. 142, Cor. 2) that it assumes these several phases according as the quantity

$$G^2 + F^2 - AC$$

is positive, zero, or negative. Now, if we apply to the Discriminant J the conditions $H = 0$, $A = B$, we find, as true for the Circle,

$$- J = A \, (G^2 + F^2 - AC).$$

And since we are always to suppose A positive, we have

$$J < 0 \; \therefore \; \text{a } \textit{real} \text{ circle.}$$
$$J = 0 \; \therefore \; \text{a } \textit{point.}$$
$$J > 0 \; \therefore \; \text{an } \textit{imaginary} \text{ circle.}$$

Remark.—In allusion to the fact that its foci do *not* in general vanish in the center, the Ellipse may be called *eccentric.*

162. The following table exhibits the analytic conditions thus far imposed upon the equation of the second degree, with their geometric consequences:

$$H^2 - AB < 0$$
$$\therefore$$
Ellipse
$$\begin{cases} H = \pm, \ A - B = \pm \therefore \text{ Eccentric Curve } . \begin{cases} \text{Real } \cdot \cdot \ \Delta < 0. \\ \text{Point } \cdot \cdot \ \Delta = 0. \\ \text{Imag. } \cdot \cdot \ \Delta > 0. \end{cases} \\ H = 0, \ A - B = 0 \therefore \text{ Circle} . \ . \ . \ . \ . \begin{cases} \text{Real } \cdot \cdot \ \Delta < 0. \\ \text{Point } \cdot \cdot \ \Delta = 0. \\ \text{Imag. } \cdot \cdot \ \Delta > 0. \end{cases} \end{cases}$$

163. Theorems of Transformation.—Before advancing further, it will be well to collect from the foregoing discussion the theorems it implies respecting transformation of co-ordinates. They are often convenient in performing the reductions to which they relate.

Theorem I.—*In transforming any equation of the second degree to parallel axes through a new origin:*

1. *The variable terms of the second degree retain their original co-efficients.*

2. *The variable terms of the first degree obtain new co-efficients, which are linear functions of the new origin.*

3. *The constant term is replaced by a new one, which is the result of substituting the co-ordinates of the new origin in the original equation.*

For, in applying this transformation to equation (1) of Art. 155, the co-efficients of x^2, xy, and y^2 continued to be A, $2H$, and B; the co-efficients of x and y respectively replaced G and F by

$$Ax' + Hy' + G, \quad By' + Hx' + F;$$

and, for the new constant term, we obtained

$$Ax'^2 + 2Hx'y' + By'^2 + 2Gx' + 2Fy' + C.$$

Theorem II.—*In transforming any equation of the second degree from one set of rectangular axes to another, the quantities* $A + B$, $H^2 - AB$ *remain unaltered.*

For the equation near the middle of p. 157 is the result of this transformation; and if we add together the co-efficients of x^2 and y^2 in it, after representing them by A' and B', we obtain

$$A' + B' = A + B.$$

In like manner, representing the co-efficient of xy by $2H'$, and performing the necessary operations, we find

$$H'^2 - A'B' = H^2 - AB.$$

Theorem III.—*If, in the process of transforming an equation of the second degree, the co-efficients of* x *and* y *vanish, the new origin is the center of the locus.*

For, in that event, the new equation will be satisfied equally by x, y and $-x$, $-y$; that is, all right lines drawn through the new origin to meet the curve will be *bisected* by that origin.

Corollary.—*If only one of these co-efficients vanish, the new origin will lie on a right line passing through the center.* For we must then have either

$$Ax' + Hy' + G = 0 \text{ or } By' + Hx' + F = 0;$$

that is, the co-ordinates of the new origin must satisfy one of the equations (Art. 155) by eliminating between which we determined the center.

Theorem IV.—*If the co-efficient of* xy *vanish, the new reference-axes, if rectangular, will be parallel to the axes of the locus.*

For when in Art. 156 * this co-efficient vanished, with the *center* as origin, the new reference-axes *coincided* with the axes of the locus; hence, if the origin is at any other point, they must be *parallel* to them.

---

* As the beginner is liable to misapprehend the argument of Art. 156, it may be well to restate it, in the form which the present connection suggests :— When we revolved the reference-axes through the angle

$$\theta = \tfrac{1}{2} \tan^{-1} \frac{2H}{A - B},$$

(*which was found by equating the co-efficient of* xy *to zero*) we produced an equation (3) identical in form with that previously obtained for the Ellipse by referring it to its axes. So far then as concerned the Ellipse, the *new* reference-axes were identical with the two lines which (Art. 146) we had described as the axes of that curve. But on account of their power to reduce the *general* equation to a fixed form, these two lines were properly assumed to have a fixed relation to its locus, analogous to that which they bore to the Ellipse ; and hence were called the "axes" of that locus.

164. These theorems not only furnish criteria for selecting such transformations as will represent required geometric conditions, but they enable us to shorten the process of transformation.

Thus, knowing Theorem I, we can henceforth *write* the result of transforming to parallel axes, without going through with the ordinary substitutions.

Knowing Theorem II, we can immediately write the *central* equation of a second order curve from its *general* equation, by merely setting down the first three given terms and adding a new constant term found as in Theorem I, 3.

By uniting Theorems II and IV, we may shorten the process of reduction to the axes. For, if such a reduction is required, we shall have $H' = 0$; and, therefore, $A' + B' = A + B$ with $A'B' = AB - H^2$: two equations from which we can easily find A' and B'. C' is found as in Theorem I, 3. It is preferable, however, to write the reduced equation at once; for its form is $A'x^2 + B'y^2 + C'' = 0$; in which A', B' are found by formulæ (*d*) and (*e*) Art. 156, and C'' is obtained as before. When the given equation is already central, this reduction becomes very brief; since we do not then have to calculate C'.

EXAMPLES.

I. NOTATION AND CONDITIONS.

1. Determine by inspection the locus of each of the equations

$$2x^2 + 3y^2 = 12.$$
$$2x^2 + 3y^2 = 0.$$
$$2x^2 + 3y^2 = -12.$$

2. Transform $3x^2 + 4xy + y^2 - 5x - 6y - 3 = 0$ to parallel axes through $(2, 3)$. Is the curve an ellipse?

3. Reduce $x^2 + 2xy - y^2 + 8x + 4y - 8 = 0$ to the center. Does this represent an ellipse?

4. If in a given equation of the second degree $H = 0$, what condition must be fulfilled in order that the equation may represent an ellipse? What, if A or B equals zero?

5. Transform $14x^2 - 4xy + 11y^2 = 60$ to the axes of the curve, by all three methods. What is the origin in the given equation? What is the locus, and is the same locus indicated by the reduced equation?

6. Find the center of $5x^2 + 4xy + y^2 - 5x - 2y = 19$, and reduce the equation to it. Show that the curve is a real ellipse, both by the original equation and the reduced one.

7. Show that $14x^2 - 4xy + 11y^2 = 0$ denotes an infinitely small ellipse; that is, an ellipse in the limiting case.

8. Show that $5x^2 + 4xy + y^2 - 5x - 2y + 19 = 0$ represents an imaginary ellipse, and verify by writing the equation as referred to the axes.

9. Find the equation to the Ellipse, the origin $x'y'$ being on the curve, and the reference-axes parallel to the axes of the curve.

10. Find the equation to an ellipse whose conjugate axis is equal to the distance between its foci, taking for axes the two lines that join the extremities of the conjugate to the left-hand focus.

II. ELLIPTIC LOCI.

1. Find the locus of the vertex of a triangle, given the base and the product of the tangents of the base angles.

Taking the base and a perpendicular through its middle point for axes, calling the given product $l^2 : n^2$, and in other respects retaining the notation of Ex. 1, p. 126, we shall have

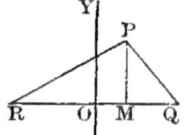

$$\tan R = \frac{y}{m + x}, \qquad \tan Q = \frac{y}{m - x}.$$

Hence, by the conditions of the problem, $\dfrac{y^2}{m^2 - x^2} = \dfrac{l^2}{n^2}$; and the equation to the required locus is

$$l^2 x^2 + n^2 y^2 = l^2 m^2.$$

Therefore, (Art. 147) *the vertex moves on an ellipse whose center is the middle point of the base, and whose foci are on the base at a distance from the center* $= m\sqrt{n^2 - l^2} : n$.

2. Find the locus of the vertex, when the base and the sum of the sides are given.

3 Find the locus of the vertex, given the base and the ratio of the sides.

4. Given the base, and the product of the tangents of the halves of the base angles: to find the locus of the vertex.

5. Two vertices of a *given* triangle move along two fixed lines which are at right angles: to find the locus of the third.

6. A right line of given length moves so that its extremities always lie one on each of two fixed lines at right angles to each other: to find the locus of a point which divides it in a given ratio.

7. In a triangle of constant base, the two lines drawn through the vertex at right angles to the sides make a constant intercept on the line of the base: find the locus of the vertex.

8. The ordinate of any circle $a^2 + \beta^2 = r^2$ is moved about its foot so as to make an oblique angle with the corresponding diameter: find the locus of its extremity in its new position.

9. The ordinate of any circle $x^2 + y^2 = r^2$ is augmented by a line equal in length to the corresponding abscissa: find the locus of the point thus reached.

10. To the ordinate of any circle there is drawn a line, from the vertex of the corresponding diameter, equal in length to the ordinate: find the locus of the point of meeting.

11. In any ellipse, find the locus of the middle point of a focal radius.

12. Find the locus of the extremity of an elliptic radius vector prolonged in a constant ratio.

13. A right line is drawn through a fixed point to meet an ellipse: find the locus of the middle point of the portion intercepted by the curve.

14. Through the focus of an ellipse, a line is drawn, bisecting the vectorial angle, and its length is a geometric mean of the radius vector and the distance from the focus to the center: find the locus of its extremity.

15. Through any point Q of an ellipse, a line is drawn parallel to the transverse axis, and upon it QP is taken equal to the corresponding focal radius: find the locus of P.

Section VI.—The Hyperbola.

I. GEOMETRIC POINT OF VIEW:—THE EQUATION TO THE HYPERBOLA IS OF THE SECOND DEGREE.

165. The Hyperbola is characterized by the following property : *The difference of the distances from the variable point of the curve to two fixed points is constant.*

Hence, we may trace the curve, and determine its figure, as follows : — Take any two points, as F' and F. At F', pivot the corner of a ruler $F'R$; at F, fasten one end of a thread, whose length is less than that of the ruler. Then, having attached the other end to the ruler at R, 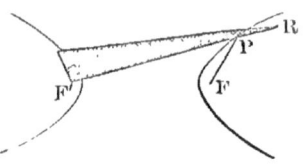 stretch the thread close against the edge of the ruler with the point of a pencil P. Move the ruler on its pivot, and slide the pencil along its edge so as to keep the thread continually stretched : the path of the pencil-point will be an hyperbola. For, in every position of P, we shall have

$$F'P - FP = (F'P + PR) - (FP + PR) = F'R - FPR.$$

That is, the difference of the distances from the variable point P to the two fixed points F' and F will always be equal to the difference between the fixed lengths of the ruler and the thread; or, we shall have

$$F'P - FP = \text{constant.}$$

By pivoting the ruler at F, and fastening the thread at F', we shall obtain a second figure similar in all respects to the former, except that it will face in the opposite direction. The complete curve therefore consists of two branches, as represented in the diagram.

166. The two fixed points, F' and F, are called the *foci* of the curve ; and the variable distances with a constant difference, $F'P$ and FP, are termed its *focal radii.*

The portion $A'A$ of the right line drawn through the foci, is called the *transverse axis.* The point O, taken midway between the foci, we shall for the present call the *focal center.*

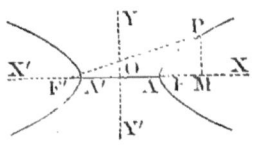

It is apparent from the diagram, that the right line $Y'Y$, drawn through the focal center at right angles to the transverse axis, does not meet the curve. We shall find, however, that a certain portion of it has a very significant relation to the Hyperbola, and is conventionally known as the *conjugate axis.* For the present, we shall speak of the whole line under that name.

167. Equation to the Hyperbola, referred to its Axes.—Putting $2c$ = the distance between the foci, and $2a$ = the constant difference of the focal radii, we shall have from the diagram above, $F'P^2 = (x+c)^2 + y^2$ and $FP^2 = (x-c)^2 + y^2$. The defining property of the Hyperbola will therefore be expressed by

$$\sqrt{\{(x+c)^2 + y^2\}} - \sqrt{\{(x-c)^2 + y^2\}} = 2a.$$

Clearing of radicals, we obtain

$$(c^2 - a^2)\,x^2 - a^2 y^2 = a^2\,(c^2 - a^2) \qquad (1);$$

and, by writing b^2 for $c^2 - a^2$ in order to abbreviate, the required equation becomes

$$b^2 x^2 - a^2 y^2 = a^2 b^2 \qquad (2):$$

which, on the analogy of the equations to the Right Line and the Ellipse, may be written

$$\frac{x^2}{a^2} - \frac{y^2}{b^2} = 1.$$

Corollary.—Hence we may regard the b^2 of the Hyperbola as the negative * of the b^2 of the Ellipse, and we infer the following principle : *Any function of* b *that expresses a property of the Ellipse, will be converted into one expressing a corresponding property of the Hyperbola by merely replacing its* b *by* b $\sqrt{-1}$.

Remark.—The relation thus suggested between these two curves will display itself completely when we come to discuss their properties. Results will continually occur, which give color to the fancy that an hyperbola is a *reversed ellipse.*

168. We must next, as in the case of the Ellipse, investigate the *general* form of the equation we have obtained.

Taking it up in the form (2), namely,

$$b^2x^2 - a^2y^2 = a^2b^2,$$

we speedily discover that a^2 and b^2 must have like signs ; for if their signs were unlike, the equation would assume one or the other of the forms

$$b^2x^2 + a^2y^2 - a^2b^2 = 0,$$
$$b^2x^2 + a^2y^2 + a^2b^2 = 0,$$

and thus would denote (Art. 148) not an *hyperbola,* but an *ellipse.* Hence, in the equation to the Hyperbola referred to its axes, the co-efficients of x^2 and y^2 must have *unlike signs.*

This condition may be fulfilled either by supposing a^2 and b^2 both positive or both negative. On the former supposition, the equation will be

$$b^2x^2 - a^2y^2 - a^2b^2 = 0 ;$$

---

* It must be remembered that b^2 is only an *abbreviation.* All that is meant, then, by the expression in the text is, that the operation for which b^2 stands in the Hyperbola is the reverse of the corresponding one in the Ellipse.

and on the latter, after changing its signs, it will become

$$b^2x^2 - a^2y^2 + a^2b^2 = 0.$$

What, now, is the geometric meaning of the variation in the sign of a^2b^2, which is thus brought into view? In the case of the Ellipse it indicated (Art. 148) the transition from the real to the imaginary state of the curve; and (as the transition to $+a^2b^2$ was made by replacing a and b by $a\sqrt{-1}$ and $b\sqrt{-1}$) we might suppose that it indicated the same thing here, were it not that a different conclusion is rendered certain by the following considerations.

Let us conceive of an hyperbola whose foci F' and F are at the same distance from their center O as those of the curve already considered, but lie upon the *conjugate* axis instead of the *transverse*. 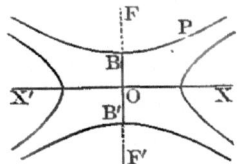 Then, retaining the same axes of reference as before, we shall evidently have, for the new positions of F' and F,

$$F'P^2 = x^2 + (c+y)^2, \quad FP^2 = x^2 + (c-y)^2.$$

Supposing now that, in addition, the constant difference of the focal radii in the new curve is $2b$ instead of $2a$, its equation will be

$$\sqrt{\{x^2 + (c+y)^2\}} - \sqrt{\{x^2 + (c-y)^2\}} = 2b;$$

or, after clearing of radicals,

$$(c^2 - b^2)\,y^2 - b^2x^2 = b^2(c^2 - b^2).$$

By substituting for $c^2 - b^2$ its value a^2, this becomes

$$a^2y^2 - b^2x^2 = a^2b^2,$$

and, by changing the signs and transposing,

$$b^2x^2 - a^2y^2 + a^2b^2 = 0:$$

which is precisely the expression we obtained above by supposing, in the equation to the original hyperbola, both a^2 and b^2 to be negative.

An hyperbola which thus has its foci on the conjugate axis of another, yet at the same distance apart, and whose a is the other's b, is said to be *conjugate* to the given one. We learn, therefore, that the equation to the Hyperbola conforms to the general type

$$A'x^2 + B'y^2 + C' = 0,$$

in which A' is *positive*, B' is *negative*, and C' is *negative* or *positive* according as the curve is primary or conjugate.

169. Theorem.—*In any hyperbola, the focal center is the center of the curve.*

For the equation obtained by taking the focal center as origin contains no variable terms except such as are of the second degree. But (Art. 163, Th. III) the origin for such an equation is the center of the curve.

170. Polar Equation to the Hyperbola, the Center being the Pole.—Changing $b^2x^2 - a^2y^2 = a^2b^2$ to polar co-ordinates, we find (Art. 57, Cor.)

$$\rho^2 = \frac{a^2b^2}{b^2 \cos^2 \theta - a^2 \sin^2 \theta} \, ;$$

or, (Trig., 838,)

$$\rho^2 = \frac{a^2b^2}{(a^2 + b^2) \cos^2 \theta - a^2} \, .$$

Dividing both terms of the second member by a^2, and putting

$$\frac{a^2 + b^2}{a^2} = e^2,$$

we obtain the usual form of the required equation, namely,

$$\rho^2 = \frac{b^2}{e^2 \cos^2 \theta - 1} \cdot$$

Remark.—Here, as well as in the case of the Ellipse, the equation implies that for any value of θ there are *two* radii vectores, numerically equal, with contrary signs.

Here, too, the equation is verified by the diagram; for the two points, P and P', obviously correspond to the same angle θ, if we fix the position of P' by the radius vector $OP' = -OP$.

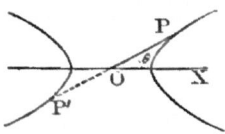

It is also worthy of notice, that this equation may be derived from the central polar equation to the Ellipse (Art. 150) by substituting $-b^2$ for b^2 in the latter.

171. The two abbreviations employed above, namely,

$$c^2 - a^2 = b^2 \text{ and } \frac{a^2 + b^2}{a^2} = e^2,$$

may evidently be derived from the two used in connection with the Ellipse (Art. 151), by substituting $-b^2$ for b^2. By combining them, however, we still obtain the relation

$$c = ae.$$

Corollary.—Hence, the central polar equation to the Hyperbola may be written

$$\rho^2 = \frac{a^2 (1 - e^2)}{1 - e^2 \cos^2 \theta} :$$

a formula which we leave the student to distinguish from that given for the Ellipse in the corollary to Art. 151.

172. Polar Equation to the Hyperbola, the Focus being the Pole.—From the annexed diagram, we have $F'P = \rho$, and $FP = \sqrt{(\rho^2 + 4c^2 - 4\rho c \cos \theta)}$. Hence, expressing the defining property of the Hyperbola,

$$\rho - \sqrt{(\rho^2 + 4c^2 - 4\rho c \cos \theta)} = 2a :$$

$$\therefore \rho = \frac{c^2 - a^2}{c \cos \theta - a}.$$

Replacing c by its equal ae, we obtain the usual form of the equation,

$$\rho = \frac{a(1 - e^2)}{1 - e \cos \theta}.$$

Remark.—The apparent identity of this equation with that of Art. 152, we leave the student to explain; and he may show that when the *right-hand* focus is the pole, the equation will be

$$\rho = \frac{a(e^2 - 1)}{1 - e \cos \theta}.$$

EXAMPLES.

1. Given the two points $(-3, 0)$ and $(3, 0)$; on the first is pivoted a ruler whose length $= 20$, and in the second is fastened a thread whose length $= 16$: form the equation to the hyperbola generated by means of this ruler and thread as in Art. 165.

2. In a given hyperbola, half the difference of the focal radii $= 2$, and half the distance between the foci $= 3$: write its equation. Why can not this example be derived from Ex. 2, p. 153, by merely substituting "difference" for "sum"?

3. Form the equation to the hyperbola whose focus is 1 foot from its center, and whose focal radii have the constant difference of 3 inches.

4. In a given hyperbola, the difference of the focal radii $= 8$, and the difference between the squares of half that difference and half the distance between the foci $= -9$: write its equation.

5. Write the equations to the hyperbolas which are the conjugates of those preceding.

6. Write the central polar equation to the hyperbola in which the squares of half the difference of the focal radii and half the distance between the foci differ by — 9, and the ratio of the distance between the foci to the difference of the focal radii = 2 : 1.

7. Write the central polar equation to the hyperbola of Ex. 2.

8. Find, in the same hyperbola, the ratio which the difference of the focal radii bears to the distance between the foci. and write the equation in the form given in the corollary to Art. 171.

9. In a given hyperbola, the difference of the focal radii = 12, and the ratio of that difference to the distance between the foci = 1 : 3. Write its polar equation, the focus being the pole.

10. The focus being the pole, form the equation to the hyperbola of Ex. 3. What would this be, if the *right-hand* focus were the pole ?

II. ANALYTIC POINT OF VIEW : — THE EQUATION OF THE SECOND DEGREE ON A DETERMINATE CONDITION REPRESENTS AN HYPERBOLA.

173. To establish this theorem, we must show (Art. 168) that the general equation of the second degree is reducible to the form

$$A'x^2 + B'y^2 + C' = 0,$$

in which A' is positive. and B' negative; and we must be able to find real conditions upon which the reduction can always be effected.

From (3) of Art. 156, we already know that, apart from the question of *signs*, the general equation is reducible to the required form. It only remains, then, to determine the condition which must be fulfilled in order that the signs of A' and B' may be such as characterize the Hyperbola.

An. Ge. 18.

174. If A' is positive, and B' negative, the product $A'B'$ must be negative. Therefore (see Art. 158) for the Hyperbola we shall have

$$AB - H^2 = \text{negative};$$

or, after changing the signs throughout,

$$H^2 - AB = positive.$$

Hence, *The equation of the second degree represents an hyperbola whenever its co-efficients fulfill the condition*

$$H^2 - AB > 0.$$

175. Let us now inquire what additional criteria the equation must satisfy, in order to distinguish between a primary hyperbola and its conjugate.

For the reduced form of the equation, (Art. 168,) the curve is primary or conjugate according as C' is negative or positive. But [Art. 155, (*b*)] we have

$$C' = -\frac{\varDelta}{H^2 - AB}:$$

in the case of the Hyperbola therefore, since $H^2 - AB$ is positive, C' will be *negative* when $\varDelta$ is *positive*, and *positive* when $\varDelta$ is *negative*. Hence,

$$H^2 - AB > 0 \text{ with } \varDelta > 0$$

is the *condition that the equation of the second degree shall represent a* primary *hyperbola;* and

$$H^2 - AB > 0 \text{ with } \varDelta < 0$$

is the *condition that it shall represent a* conjugate *one.*

176. The limit separating the two conditions just determined, is evidently

$$H^2 - AB > 0 \text{ with } \Delta = 0.$$

But (Arts. 131, 132; cf. 160) this is the condition that the equation of the second degree shall represent two real right lines intersecting in the center of its locus.

Hence, *Two real right lines, intersecting in its center, form the limiting case of the Hyperbola.*

Remark.—The student may verify this by applying the condition $\Delta = 0$ to the equation $A'x^2 - B'y^2 + C' = 0$.

177. In Art. 161, we found the Circle to be a particular case of the Ellipse. We shall now see that the Hyperbola has an analogous case.

One way of satisfying the criterion of the Hyperbola is, to have in the general equation the condition

$$A + B = 0.$$

For then $H^2 - AB$ will become $H^2 + A^2$: which is necessarily positive. But if $A + B = 0$, then (Art. 163, Th. II) $A' + B' = 0$; and, after dividing through by A', the equation referred to the axes will become

$$x^2 - y^2 = \text{constant}:$$

an expression denoting an hyperbola, by the condition just established, and strongly resembling the equation to the Circle.

$$x^2 + y^2 = \text{constant}.$$

Corollary.—Suppose now we push *this* hyperbola to its limiting case. Its equation will of course continue to fulfill the condition $A + B = 0$, and the corresponding pair of right lines will therefore (Art. 128, Cor.) intersect *at right angles.* Accordingly, this curve is known as the *Rectangular Hyperbola.*

178. The following table presents, in their proper subordination, the several conditions by which we may distinguish the varieties of the Hyperbola in the equation of the second degree:

$$H^2 - AB > 0 \; \therefore \; \text{Hyperbola} \begin{cases} A + B = \pm \;\; .. \;\; \text{Oblique} \; . \; . \begin{cases} \text{Primary} \; \cdot \cdot \; \Delta > 0. \\ \text{Two Intersect. Lines} \; \cdot \cdot \; \Delta = 0. \\ \text{Conjugate} \; \cdot \cdot \; \Delta < 0. \end{cases} \\ A + B = 0 \;\; \therefore \;\; \text{Rectangular} \begin{cases} \text{Primary} \; \cdot \cdot \; \Delta > 0. \\ \text{Two Perpendiculars} \; \cdot \cdot \; \Delta = 0. \\ \text{Conjugate} \; \cdot \cdot \; \Delta < 0. \end{cases} \end{cases}$$

By comparing this table with that of Art. 162, the student will see the truth of the Remark under Art. 167.

EXAMPLES.

I. NOTATION AND CONDITIONS.

1. Determine whether the following equations represent hyperbolas:

$$3x^2 - 8xy + 5y^2 - 6x + 4y - 2 = 0,$$
$$x^2 + 2xy - y^2 + x - 3y + 7 = 0,$$
$$5x^2 - 12xy - 7y^2 + 8x - 10y + 3 = 0.$$

2. Show that $2x^2 - 12xy + 5y^2 - 6x + 8y - 9 = 0$ represents an oblique primary hyperbola.

3. Show that $3x^2 + 8xy - 3y^2 + 6x - 10y + 5 = 0$ represents a rectangular conjugate hyperbola.

4. Show that $3x^2 + 8xy - 3y^2 + 6x + 10y - 5 = 0$ represents a rectangular primary hyperbola.

5. Verify the proposition of Ex. 2 by reducing the equation to the axes of the curve.

6. Verify the propositions of Ex. 3 and 4 in the same manner.

7. Given the hyperbola $5x^2 - 6y^2 = 30$: form the equation to its conjugate, and find the quantities a, b, c, and e.

8. Show that $2x^2 + xy - 15y^2 - x + 19y - 6 = 0$ denotes an oblique hyperbola in its limiting case, and find the corresponding center.

9. Show that $3x^2 - 8xy - 3y^2 + x + 17y - 10 = 0$ denotes a rectangular hyperbola in its limiting case, and find the center.

10. Transform the two equations last given, to the centers of their respective curves.

II. HYPERBOLIC LOCI.

1. Given the base of a triangle, and the difference of the angles at the base: to find the locus of the vertex.

Since the difference of the base angles is given, the tangent of their difference is given. Let us call it $= \dfrac{a}{h}$. Then, using the axes and notation of Ex. 1, p. 126, we shall have

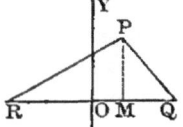

$$\frac{\dfrac{y}{m-x} - \dfrac{y}{m+x}}{1 + \dfrac{y^2}{m^2 - x^2}} = \frac{a}{h}\; ;\; \text{or,}\; \frac{2xy}{m^2 - x^2 + y^2} = \frac{a}{h}.$$

Hence, the equation to the required locus is

$$ax^2 + 2hxy - ay^2 = am^2\;;$$

and the vertex moves (Art. 177) on *a rectangular hyperbola whose center* [Art. 155, (2)] *is the middle point of the base, and whose transverse axis* [Art. 156, (c)] *is inclined to the base at an angle* $\theta = \dfrac{1}{2}\tan^{-1}\dfrac{h}{a}$: *that is, at an angle = half the complement of the given difference between the base angles.*

2. Given the base of a triangle, and the difference between the tangents of the base angles: to find the locus of the vertex.

3. Find the locus of the vertex, given the base of a triangle, and that one base angle is double the other.

4. Find the locus of the vertex in an isosceles triangle, when the extremity of one equal side is fixed, and the other equal side passes through a fixed point.

5. Given the vertical angle of a triangle, and also its area: find the locus of the point where the base is cut in a given ratio.

6. Find the locus of a point so situated in a given angle, that, if perpendiculars be dropped from it upon the sides of the angle, the quadrilateral thus formed will be of constant area.

7. Given a fixed point and a fixed right line: to find the locus of P, from which if there be drawn a right line to the fixed point and a perpendicular to the fixed line, they will make a constant intercept on the latter.

8. In the annexed diagram, QP is perpendicular to OQ, and RP to OR: find the locus of P, on the supposition that QR is constant.

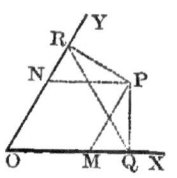

9. Supposing that QR in the same diagram passes through a fixed point, find the locus of the intersection of two lines drawn through Q and R parallel respectively to OR and OQ.

10. QR is a line of variable length, revolving upon the fixed point $\alpha\beta$: find the locus of the center of the circle described about the triangle ORQ.

11. QR moves between OQ and OR so that the area of the triangle ORQ is constant: find the locus of the center of the circumscribed circle.

12. A circle cuts a constant chord from each of two intersecting right lines: find the locus of its center.

13. Find the locus of the middle point of any hyperbolic focal radius.

14. From the extremity of any hyperbolic focal radius a line is drawn, parallel to the transverse axis and equal in length to the radius: find the locus of its extremity.

15. The ordinate of an hyperbola is prolonged so as to equal the corresponding focal radius: find the locus of the extremity of the prolongation.

Section VII.—The Parabola.

I. GEOMETRIC POINT OF VIEW:—THE EQUATION TO THE PARABOLA IS OF THE SECOND DEGREE.

179. The Parabola may be defined by the following property: *The distance of the variable point of the curve from a fixed point is equal to its distance from a fixed right line.*

We may therefore trace the curve and find its figure as follows: — Take any point F, and draw any right line $D'D$. Along the latter, fix the edge of a ruler; and in the former, fasten one end of a thread whose length is equal to that of a second ruler RD, which is right-angled at D. Then, having attached the other end to this ruler at R, keep the thread stretched against the edge RD with the point P of a pencil, while the ruler is slid on its edge QD along $D'D$ toward F: the path of P will be a parabola. For, in every position of P, we shall have

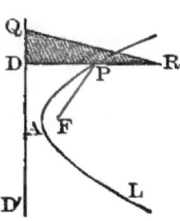

$$FP = PD,$$

as these distances will in all cases be formed by subtracting the same length RP from the equal lengths of the thread and ruler.

180. The fixed point F is called the *focus* of the parabola, and the fixed line $D'D$ its *directrix*.

The line OF, drawn through the focus at right angles to the directrix, and extending to infinity, is called the *axis* of the curve. The point A, where the axis cuts the curve, is termed the *vertex*.

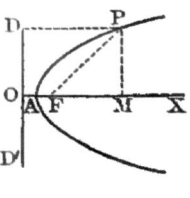

We shall refer to the distance FP under the name of the *focal radius*.

181. Equation to the Parabola, referred to its Axis and Directrix.—By putting $2p =$ the constant distance of the focus from the directrix, we shall have, in the above diagram, $FP = \sqrt{\{(x - 2p)^2 + y^2\}}$ and $PD = x$. The algebraic expression for the defining property of the Parabola will therefore be

$$\sqrt{\{(x - 2p)^2 + y^2\}} = x.$$

Clearing of radicals, and reducing, we find the required equation,

$$y^2 = 4p \, (x - p).$$

182. Let us now investigate the *general* type to which this equation conforms. It may evidently be written

$$y^2 - 4px + 4p^2 = 0 \tag{1},$$

and is therefore a particular case of the general equation

$$B'y^2 + 2G'x + C' = 0 \tag{2},$$

in which B', G', C' are any three constants whatever. Accordingly our real object is, to determine whether *every* equation in the type of (2) represents a parabola. We may settle this point as follows:

Let us transform (1) to parallel axes whose origin is somewhere on the primitive axis of x, say at the distance x' from the given origin. To effect this, we merely replace the x of (1) by $x' + x$, and thus obtain

$$y^2 - 4px + 4p \, (p - x') = 0 \tag{3}.$$

Now, since (1) represents a parabola, (3) also does. But in (3), since x' is arbitrary, the absolute term may have any ratio whatever to the co-efficient of x. Moreover, by taking x' of the proper value, we can render the absolute term positive or negative at pleasure; and, by supposing $2p$ susceptible of the double sign, we shall accomplish the same with respect to the co-efficient of x. By carrying out these suppositions, and then multiplying the whole equation by some arbitrary constant, we can give it three co-efficients which will be entirely arbitrary, and may therefore write it

$$B'y^2 + 2G'x + C' = 0.$$

To show, then, that *every* equation in the type of (2) represents a parabola, we have only to prove that $2p$ may be either positive or negative, without affecting the form of the curve represented by (1).

Now this supposition is clearly correct; for (see diagram, Art. 179) a negative value of $2p$ merely indicates that the focus is taken on the *left* of $D'D$, instead of on the *right:* while, by using the thread and ruler on the left of the directrix, we can certainly describe a curve similar in all respects to PAL, except that it will face in the opposite direction.

Hence we conclude that the equation to the Parabola may always be written in the form

$$B'y^2 + 2G'x + C' = 0,$$

B', G', C' being any three constants whatever: and, conversely, that every equation of this form represents a parabola.

183. Polar Equation to the Parabola, the Focus being the Pole.—From the annexed diagram, we have $FP = \rho$ and $OM = OF + FM = 2p + \rho \cos \theta$. Accordingly, the polar expression for the fundamental property of the Parabola will be

$$\rho = 2p + \rho \cos \theta.$$

The required equation is therefore

$$\rho = \frac{2p}{1 - \cos \theta}.$$

Remark.—This expression implies that the angle θ is measured from FX toward the *left*. The student may show that, supposing θ to be estimated from FO toward the *right*, the equation to the Parabola will be

$$\rho = \frac{2p}{1 + \cos \theta}.$$

An. Ge 19.

184. To exhibit in part the analogy of the equation just obtained to those of the Ellipse and Hyperbola in Arts. 152, 172, let us agree to write

$$e = 1$$

as an abbreviation characteristic of the Parabola, and analogous to those adopted in Arts. 150, 170 for the other two curves. We may then write (see also Art. 627)

$$\rho = \frac{2p}{1 - e \cos \theta}.$$

Corollary.—Adopting the convention last suggested, we may arrange the abbreviations referred to, according to their numerical order, thus:

Ellipse   . . . . $e < 1.$

Parabola   . . . . $e = 1.$

Hyperbola . . . . $e > 1.$

EXAMPLES.

1. Given the points $(4, 0)$, $(1, 0)$, $(3, 0)$: write the rectangular equations to the three parabolas of which they are the foci.

2. Write the rectangular equation to the parabola whose focus is the point $(-3, 0)$.

3. Transform the equations just found to parallel axes passing through the foci of their respective curves.

4. What are the positions of the foci with respect to the directrices, in the parabolas $y^2 = 4(x - 4)$, $4y^2 = -3(4x + 3)$, and $5y^2 - 6x + 9 = 0$?

5. Write the focal polar equation to the parabola whose focus is 2 feet distant from the directrix, and find the length of its radius vector when $\theta = 90°$. Also, write the polar equation to any parabola, the pole being at the intersection of the axis and directrix.

II. ANALYTIC POINT OF VIEW: — THE EQUATION OF THE
SECOND DEGREE ON A DETERMINATE CONDITION
REPRESENTS A PARABOLA.

185. To establish this theorem, we must show that
there are real conditions upon which the general equation
of the second degree may always be reduced to the form
(Art. 182)

$$B'y^2 + 2G'x + C' = 0.$$

186. In the investigation on which we are about to
enter, *we must confine our attention to those equations of
the second degree whose co-efficients fulfill the condition*

$$H^2 - AB = 0.$$

For we have already proved (Arts. 158, 174) that
every equation of the second degree in which $H^2 - AB$
is *not* equal to zero, represents either an ellipse or an
hyperbola.

187. Further: The restriction just established carries
with it the additional one, that, *in the equations we are
permitted to consider, the condition*

$$\varDelta > 0$$

can not occur.

For, reverting to the general value of the Discriminant (Art. 159), we have

$$\varDelta = ABC + 2FGH - AF^2 - BG^2 - CH^2.$$

Whence, multiplying the second member by $B : B$,
adding $H^2F^2 - H^2F^2$ to the numerator of the result,
and factoring, we may write

$$\varDelta = \frac{(H^2 - AB)(F^2 - BC) - (HF - BG)^2}{B}.$$

But, by the preceding article, we must assume

$$H^2 - AB = 0;$$

hence, for the purposes of the present inquiry,

$$\varDelta = -\frac{(HF - BG)^2}{B}.$$

Now the condition $H^2 - AB = 0$ obviously requires that A and B shall have like signs; and we have agreed (see foot-note, p. 160) to write all our equations so that A shall be *positive:* therefore B, in the present inquiry, is positive. Whence it follows, that the foregoing expression for $\varDelta$ is essentially *negative;* unless $HF - BG = 0$, when it will *vanish.* Our proposition is therefore established.

188. The restrictions of the two preceding articles being accepted, our actual problem is, to determine whether the equation

$$Ax^2 + 2Hxy + By^2 + 2Gx + 2Fy + C = 0 \quad (1),$$

in which we suppose $H^2 - AB = 0$, can be reduced to the form

$$B'y^2 + 2G'x + C' = 0;$$

that is, whether it can be subjected to such a transformation of co-ordinates as will destroy the co-efficients of x^2, xy, and y.

189. In the first place, then, we can certainly destroy the co-efficient of xy. For, to effect this, we need only revolve the axes through an angle θ, such that [Art. 156, (c)] we may have

$$\tan 2\theta = \frac{2H}{A - B}:$$

a condition compatible with *any* values of A, H, and B.

Making this transformation, therefore, we shall get an equation of the form

$$A'x^2 + B'y^2 + 2G'x + 2F'y + C = 0:$$

in which (see the third equation in Art. 156)

$$A' = \tfrac{1}{2}\{(A + B) + (A - B)\cos 2\theta + 2H \sin 2\theta\},$$

$$B' = \tfrac{1}{2}\{(A + B) - (A - B)\cos 2\theta - 2H \sin 2\theta\},$$

and (Art. 56, Cor. 3)

$$G' = G\cos\theta + F\sin\theta, \quad F' = F\cos\theta - G\sin\theta.$$

Now, from the value of $\tan 2\theta$ above, (see foot-note, p. 158,) we know that

$$\sin 2\theta = \frac{2H}{\sqrt{\{(A-B)^2 + (2H)^2\}}}, \quad \cos 2\theta = \frac{A-B}{\sqrt{\{(A-B)^2 + (2H)^2\}}},$$

or, by applying the condition $H^2 = AB$, and taking the radical as negative, that

$$\sin 2\theta = -\frac{2H}{A+B}, \quad \cos 2\theta = -\frac{A-B}{A+B};$$

and therefore, by Trig., 847, IV, and by again applying the condition $H^2 = AB$, that

$$\sin\theta = -\frac{H}{\sqrt{(H^2 + B^2)}}, \quad \cos\theta = \frac{B}{\sqrt{(H^2 + B^2)}}$$

Substituting these values in the expressions for A', B', G', and F', we obtain

$$A' = 0; \ B' = A + B; \ G' = \frac{BG - HF}{\sqrt{(H^2 + B^2)}}; \ F' = \frac{HG - BF}{\sqrt{(H^2 + B^2)}}$$

so that the proposed transformation has destroyed the co-efficient of x^2 as well as that of xy, and (1) becomes

$$B'y^2 + 2G'x + 2F'y + C = 0 \qquad (2).$$

190. In the second place, we can certainly destroy the co-efficient of y. For our ability to do so depends on finding some new origin, to which if we transform (2) in parallel axes, the new co-efficient of y shall be zero; and that we can find such an origin is easily proved. For, if the new origin be $x'y'$, the new equation (Art. 163, Th. I) will assume the form

$$B'y^2 + 2G'x + 2(B'y' + F')y + C' = 0:$$

in which the co-efficient of y will vanish, if

$$y' = -\frac{F'}{B'} = -\frac{B(HG + BF)}{\sqrt{(H^2 + B^2)^3}};$$

and y' being thus necessarily finite and real, while x' is indeterminate, there is an infinite number of points, lying on one right line, to any of which if we reduce (2) by parallel transformation, the co-efficient of y will disappear, and (2) will become

$$B'y^2 + 2G'x + C' = 0 \qquad\qquad (3).$$

191. We see, then, that we *can* reduce (1) to the required form; and that, too, without imposing any condition upon it other than the original one, that $H^2 - AB$ shall be equal to zero.

Hence, *The equation of the second degree will represent a parabola whenever its co-efficients fulfill the condition*

$$H^2 - AB = 0.$$

Corollary.—It follows from this, that *whenever the equation of the second degree denotes a parabola, its first three terms form a perfect square.*

192. From the restriction established in Art. 187, we conclude that the Parabola presents only two varieties of the condition just determined. They are,

I. $H^2 - AB = 0$ with $J < 0$.

II. $H^2 - AB = 0$ with $J = 0$.

By referring to Arts. 131, 132, it will be seen that the second of these is identical with the condition upon which the equation of the second degree represents *two parallels*.

Hence, *Two parallels constitute a particular case of the Parabola.*

193. If we apply the criterion of the Parabola to the two lines

$$Ax + Hy + G = 0, \quad Hx + By + F = 0,$$

at whose intersection (Art. 155) the center of the second order curve is found, we shall have A equal to the quotient of H^2 by B: and the two lines will become

$$Hx + By + \frac{BG}{H} = 0, \quad Hx + By + F = 0 \quad (n):$$

which (Art. 98, Cor.) are evidently parallel. Hence, since we may always suppose that parallels intersect at infinity, *the center of a parabola is in general situated at infinity.*

194. To this general law, however, the case brought out in Art. 192 presents a striking exception. For, when the Parabola passes into two parallels, J vanishes; and we obtain (Art. 187)

$$HF - BG = 0 \ \therefore \ \frac{BG}{H} = F :$$

so that, in this case, the lines in (n) become coincident, and the center is any point on the line

$$Hx + By + F = 0.$$

We thus arrive at the conception of the *Right Line as the Center of Two Parallels.*

Remark.—The result of this article is fully corroborated by the equations to the two parallels themselves. For (Art. 160) these are

$$Hx + By + F + \sqrt{F^2 - BC} = 0,$$
$$Hx + By + F - \sqrt{F^2 - BC} = 0:$$

which obviously represent two lines equally distant from

$$Hx + By + F = 0.$$

195. Two parallels, considered as a variety of the Parabola, present three subordinate cases, each of which has its proper criterion.

For, since the equations to the parallels are

$$Hx + By + F + \sqrt{F^2 - BC} = 0,$$
$$Hx + By + F - \sqrt{F^2 - BC} = 0,$$

we shall evidently have the following series of conditions:

I. $F^2 - BC > 0$ ∴ Two parallels, *separate* and *real.*

II. $F^2 - BC = 0$ ∴ Two *coincident* parallels.

III. $F^2 - BC < 0$ ∴ Two parallels, *separate* but *imaginary.*

Corollary.—Hence, *The Right Line, as the limit of two parallels, is the limiting case of the Parabola.*

Remark.—It is noticeable that the limit into which the two parallels vanish when $F^2 - BC = 0$, is the line

$$Hx + By + F = 0,$$

which we have just shown to be the center of the rectilinear case of the Parabola.

196. The results of the foregoing articles, as fixing the varieties of the Parabola and their corresponding analytic conditions, may be summed up in the following table :

$H^2-AB=0$
$\therefore$
Parabola.
$\begin{cases} \Delta<0 \ . \ . \ \text{Center at Infinity.} \\[2ex] \Delta=0 \ . \ . \ \text{Center a R't Line} \begin{cases} \text{Two Real Parallels } \cdot\cdot \ F^2-BC>0. \\ \text{Single Right Line } \cdot\cdot \ F^2-BC=0. \\ \text{Two Imag. Parallels} \cdot\cdot F^2-BC<0. \end{cases} \end{cases}$

EXAMPLES.

I. NOTATION AND CONDITIONS.

1. Show why the following equations represent parabolas :

$$4x^2 + 12xy + 9y^2 + 6x - 10y + 5 = 0,$$
$$(2x - 5y)^2 = 3x + 4y - 5,$$
$$5y^2 - 6x - 2y - 7 = 0.$$

2. Show that the preceding equations represent true parabolas, having their centers at infinity ; but that

$$4x^2 - 12xy + 9y^2 = 25$$

denotes two parallels, whose center is the line

$$2x = 3y.$$

3. Show that $4x^2 - 12xy + 9y^2 + 8x - 12y + 5 = 0$ denotes two imaginary parallels, whose center is the real line $2x - 3y - 2 = 0$: and that $4x^2 - 12xy + 9y^2 - 8x + 12y + 4 = 0$ denotes the limit of these parallels.

4. Reduce $9x^2 - 24xy + 16y^2 + 4x - 8y - 1 = 0$ to the form $B'y^2 + 2G'x + C' = 0.$

5. Show that when a parabola breaks up into two parallels, the line $Hx + By + F = 0$ becomes the axis of the curve.

II. PARABOLIC LOCI.

1. Given the base of a triangle, and the sum of the tangents of the base angles: to find the locus of the vertex.

Let us take the base for the axis of y, and a perpendicular through its middle point for the axis of x. Then, in the annexed diagram, OM will be the ordinate, and MP the abscissa of the variable vertex P. Therefore, supposing the length of the base $= 2m$, and the given sum of tangents $= n$, we shall have

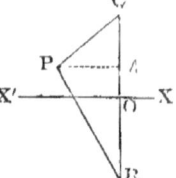

$$\frac{-x}{m+y} + \frac{-x}{m-y} = n;$$

and the equation to the locus sought will be $ny^2 - 2mx - nm^2 = 0.$

Hence, (see close of Art. 182,) the vertex moves on *a parabola whose axis is the perpendicular through the middle of the base.*

By comparing this equation with (3) of Art. 182, it will be seen that the distances of the directrix and focus from the base are respectively

$$\frac{m}{2n}(n^2+1), \quad \frac{m}{2n}(n^2-1).$$

2. Given the base and altitude of a triangle: to find the locus of the intersection of perpendiculars drawn from the extremities of the base to the opposite sides.

3. Given a fixed line parallel to the axis of x, and a movable line passing through the origin: to find the locus of a point on the latter, so taken that its ordinate is always equal to the portion of the former included between the axis of y and the moving line.

4. Lines are drawn, through the point where the axis of a parabola meets the directrix, so as to intersect the curve in two points: to find the locus of the points midway between the intersections.

5. Through any point Q of a circle, OQ is drawn from the center O, and QR made a chord parallel to the diameter EOI and bisected in S: to find the locus of P, where OQ and ES intersect.

6 Find the locus of the center of a circle, which passes through a given point and touches a given right line.   [Take given line for axis of y, and its perpendicular through given point for axis of x.]

7. Given a right line and a circle: to find the locus of the center of a circle which touches both.   [Take perpendicular to given line, through center of given circle, for axis of x.]

8. *OA* is a fixed right line, whose length $= a$; about *O*, a second line *POP'* revolves in such a manner that the product of the areas *P.1O*, *P'AO* $= a^4$, and their quotient $= \cot^2 \frac{1}{2} POA$. to find the locus of *P* or *P'*. [Polars.]

9. Given the base of a triangle, and that the tangent of one base angle is double the cotangent of half the other: to find the locus of the vertex.

10. Find the locus of the center of a circle inscribed in a sector of a given circle, one of the bounding radii of the sector being fixed.

SECTION VIII.—THE LOCUS OF THE SECOND ORDER IN GENERAL.

197. We have now seen that the equations to the Pair of Right Lines, the Circle, the Ellipse, the Hyperbola, and the Parabola, are all of the second degree. We have proved, too, that the *general* equation of the second degree may be made to represent either of these loci, by giving it co-efficients which fulfill the proper conditions; and, in the course of the argument, it has come to light that the Point. the Pair of Lines in their various states of intersection, parallelism, and coincidence, and the Circle, are phases of the three curves mentioned last.

The latter result suggests the question, *Is not the general equation, considered without reference to any of these conditions. the symbol of some locus still more generic than either the Ellipse. the Hyperbola, or the Parabola, of which these three curves are themselves successive phases?* It is the object of the present section, to show that this question, in a certain important sense. is to be answered in the affirmative; and to aid the student in forming an exact conception of what is meant by the phrase *Locus of the Second Order in General.*

198. We proceed, then, to show that such a locus exists; and to explain the peculiar nature of the existence which belongs to it.

In the first place, a moment's reflection upon the discussions in the preceding pages will convince us that hitherto we have not regarded the equation

$$Ax^2 + 2Hxy + By^2 + 2Gx + 2Fy + C = 0 \qquad (1)$$

in the strictly *general* aspect at all. For we have supposed its co-efficients to be subject to some one of the three conditions

$$H^2 - AB < 0, \quad H^2 - AB = 0, \quad H^2 - AB > 0,$$

and therefore to be *actual numbers*, since it is only in actual numbers that the existence of such conditions can be tested. But, obviously, we can conceive of equation (1) as not yet subjected to any such conditions, the constants A, B, C, F, G, H not being *actual* numbers, but symbols of *possible* ones; and, in fact, we *must* so conceive of it, if we would take it up in pure generality.

In the second place, not only does the equation await this purely general consideration, but when so considered it still has geometric meaning. For, though its *co-efficients* are indeterminate, its *exponents* are numerical and fixed: it therefore still holds its variables under a constant law, not so explicit as before, but certainly as real. It is still impossible to satisfy it by the co-ordinates of points taken at random; it will accept only such as will combine to form an equation *of the second degree.*

Since, then, we must consider (1) in its purely general aspect as well as under special conditions; since, even in this aspect, it still expresses a *law of form;* and since this law, consisting as it does in the mere fact that the equation is of the second degree, must pervade all the curves of the Second order: it follows that this law may be regarded as a *generic locus*, whose properties are shared alike by the Ellipse, the Hyperbola, and the Parabola. •

199. By the phrase *Locus of the Second Order in General* we therefore mean not a *figure* but an abstract *law of form*. It exists to abstract thought, but can not be drawn or imagined. To illustrate the nature of its existence by a more familiar case, we may compare it to that of the generic conception of a parallelogram. We can *define* a parallelogram ; but if we attempt to *imagine* or *draw* one, we invariably produce some particular *phase* of the conception — either a rhomboid, a rhombus, a rectangle, or a square. In the same way, the Locus of the Second Order exists so as to be defined ; but not otherwise, except in its special phases. In short, when speaking of it, we are dealing with a purely analytic conception ; and the beginner should avoid supposing that it is any thing else.

200. Further: This common *law of form* not only manifests itself in all the three curves we have been considering, but they may be regarded as *successive* phases of it, whose order is predetermined. For, as we have seen, they may be supposed to arise out of the general locus whenever the condition characteristic of each is imposed on the general equation. Now these conditions may be summed up as follows:

$$H^2 - AB < 0, \quad H^2 - AB = 0, \quad H^2 - AB > 0.$$
$$\Delta < 0. \qquad\qquad \Delta = 0. \qquad\qquad \Delta > 0.$$
$$F^2 - BC < 0, \quad F^2 - BC = 0, \quad F^2 - BC > 0.$$

Hence, since 0 lies between — and +, while the conditions in $\varDelta$ actually entered our investigations as subordinates of those in $H^2 - AB$, and the conditions in $F^2 - BC$ as subordinates of $\varDelta = 0$, it follows that the three curves have a natural order corresponding to the analytic order of their criteria, and that their several varieties have a similar order corresponding to theirs. Accordingly, we should expect the curves to occur thus: Ellipse; Parabola; Hyperbola. In due time hereafter, this order will be verified geometrically.

201. Our three curves and their several varieties are thus shown to be species of the Locus of the Second Order: are there any others? We shall now show that there are not, by proving that *every* equation of the second degree must represent one of the three curves already considered.

We have just seen that all the conditions hitherto imposed on the general equation are subordinate to the three,

$$H^2 - AB < 0, \quad H^2 - AB = 0, \quad H^2 - AB > 0;$$

and we have proved that any equation of the second degree fulfilling either of these must represent one of the three curves. But no equation of the second degree can exist without being subject to one of these conditions; for, whatever be the numerical values of A, H, and B, we can always form the function $H^2 - AB$, which can not but be less than, equal to, or greater than zero. Hence, the series of conditions already imposed on the general equation exhaust the possible varieties of its locus, and we have the proposition at the head of this article.

202. We mentioned in Art. 47, that the term *Conic Section* or *Conic* is used to describe a curve of the

Second order. From what has now been shown, we may define the **Conic in General** as *the embodiment of that general law of form which is expressed by the unconditioned equation of the second degree.*

It also follows that there are three species of the Conic, corresponding to the three leading conditions which have become so familiar.

203. Let us now recapitulate the argument by which we have thus gradually established the theorem :—*Every equation of the second degree represents a conic.*

I. We proved (Arts. 153—162) that every equation of the second degree whose co-efficients fulfill the condition $H^2 - AB < 0$, represents an ellipse, and showed that the Point, the Pair of Imaginary Lines, and the Circle, are particular cases of that curve.

II. We proved (Arts. 173—178) that every equation of the second degree whose co-efficients fulfill the condition $H^2 - AB > 0$, represents an hyperbola, and showed that the Pair of Real Intersecting Lines are a case of *that* curve.

III. We proved (Arts. 185—196) that every equation of the second degree whose co-efficients fulfill the condition $H^2 - AB = 0$, represents a parabola, and showed that Two Parallels, whether separate, coincident, or imaginary, are a case of *that* curve.

IV. We combined (Art. 200) the results of the three preceding steps, and inferred that, of the conditions previously imposed, the three

$$H^2 - AB < 0, \quad H^2 - AB = 0, \quad H^2 - AB > 0$$

were all that we needed to consider in testing the leading signification of the equation of the second degree, since all the others had proved to be subordinates of these.

V. We showed (Art. 201) that these three conditions are of such a nature that *every* equation of the second degree must be subject to one of them; and thence inferred the theorem.

204. The existence of *three* conditions by which the signification of the general equation may be varied, indicates, as we have already noticed, three species of the Conic. But we must not overlook a previous subdivision of the locus. The three conditions are themselves subject to classification : two of them are *finite*, while the third is *infinitely small*. This classification, too, has its geometric counterpart : for the Ellipse and Hyperbola, in which $H^2 - AB$ is finite, have each a finite point as their center ; while the Parabola, in which $H^2 - AB$ is infinitely small, has no such center. In order, then, to include all the facts, we must say that the Conic consists of two families of curves, one *central* and the other *non-central;* and that these two families break up into the three species which we have already described.

205. The entire Locus of the Second Order, with its subdivisions arranged according to their mutual relationships as fixed by their analytic conditions, may be presented as follows:

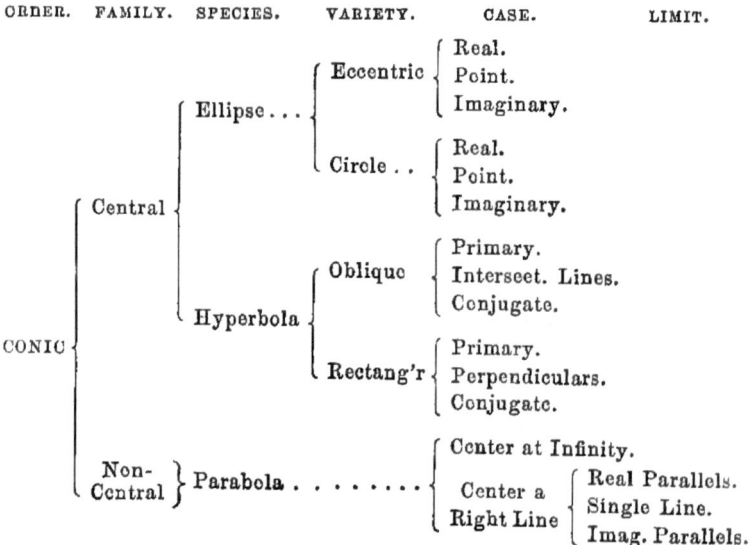

ORDER.	FAMILY.	SPECIES.	VARIETY.	CASE.		LIMIT.
CONIC	Central	Ellipse	Eccentric	Real. Point. Imaginary.		
			Circle	Real. Point. Imaginary.		
		Hyperbola	Oblique	Primary. Intersect. Lines. Conjugate.		
			Rectang'r	Primary. Perpendiculars. Conjugate.		
	Non-Central	Parabola		Center at Infinity. Center a Right Line	Real Parallels. Single Line. Imag. Parallels.	

CHAPTER SECOND.

THE MODERN GEOMETRY: TRILINEAR AND TANGENTIAL CO-ORDINATES.

SECTION I.—TRILINEAR CO-ORDINATES.

206. Modern geometers frequently employ the following method of representing a point:

Any three right lines that *form a triangle*, as AB, BC, CA, are assumed as the Fixed Limits to which all positions shall be referred. The position of any point P in the plane of the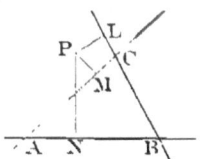
triangle is then determined by finding the lengths PL, PM, PN of three perpendiculars dropped from it upon the three fixed lines.

The triangle whose sides are thus employed as limits, is called the *triangle of reference*. The three perpendiculars let fall upon its sides from any point, are termed the *trilinear co-ordinates* of the point, and are designated by the Greek letters a, β, γ.

207. On a first glance, this system of co-ordinates seems redundant; for, in the Cartesian system, we have seen that *two* co-ordinates are sufficient to determine a point: and it is obvious from the diagram that P is determinable by any *two* of the perpendiculars PL, PM, PN.

The reader will therefore not be surprised to learn that the new method came into use as an unexpected consequence of abridging Cartesian equations. The process

An. Ge. 20.

by which the trilinear system thus grows out of the bilinear, will now be explained.

208. In Art. 108, we have already hinted at the Abridged Notation, which gives rise to the system of which we are speaking. We will now present the subject in detail.

If the equation to any right line is written in terms of the *direction-cosines* of the line, namely, in the form

$$x \cos a + y \sin a - p = 0,$$

we may use a as a convenient abbreviation for the whole member equated to zero; for it naturally recalls the expression into which it enters as so prominent a constant. Similarly, in the equations

$$x \cos \beta + y \sin \beta - p' = 0,$$
$$x \cos \gamma + y \sin \gamma - p'' = 0,$$

we may represent the first members by β and γ. Thus the equations to any three right lines may be written

$$a = 0, \quad \beta = 0, \quad \gamma = 0.$$

The brevity of these expressions is advantageous, even when they are taken separately; but it is not until we combine them, that the chief value of the abridgment appears. We then find that *it enables us to express, by simple and manageable symbols, any line of a given figure in terms of three.others.*

209. It is this last named fact, which constitutes the fundamental principle of trilinear co-ordinates. That we *can* so express a line, follows from our being able (see Art. 108) to write, in terms of the equations to two given right lines, the equation to any line passing through their intersection.

210. We may convince ourselves of this, by a few simple examples. If $a = 0$, $\beta = 0$ represent any two right lines, then (Art. 108)

$$a + k\beta = 0$$

is the equation to *any* right line passing through their intersection, provided k is indeterminate. Now k, in this equation, (Art. 108, Cor. 2) is the negative of the ratio of the perpendiculars dropped from any point in the line $a + k\beta = 0$ upon the two lines $a = 0$ and $\beta = 0$. Therefore, writing k so as to display its *intrinsic* sign,

$$a + k\beta = 0$$

denotes a right line passing through the intersection of $a = 0$ and $\beta = 0$, and lying in that angle of the two lines which is *external* to the origin : but

$$a - k\beta = 0$$

denotes one lying in the *same* angle as the origin. Moreover, when the perpendiculars mentioned are equal, the value of $k = \pm 1$; and we have (Art. 109, Cor. 3)

$$a + \beta = 0$$

the equation to the bisector of the *external* angle between two given lines, and

$$a - \beta = 0$$

the equation to the bisector of the *internal* angle.

Suppose, then, that we have a given triangle, whose sides are the three lines

$$a = 0, \quad \beta = 0, \quad \gamma = 0.$$

Granting, as we may, that *the origin is within the triangle*, the equations to the three bisectors of the angles will be

$$a - \beta = 0, \quad \beta - \gamma = 0, \quad \gamma - a = 0.$$

The equations to the three lines which bisect the three external angles of the triangle, will be

$$\alpha + \beta = 0, \quad \beta + \gamma = 0, \quad \gamma + \alpha = 0.$$

Thus, these six lines of the triangle are all expressed in terms of its three sides.

211. We can of course extend this system of abbreviations to the case of lines whose equations are in the general form

$$Ax + By + C = 0,$$

by representing the member equated to zero by a single English letter, such as L or v. Thus,

$$L + kL' = 0 \text{ or } v + kv' = 0$$

denotes a line passing through the intersection of the lines

$$Ax + By + C = 0, \quad A'x + B'y + C' = 0.$$

It must be borne in mind, however, that in these equations k does not denote the negative of the ratio of the perpendiculars mentioned above, and in consequence does not become ± 1 when those perpendiculars become equal. Hence, in these cases, the equations to the external and internal bisectors of the angle between two given lines, are respectively (see Art. 109, Cor. 4)

$$L + rL' = 0 \text{ or } lv + mv' = 0,$$
$$L - rL' = 0 \text{ or } lv - mv' = 0,$$

in which r, or $m : l$, is to be determined by the formula on p. 116.

In this notation, if the three sides of a triangle are

$$u = 0, \quad v = 0, \quad w = 0,$$

the three bisectors of its interior angles may be represented by

$$lu - mv = 0, \quad mv - nw = 0, \quad nw - lu = 0;$$

and the three bisectors of its exterior angles, by

$$lu + mv = 0, \quad mv + nw = 0, \quad nw + lu = 0.$$

Here, too, the six lines are all expressed in terms of the three sides.

212. Having thus learned how to interpret the equations

$$a \pm \beta = 0, \quad \beta \pm \gamma = 0, \quad \gamma \pm a = 0$$

when $a = 0$, $\beta = 0$, $\gamma = 0$ are given as forming a triangle, we next advance to the interpretation of

$$la + m\beta + n\gamma = 0:$$

an equation of which the preceding six are evidently particular cases, and in which l, m, n are any three constants whatever.

On the surface, this looks like the condition (Art. 114) that the three lines $a = 0$, $\beta = 0$, $\gamma = 0$ shall meet in one point. But the terms of that condition are, that three lines will meet in one point whenever three constants can be found *such that, if the equations to the lines be each multiplied by one of them, the sum of the products will be zero.* Hence, the l, m, and n of that condition are not *absolutely* arbitrary, but only arbitrary within the limits consistent with causing the function $la + m\beta + n\gamma$ to vanish identically. On the contrary, the l, m, and n of the present equation *are* absolutely arbitrary.

With this fact premised, let us now investigate the meaning of the equation.

213. If we replace α, β, and γ by the functions for which they stand, and reduce the equation on the supposition that x and y have but one signification in all its three branches, we obtain

$$\left.\begin{array}{c}(l \cos \alpha + m \cos \beta + n \cos \gamma)\, x \\ +\ (l \sin \alpha + m \sin \beta + n \sin \gamma)\, y \\ +\ lp + mp' + np'' \end{array}\right\} = 0:$$

which is evidently the equation to some *right line*, since it conforms to the type

$$Ax + By + C = 0.$$

Its full significance, however, will not appear until we discuss it under each of three hypotheses concerning the relative position of the three lines whose equations enter it. To this discussion, we devote the next three articles.

214. Let the three lines $\alpha = 0$, $\beta = 0$, $\gamma = 0$ meet in one point.

When this is the case, the equation

$$l\alpha + m\beta + n\gamma = 0$$

denotes a right line *passing through the point of triple intersection*. For the co-ordinates of this point will render α, β, and γ equal to zero simultaneously, and will therefore satisfy the equation just written.

215. Let the three given lines be parallel.

In this case, their equations (Art. 98, Cor.) may be written

$$\alpha = 0, \quad \alpha + c' = 0, \quad \alpha + c'' = 0,$$

and the equation we are discussing will thus become

$$(l + m + n)\, \alpha + (mc' + nc'') = 0;$$

that is, it will assume the form

$$\alpha + c = 0,$$

and will therefore denote a right line *parallel to the three given ones.*

216. Let the three given lines form a triangle.

In this case the preceding results are avoided; hence, the equation of Art. 213 assumes the form

$$Ax + By + C = 0$$

without imposing any restriction upon the values of A, B, and C. In this case, therefore, it denotes *any right line whatever.*

This result is simply the extension of that obtained in Arts. 210, 211; and shows that we can (as was stated in Art. 208) express any line of a figure in terms of three given ones.

217. From the results of the last three articles, we conclude that

$$la + m\beta + n\gamma = 0$$

is the equation to any right line; provided, however, that the lines

$$a = 0, \quad \beta = 0, \quad \gamma = 0$$

are so situated as to *form a triangle.* This proviso, we can not too carefully remember.

218. The examination of an example somewhat more complex than any yet presented, will convince us that this abridged method of writing equations is in effect a *new system of co-ordinates,* applicable to lines of any order. Before commencing this example, it may be necessary to remind the student that the use of a *Greek* letter for an abbreviation, always implies that the equation to the line so represented is in the form

$$x \cos a + y \sin a - p = 0,$$

and the like; and that, when the fundamental equations

are in any other form, they will be abridged by means of *English* characters.

For the sake of still greater brevity, the line $a = 0$ is often cited as the line a, the line $\beta = 0$ as the line β, etc. The point of intersection of two lines is frequently spoken of as the point $a\beta$, etc. The last notation should be carefully distinguished from that of the *co-ordinates* of the same point.

219. Example.—*Any line of a quadrilateral in terms of any three.*

Let $ABEF$ be any quadrilateral, and let

$$a = 0, \quad \beta = 0, \quad \gamma = 0$$

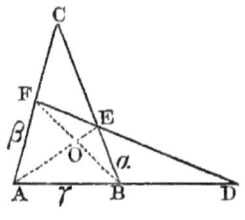

be the equations to its three sides BC, CA, AB. We can now represent any other line in the figure, in terms of a, β, and γ.

Suppose the origin of bilinear co-ordinates to be within the triangle ABC, and, in fact, within the triangle EOB. The equation to AE, which passes through the intersection of β and γ, and lies in their internal angle, will be of the form

$$m\beta - n\gamma = 0 \qquad\qquad (AE).$$

The equation to BF, which passes through the intersection of γ and a, and lies in their internal angle, will be of the form

$$n\gamma - la = 0 \qquad\qquad (BF).$$

The equation to EF, which joins the intersections of (a, AE) and (β, BF), and lies in the external angle of the first two lines, but in the internal angle of the second pair, must be formed (Art. 108, Cor. 2) so as to equal either the sum of la and $m\beta - n\gamma$ or the difference of $m\beta$ and $n\gamma - la$. It is therefore

$$la + m\beta - n\gamma = 0 \qquad\qquad (EF).$$

The equation to CD, which joins $a\beta$ to the intersection of (γ, EF), and lies in the external angle of both pairs of lines, must be the sum of either la and $m\beta$, or $n\gamma$ and $la + m\beta - n\gamma$. Consequently, it is

$$la + m\beta = 0 \qquad\qquad (CD).$$

The equation to OC, which joins $a\beta$ to the intersection of (AE, BF), and lies in the internal angle of $a\beta$, but in the external angle of the other two lines, must be equal either to the difference of la and $m\beta$ or the sum of $m\beta - n\gamma$ and $n\gamma - la$. Accordingly, it is

$$la - m\beta = 0 \qquad\qquad (OC).$$

Finally, the equation to OD, which joins the intersection of (γ, EF) to that of (AE, BF), and lies in the internal angle of both pairs of lines, must be formed so as to equal the difference of either $la + m\beta - n\gamma$ and $n\gamma$, or $m\beta - n\gamma$ and $n\gamma - la$. Therefore, it is

$$la + m\beta - 2n\gamma = 0 \qquad\qquad (OD).$$

We have thus expressed all the lines of the quadrilateral in terms of the three lines a, β, and γ. We can do more: we can solve problems involving the properties of the figure, by means of these equations, and test the relative positions of its lines without any direct reference to the x and y which the symbols a, β, γ conceal. Thus, the form of the equations

$$la - m\beta = 0, \quad m\beta - n\gamma = 0, \quad n\gamma - la = 0$$

shows (Art. 114) that the three lines OC, AE, BF meet in one point, and the same relation between OD, AE, BF is shown in the form of their equations.

220. We see, then, that by introducing this abridged notation we can replace the Cartesian equations in x and y by a set of equations in a, β, γ. Moreover, it is noticeable that all these abridged equations to right lines are *of the first degree* with respect to a, β, and γ. Since, therefore, we operate upon these symbols (as the last example shows) just as if they were variables, and since they combine in equations which satisfy the condition that an equation to a right line must be of the first degree, it appears that we may use a, β, γ as *co-ordinates*. Thus, we may say that the equation

$$la + m\beta + n\gamma = 0$$

is the equation to any right line, and that a, β, γ are the co-ordinates of any point in the line.

An. Ge. 21.

What now is the system of reference to which these co-ordinates belong? We have seen that, in order to the interpretation we have given of the last-named equation, the *lines* a, β, γ must *form a triangle.* We know, too, (Art. 105) that a is the length of a perpendicular let fall from any point to the line a; that β is the length of a perpendicular from the same point to the line β; and that γ is the length of a perpendicular from the same point to the line γ. Hence, we see that if a, β, γ *are* taken as co-ordinates, they are referred to the triangle formed by

$$a = 0, \quad \beta = 0, \quad \gamma = 0,$$

and that they signify the three perpendiculars dropped from any point in its plane upon its three sides. In short, we have come out upon the system of trilinear co-ordinates described in Art. 206.

PECULIAR NATURE OF TRILINEAR CO-ORDINATES.

221. Before making any further application of the new system, it is important to notice that *trilinear co-ordinates are in one respect essentially different from bilinear.* In the Cartesian system, the x and y are independent of each other, unless connected by the equation to some locus. In the trilinear system, on the contrary, the a, β, and γ are each of them determined by the other two; that is, there is a certain equation between them, which holds true in all cases, whether the point which they represent be restricted to a locus or not. In the language of analysis, we express this peculiarity by saying that each of these co-ordinates is a *determined* function of the other two.

That trilinear co-ordinates are subject to such a condition, is proved by the fact, that *two* co-ordinates completely fix a point, and must therefore determine the value of any third. The equation expressing the exact limits of the condition, we now proceed to develop.

222. Let ABC be the triangle of reference, $a = 0$ being the equation to the side BC, $\beta = 0$ the equation to CA, and $\gamma = 0$ the equation to AB. Then, if P be any point in the plane of the triangle, the three perpendiculars PL, PM, 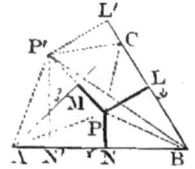 PN will be represented by a, β, γ. Supposing now that a, b, and c denote the *lengths* of the three sides BC, CA, AB, we shall have

$$aa = \text{twice the area of } BPC,$$
$$b\beta = \qquad \text{``} \qquad \text{``} \qquad CPA,$$
$$c\gamma = \qquad \text{``} \qquad \text{``} \qquad APB.$$

But the sum of these areas is constant, being equal to the area of the triangle of reference : therefore, representing the double area of this triangle by M, we obtain

$$aa + b\beta + c\gamma = M :$$

which is the constant relation connecting the trilinear co-ordinates of any point.

Remark.—When we say that the sum of the areas of APB, BPC, CPA is equal to the area of ABC, we of course mean their *algebraic* sum. For, if we take a point *outside* of the triangle of reference, such as P', we shall evidently have

$$ABC = AP'B + BP'C - CP'A :$$

that is, in such a case, one of the three areas becomes negative. And this is as it should be : for, *as we have agreed to suppose the Cartesian origin to be* WITHIN *the triangle of reference*, the perpendicular $P'M$ is negative, according to the third corollary of Art. 105.

223. This peculiarity of trilinear co-ordinates may be made to promote the advantages of the system.

In the first place, it gives rise to a very symmetrical expression for *an arbitrary constant.* For if k be arbitrary, kM will also be; and we shall have, for the symbol alluded to,

$$k (a\alpha + b\beta + c\gamma) \tag{1}.$$

We can also modify this symbol, and render it still more useful. Let $1 : r =$ the ratio of the side a, in the triangle of reference, to the sine of the opposite angle A. Then, by Trig., 867, we shall have

$$\frac{\sin A}{a} = \frac{\sin B}{b} = \frac{\sin C}{c} = r.$$

Now, by the principle of Art. 222,

$$r (a\alpha + b\beta + c\gamma) = \text{constant};$$

hence, substituting for ra, rb, rc from the equal ratios above,

$$a \sin A + \beta \sin B + \gamma \sin C = \text{constant};$$

or, the symbol for an arbitrary constant may be written

$$k (a \sin A + \beta \sin B + \gamma \sin C) \tag{2}.$$

224. In the second place, the peculiarity of trilinears enables us to *use homogeneous equations* in all cases. For, if a given trilinear equation is not homogeneous, we can at once render it so by means of the relation in Art. 222. Thus, if the given equation were $\beta = p$, we might write

$$M\beta = p (a\alpha + b\beta + c\gamma).$$

225. In the third place, instead of the actual trilinear co-ordinates of a point, we may employ any three quantities that are in the same ratio, without affecting the

equation to the locus of the point. For, since all trilinear equations are homogeneous, the effect of replacing
a, β, γ by ρa, $\rho\beta$, $\rho\gamma$ will only be to multiply the given
equation by the constant ρ: which of course leaves it
essentially unchanged.

This principle will often prove of great convenience.

Remark.—In case it becomes desirable to find the actual trilinear co-ordinates when they have been displaced in the manner
described, we may proceed as follows: — Let l, m, n be the three
quantities in a constant ratio to a, β, γ: then, supposing the ratio
to be $1 : r$,

$$a = rl, \quad \beta = rm, \quad \gamma = rn.$$

Hence. (Art. 222,)

$$r(la + mb + nc) = M:$$

from which r is readily found.

226. Finally, the property in question enables us to
pass from trilinear co-ordinates to Cartesian. For, by
means of the relation $aa + b\beta + c\gamma = M$, we can convert
any trilinear equation into one which shall contain only
β and γ. Then, supposing the side γ of the triangle of
reference to be the axis of x, and the side β the axis of y,
we shall have

$$\beta = x \sin A, \quad \gamma = y \sin A.$$

Corollary.—If the triangle of reference should be right-
angled at A. the reduction-formulæ will become

$$\beta = x. \quad \gamma = y.$$

But, in general, to pass from trilinears to rectangulars
when the side γ is taken for the axis of x, we must use

$$\beta = x \sin A - y \cos A. \quad \gamma = y.$$

The student may draw a diagram. and verify the last
formulæ.

227. The advantage of the trilinear system consists in this: Its equations may all be referred to *three* of the most prominent lines in the figure to which they belong, and hence become shorter and more expressive than those of the Cartesian system, which can go no farther in the process of simplification than the use of *two* prominent lines. The student will see a good illustration of this in the equations to the bisectors of the internal angles of a triangle. The trilinear equations are

$$\alpha - \beta = 0, \quad \beta - \gamma = 0, \quad \gamma - \alpha = 0$$

which are much simpler than the Cartesian ones given in Ex. 32, p. 124.

TRILINEAR EQUATIONS IN DETAIL.

228. Equation to any Right Line.—This we have already (Arts. 212—217) found to be

$$l\alpha + m\beta + n\gamma = 0.$$

229. Equation to a right line parallel to a given one.—It is obvious geometrically, that the α, β, γ of the parallel line will each differ from the α, β, γ of the given one by some constant. Hence, the given line being $l\alpha + m\beta + n\gamma = 0$, the required equation [Art. 223, (2)] will be

$$l\alpha + m\beta + n\gamma + k\,(\alpha \sin A + \beta \sin B + \gamma \sin C) = 0.$$

230. Equation to a right line situated at infinity.—The Cartesian equation to this is $C = 0$, in which (Art. 110) C is a finite constant. Therefore (Art. 223) the trilinear equation is

$$\alpha \sin A + \beta \sin B + \gamma \sin C = 0.$$

231. Condition that two right lines shall be mutually perpendicular.—Let the two lines be

$$la + m\beta + n\gamma = 0, \quad l'a + m'\beta + n'\gamma = 0.$$

By writing a, β, γ in full, collecting the terms, and applying (Art. 96, Cor. 3) the criterion $AA' + BB' = 0$, we obtain, as the required condition,

$$ll' + mm' + nn' + \left\{ \begin{matrix} (mn' + m'n) \cos (\beta - \gamma) \\ (nl' + n'l) \cos (\gamma - a) \\ (lm' + l'm) \cos (a - \beta) \end{matrix} \right\} = 0.$$

Now, in this expression, a, β, γ are the angles made with the axis of x by perpendiculars from the origin on the *lines* a, β, γ. Supposing the latter to form the triangle of reference, and to inclose the origin, it is evident from the diagram, that $\beta - \gamma$ is the angle between the *perpendiculars* β and γ: that $\gamma - a$ is the angle between the perpendiculars γ and a: and that $a - \beta$ is the angle between the perpendiculars a and β. From the properties of a quadrilateral it then follows, that $\beta - \gamma$ is the supplement of A, $\gamma - a$ of B, and $a - \beta$ of C. Hence, the required condition may be otherwise written

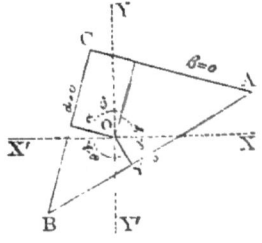

$$ll' + mm' + nn' - \left\{ \begin{matrix} (mn' + m'n) \cos A \\ (nl' + n'l) \cos B \\ (lm' + l'm) \cos C \end{matrix} \right\} = 0.$$

Corollary.—The condition that $la + m\beta + n\gamma = 0$ shall be perpendicular to $\gamma = 0$, is

$$n = m \cos A + l \cos B.$$

232. Length of a perpendicular from any point to a given right line.—Let the point be $\alpha\beta\gamma$, and the line $l\alpha + m\beta + n\gamma = 0$. Write the latter equation in full, collect its terms, and apply the formula of Art. 105, Cor. 2. We thus find

$$P = \frac{l\alpha + m\beta + n\gamma}{\sqrt{(l^2 + m^2 + n^2 - 2mn\cos A - 2nl\cos B - 2lm\cos C)}}.$$

233. Equation to a right line passing through Two Fixed Points.—The *form* of this will of course be

$$l\alpha + m\beta + n\gamma = 0;$$

and our problem is, to determine the ratios $l : m : n$ so that the line shall pass through the two points $\alpha_1\beta_1\gamma_1$, $\alpha_2\beta_2\gamma_2$.

Since the two points are to be on the line, we shall have

$$l\alpha_1 + m\beta_1 + n\gamma_1 = 0,$$
$$l\alpha_2 + m\beta_2 + n\gamma_2 = 0.$$

Solving for $l : n$ and $m : n$ between these conditions, we find

$$l : m : n = (\beta_1\gamma_2 - \gamma_1\beta_2) : (\gamma_1\alpha_2 - \alpha_1\gamma_2) : (\alpha_1\beta_2 - \beta_1\alpha_2).$$

The required equation is therefore

$$\alpha\,(\beta_1\gamma_2 - \gamma_1\beta_2) + \beta\,(\gamma_1\alpha_2 - \alpha_1\gamma_2) + \gamma\,(\alpha_1\beta_2 - \beta_1\alpha_2) = 0.$$

Corollary.—Hence, in trilinears, the *condition that three points shall lie on one right line* is

$$\alpha_3\,(\beta_1\gamma_2 - \gamma_1\beta_2) + \beta_3\,(\gamma_1\alpha_2 - \alpha_1\gamma_2) + \gamma_3\,(\alpha_1\beta_2 - \beta_1\alpha_2) = 0.$$

Expanding, and re-collecting the terms, we may write this more symmetrically (see Art. 112),

$$\alpha_1\,(\beta_2\gamma_3 - \gamma_2\beta_3) + \alpha_2\,(\beta_3\gamma_1 - \gamma_3\beta_1) + \alpha_3\,(\beta_1\gamma_2 - \gamma_1\beta_2) = 0.$$

234. **Theorem.**—*Every trilinear equation of the second degree represents a conic.*

For, since a, β, γ are linear functions of x and y, every such equation is reducible to an equation of the second degree in x and y. But the latter (Art. 203) will represent a conic.

Remark.—In passing now to the trilinear expressions for curves of the Second order, we shall at first suppose the triangle of reference to have a special position, such as will tend to simplify the resulting equations. The more general equations, corresponding to *any* position of the triangle, will be investigated afterward.

235. Equation to the Conic, referred to the Inscribed Triangle.—To obtain this, we must form an equation of the second degree in a, β, γ, such as the co-ordinates of the vertices of the reference-triangle will satisfy.

Now, at the vertex A, we have $\beta = 0$, $\gamma = 0$; at the vertex B, $\gamma = 0$, $a = 0$; and, at the vertex C, $a = 0$, $\beta = 0$. Hence, the required equation may be written

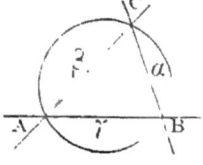

$$l\beta\gamma + m\gamma a + na\beta = 0;$$

for this expression is obviously satisfied by either of the three suppositions

$$\beta = 0, \ \gamma = 0; \quad \gamma = 0, \ a = 0; \quad a = 0, \ \beta = 0.$$

Corollary.—Dividing through by $a\beta\gamma$, we may write the equation just found in the more symmetrical form

$$\frac{l}{a} + \frac{m}{\beta} + \frac{n}{\gamma} = 0.$$

236. Equation to the Circle, referred to the Inscribed Triangle.—Our problem here is, to determine $l : m : n$ so that

$$\frac{l}{\alpha} + \frac{m}{\beta} + \frac{n}{\gamma} = 0 \tag{1}$$

shall represent a circle.

Write α, β, γ each in full, expand the equation, collect the terms with reference to x and y, and apply the criterion (Art. 140) $H = 0$, $A - B = 0$. The conditions in order that (1) may represent a circle will thus be found to be

$$l \cos (\beta + \gamma) + m \cos (\gamma + \alpha) + n \cos (\alpha + \beta) = 0,$$
$$l \sin (\beta + \gamma) + m \sin (\gamma + \alpha) + n \sin (\alpha + \beta) = 0.$$

Solving these for $l : n$ and $m : n$, and applying Trig., 845, III, we obtain

$$l : m : n = \sin (\beta - \gamma) : \sin (\gamma - \alpha) : \sin (\alpha - \beta).$$

But (see Art. 231) $\beta - \gamma = 180° - A$, $\gamma - \alpha = 180° - B$, $\alpha - \beta = 180° - C$. Hence, the equation sought is

$$\frac{\sin A}{\alpha} + \frac{\sin B}{\beta} + \frac{\sin C}{\gamma} = 0.$$

Corollary.—The *equation to any circle concentric with the one circumscribed about the triangle of reference,* will only differ from the foregoing (Art. 142, Cor. 1) by some constant. Hence, it may be written

$$\frac{\sin A}{\alpha} + \frac{\sin B}{\beta} + \frac{\sin C}{\gamma} = k \left(\alpha \sin A + \beta \sin B + \gamma \sin C \right).$$

237. General equation to the Circle.—The equation of the preceding article applies only to the circle described about the triangle of reference. We are now to seek an equation which will represent *any* circle.

The rectangular equation to a circle (Art. 140) may be put into the form

$$x^2 + y^2 + Ax + By + C = 0.$$

In this, the only arbitrary constants are A, B, C. Hence, rectangular equations to different circles will vary only in the linear part, and the equation to *any* circle may be formed from that of a *given* one by merely adding to the latter an arbitrary linear function. Now this property is as true of trilinear as of rectangular equations, since a trilinear equation is only a rectangular one written in a peculiar way. We can therefore form the equation to *any* circle from that of the circle described about the triangle of reference, by adding to the latter, terms in the type of $la + m\beta + n\gamma$.

By clearing of fractions, and replacing $\sin A$, $\sin B$, $\sin C$ by a, b, c, which are in the same ratio, the equation to the circumscribed circle becomes

$$a\beta\gamma + b\gamma a + ca\beta = 0.$$

Hence, the required equation is

$$a\beta\gamma + b\gamma a + ca\beta - M(la + m\beta + n\gamma) = 0:$$

in which M is the fixed constant $aa + b\beta + c\gamma$, and is multiplied into the linear function in order to render the equation homogeneous.

Remark.—When convenience requires it, we can replace M (Art. 223) by the constant $a\sin A + \beta\sin B + \gamma\sin C$, and write the equation

$$\beta\gamma\sin A + \gamma a\sin B + a\beta\sin C - (la + m\beta + n\gamma)(a\sin A + \beta\sin B + \gamma\sin C) = 0.$$

238. Trilinear equation to the Conic in General.—From Arts. 224, 234, it follows that this is simply the general homogeneous equation of the second degree in a, β, γ.   We therefore write it

$$Aa^2 + B\beta^2 + C\gamma^2 + 2F\beta\gamma + 2G\gamma a + 2Ha\beta = 0.$$

Remark.—It is worthy of notice, that, if we suppose $\gamma = 1$, this expression becomes

$$Aa^2 + 2Ha\beta + B\beta^2 + 2Ga + 2F\beta + C = 0,$$

and is, in *form*, identical with

$$Ax^2 + 2Hxy + By^2 + 2Gx + 2Fy + C = 0,$$

the Cartesian equation to the Conic in General.   This fact has led Salmon to the opinion that Cartesian co-ordinates are a case of trilinear.*

239. Problem.—*To determine the condition in order that the general trilinear equation of the second degree may represent a circle.*

From the equation of Art. 222, it is evident that we have the following relations :

$$aa^2 = Ma - ba\beta - c\gamma a,$$
$$b\beta^2 = M\beta - c\beta\gamma - aa\beta,$$
$$c\gamma^2 = M\gamma - a\gamma a - b\beta\gamma.$$

Substituting from these for a^2, β^2, γ^2 in the equation of Art. 238, we find that it may be written

$$\left. \begin{array}{l} \left(2F - \dfrac{Bc}{b} - \dfrac{Cb}{c}\right)\beta\gamma \\[2mm] + \left(2G - \dfrac{Ca}{c} - \dfrac{Ac}{a}\right)\gamma a \\[2mm] + \left(2H - \dfrac{Ab}{a} - \dfrac{Ba}{b}\right)a\beta \end{array} \right\} + M\left(\dfrac{A}{a}a + \dfrac{B}{b}\beta + \dfrac{C}{c}\gamma\right) = 0.$$

Comparing this with the equation to the circle,

$$a\beta + b\gamma a + ca\beta + M(la + m\beta + n\gamma) = 0,$$

we see that it will represent a circle, provided

$$\frac{2F - \dfrac{Bc}{b} - \dfrac{Cb}{c}}{a} = \frac{2G - \dfrac{Ca}{c} - \dfrac{Ac}{a}}{b} = \frac{2H - \dfrac{Ab}{a} - \dfrac{Ba}{b}}{c}$$

Hence, the required condition is

$$2Fbc - Bc^2 - Cb^2 = 2Gca - Ca^2 - Ac^2 = 2Hab - Ab^2 - Ba^2.$$

240. Equations to the Chord and the Tangent of any Conic.—For the sake of simplicity, we shall suppose the inscribed triangle to be the triangle of reference.

I. The equation to the chord is the equation to the right line which joins any two points on the curve, as $a'\beta'\gamma'$, $a''\beta''\gamma''$. Now since these points are on the conic, we shall have (Art. 235, Cor.)

$$\frac{l}{a'} + \frac{m}{\beta'} + \frac{n}{\gamma'} = 0, \qquad \frac{l}{a''} + \frac{m}{\beta''} + \frac{n}{\gamma''} = 0.$$

We can now easily see that the equation to the chord may be written

$$\frac{la}{a'a''} + \frac{m\beta}{\beta'\beta''} + \frac{n\gamma}{\gamma'\gamma''} = 0.$$

For this is the equation to a right line, since it is of the first degree: and to the line that passes through $a'\beta'\gamma'$, $a''\beta''\gamma''$, since it is satisfied by the co-ordinates of either point, inasmuch as they convert it into one of the conditions just found above.

II. The point $a''\beta''\gamma''$ may be supposed to approach the point $a'\beta'\gamma'$ as closely as we please. At the moment of coincidence, the chord becomes the *tangent* at $a'\beta'\gamma'$, and we also have $a''=a'$, $\beta''=\beta'$, $\gamma''=\gamma'$. By making the corresponding substitutions in the equation to the chord, we shall therefore obtain the equation to the tangent. It is

$$\frac{la}{a'^2}+\frac{m\beta}{\beta'^2}+\frac{n\gamma}{\gamma'^2}=0.$$

241. There are many other equations, more or less symmetrical, representing the Conic or the Circle, but our limits forbid their presentation. Those already developed have the widest application, and are sufficient to illustrate the trilinear method. The student who wishes further information on the subject, may consult SALMON'S *Conic Sections.*

EXAMPLES.

1. When will the locus of a point be a circle, if the product of its perpendiculars upon two sides of a triangle is in a constant ratio to the square of its perpendicular on the third?

Take the triangle mentioned, as the triangle of reference; and represent the constant ratio spoken of, by k. The conditions then give us

$$a\beta = k\gamma^2;$$

so that the locus is, in general, a conic of *some* form. In order that it may be a *circle*, the equation just found must satisfy the condition of Art. 239. Applying this, and observing the fact that in the present equation $C=-k$, $2H=1$, and A,B,F,G each $=0$, we find that the locus will be a circle if

$$kb^2 = ka^2 = ab.$$

From the first of these conditions, we obtain $a=b$; from the second and third, $k=1$. The required condition then is, that *the triangle shall be isosceles, and the constant ratio unity.*

It is interesting to notice how clearly the equation above written expresses the position of the locus with respect to the triangle. To find where the side a cuts the conic, we simply make (Art. 62, Cor. 1) $a=0$ in the equation

$$a\beta = k\gamma^2.$$

The result is $\gamma^2 = 0$; that is, the equation whose roots are the co-ordinates of the points of intersection is a quadratic with *equal* roots. Hence,

(Art. 62, 3d paragraph of Cor. 2.) the line a meets the curve in two *coincident* points ; in other words, it *touches* the conic. Moreover, since the quadratic of intersection is $\gamma^2 = 0$, the two points coincide on the line γ. We thus learn that the conic touches the side a of the triangle at the point γa.

If we make $\beta = 0$ in the equation, we again obtain $\gamma^2 = 0$. Hence, the conic touches the side β of the triangle at the point $\beta\gamma$. If we make $\gamma = 0$, we get either $a = 0$ or $\beta = 0$; the side γ therefore cuts the conic in two points : one on the side a, the other on β; or, one the point γa, the other, the point $\beta\gamma$.

We may sum up the whole result by saying that $a\beta = k\gamma^2$ denotes *a conic, to which two of the sides of the triangle of reference are tangents, while the third is a chord uniting their points of contact.*

2. Form the trilinear equations to the right lines joining the vertices of a triangle to the middle points of the opposite sides.

Take the triangle itself as the triangle of reference, and suppose the Cartesian origin within it. The required equations (Art. 108, Cor. 2) will be of the form
$$a - k\beta = 0, \quad \beta - k'\gamma = 0, \quad \gamma - k''a = 0 :$$
in which we are to determine k, k', k'' so that the lines shall bisect the sides of the triangle.

We know that k is = the ratio of the perpendiculars dropped from the middle of γ on a and β respectively. But this ratio is evidently $= \sin B : \sin A$. Similarly, $k' = \sin C : \sin B$; and $k'' = A : \sin C$. Hence, the equations sought are
$$a \sin A - \beta \sin B = 0, \quad \beta \sin B - \gamma \sin C = 0, \quad \gamma \sin C - a \sin A = 0.$$

3. Form the trilinear equations to the three perpendiculars which fall from the vertices of a triangle upon the opposite sides.

We begin, as before, with the *forms* of the equations, namely, $a - k\beta = 0$, $\beta - k'\gamma = 0$, $\gamma - k''a = 0$. Now the condition that the first line shall be perpendicular to γ (Art. 231, Cor.) gives us
$$\cos B - k \cos A = 0.$$
Hence, $k = \cos B : \cos A$; and, similarly, $k' = \cos C : \cos B$, $k'' = \cos C : \cos A$. The required equations are therefore
$$a \cos A - \beta \cos B = 0, \quad \beta \cos B - \gamma \cos C = 0, \quad \gamma \cos C - a \cos A = 0.$$

4. Form the trilinear equations to the three perpendiculars through the middle points of the sides of a triangle.

The middle points of the sides may be regarded as the intersections of γ, a, and β with the lines of Ex. 2. Hence, the *form* of the equations will be
$$a \sin A - \beta \sin B + n\gamma = 0, \quad \beta \sin B - \gamma \sin C + la = 0, \quad \gamma \sin C - a \sin A + m\beta = 0.$$

But the condition of Art. 231, Cor., gives us $n = \sin (A-B)$, $l = \sin (B-C)$, $m = \sin (C-A)$. Therefore, the equations sought are

$$a \sin A - \beta \sin B + \gamma \sin (A - B) = 0,$$
$$\beta \sin B - \gamma \sin C + a \sin (B - C) = 0,$$
$$\gamma \sin C - a \sin A + \beta \sin (C - A) = 0.$$

If the student will now compare the equations of the last three examples with those of Exs. 4, 6, 7 on page 121, he will at once see how much simpler the trilinear expressions are than the Cartesian.

5. Show that the equation representing a perpendicular to the base of a triangle at its extremity, is $a + \gamma \cos B = 0$.

6. Show that the lines $a - k\beta = 0$, $\beta - ka = 0$ are equally inclined to the bisector of the angle between a and β.

7. Prove that the equation to the line joining the feet of two perpendiculars from the vertices of a triangle on the opposite sides is

$$a \cos A + \beta \cos B - \gamma \cos C = 0.$$

Also, that the equation to a line passing through the middle points of two sides is

$$a \sin A + \beta \sin B - \gamma \sin C = 0.$$

8. Show that the equation to a line through the vertex of a triangle, parallel to the base, is $a \sin A + \beta \sin B = 0$.

9. Find the equation to the line which joins the centers of the inscribed and circumscribed circles belonging to any triangle.

By the principle of Art. 225, we may take for the co-ordinates of the first point 1, 1, 1; and, of the second, $\cos A$, $\cos B$, $\cos C$. Hence, (Art. 233,) the equation is

$$a (\cos B - \cos C) + \beta (\cos C - \cos A) + \gamma (\cos A - \cos B) = 0.$$

10. What is the locus of a point, the sum of the squares of the perpendiculars from which on the sides of a triangle is constant? Show that when the locus is a circle, the triangle is equilateral.

11. Find, by a method similar to that of Art. 231, the tangent of the angle contained by the lines

$$la + m\beta + n\gamma = 0, \quad l'a + m'\beta + n'\gamma = 0.$$

12. Write the equation to the circle circumscribing the triangle whose sides are 3, 4, 5.

13. A conic section is described about a triangle ABC; lines bisecting the angles A, B, and C meet the conic in the points A', B', and C': form the equations to $A'B$, $A'C$, $A'B'$

14. Find an equation to the conic which touches the three sides of a triangle.

If the equations to the three sides of the triangle are $a = 0$, $\beta = 0$, $\gamma = 0$, the required equation may be written

$$1\,\overline{la} + 1\,\overline{m\beta} + 1\,\overline{n\gamma} = 0.$$

Verify this, by clearing of radicals, and showing that γ, a, and β are all tangents to the curve. (See Art. 240.)

Note.—The preceding equation is of great importance in some investigations, and it will be found upon expansion to involve four varieties of sign, in the terms containing $\beta\gamma$, γa, $a\beta$. This agrees with the fact that there are *four* conics which touch the sides of a triangle, namely, one *inscribed*, and three *escribed*—that is, tangent to one side externally, and to the prolongations of the other two internally. If we suppose a, β, γ all *positive*, therefore, the equation will represent the *inscribed* conic; thus,

$$\sqrt{la} + 1\,\overline{m\beta} + 1\,\overline{n\gamma} = 0.$$

The equations to the *escribed* conics will be

$$1\,\overline{-la} + \sqrt{m\beta} + 1\,\overline{n\gamma} = 0, \quad 1\,\overline{la} + 1\,\overline{-m\beta} + 1\,\overline{n\gamma} = 0, \quad \sqrt{la} + \sqrt{m\beta} + 1\,\overline{-n\gamma} = 0,$$

since, in each, one set of perpendiculars must fall on the triangle externally.

15. Find the equation to the circle inscribed in any triangle.

We may derive this from $1\,\overline{la} + 1\,\overline{m\beta} - 1\,\overline{n\gamma} = 0$, by clearing of radicals on the assumption that a, β, γ are positive, and then taking l, m, n so as to satisfy the condition of Art. 239. Or, we may develop the equation from that of the *circumscribed* circle, as follows[*]:—Let the sides of the triangle formed by joining the points of contact of the inscribed circle be a', β', γ'. Its equation (Art. 236) will then be

$$\frac{\sin A'}{a'} + \frac{\sin B'}{\beta'} + \frac{\sin C'}{\gamma'} = 0.$$

But, with respect to the triangles $A'B'C'$, etc., we have (Ex. 1) $a'^2 = \beta\gamma$, $\beta'^2 = \gamma a$, $\gamma'^2 = a\beta$; and $A' = 90° - \tfrac{1}{2}A$, $B' = 90° - \tfrac{1}{2}B$, $C' = 90° - \tfrac{1}{2}C$. Substituting these values, and multiplying the resulting expression throughout by $1\,\overline{a\beta\gamma}$, the required equation is found to be

$$\cos\tfrac{1}{2}A\,1\,\overline{a} + \cos\tfrac{1}{2}B\,1\,\overline{\beta} + \cos\tfrac{1}{2}C\,1\,\overline{\gamma} = 0.$$

The student may now investigate the equations to the three escribed circles. By the principle developed in Ex. 14, Note, these may be written down at once, from the equation just found.

---

[*] Hart's solution, quoted by Salmon. An. Ge. 22.

Section II. — Tangential Co-ordinates.

242. We now come to a method of representing lines and points, which, in connection with the trilinear, plays a very important part in the Modern Geometry. It is known as the method of Tangential Co-ordinates, and was first employed by Möbius, in his *Barycentrische Calcul*, which appeared in 1827.

We can here give only an outline of the system. For a full exhibition of it in its most important applications, the student is referred to the work just cited, to Plücker's *System der analytischen Geometrie*, and to Salmon's *Conic Sections*.

243. The following considerations will bring into view the relations of the new system to the Cartesian and the trilinear.

Let $\lambda x + \mu y + \nu = 0$ be the equation to a right line. Since the position of the line is determined by fixing the values of λ, μ, ν, it is evident that we may regard these co-efficients as the *co-ordinates of a right line*.

Suppose then we take λ, μ, ν as variables, and connect them by any equation of the first degree, $a\lambda + b\mu + c\nu = 0$. Such an equation (Art. 115, Cor.) is the condition that the whole system of lines denoted by $\lambda x + \mu y + \nu = 0$ shall pass through a fixed point, whose co-ordinates are $a : c$, $b : c$. Since, then, the point becomes known whenever $a\lambda + b\mu + c\nu = 0$ has known co-efficients, we must regard the condition just written as the *equation to a point*.

Accepting this result, and putting α, β, etc., as abbreviations for the expressions equated to zero in the equations to different points, we shall have, by the analogy of Art. 108, $l\alpha + m\beta = 0$ as the equation to a point dividing in a given ratio the distance between the points $a = 0$, $\beta = 0$. Similarly,

$$la - m\beta = 0, \quad m\beta - n\gamma = 0, \quad n\gamma - la = 0$$

denote three points which lie on one right line.

244. From what has just been said, we can see that we may have a system of notation in which *co-ordinates represent right lines, while points are represented by equations of the first degree between the co-ordinates.* This is what we mean by a system of tangential co-ordinates.

The tangential method is in a certain sense the reciprocal of the Cartesian. It begins with the *Right Line*, and, by means of an infinite number of right lines all passing through the same point, determines the *Point*; the Cartesian method, on the contrary, begins with the *Point*, and determines the Right Line as the assemblage of an infinite number of points. In the tangential system, accordingly, the Right Line fulfills the office assigned to the Point in the Cartesian: it is the determinant of all forms, which are conceived to be obtained by causing an infinite number of right lines to intersect each other in points infinitely close together.

245. Article 243 has shown us that Cartesian *co-efficients* are tangential *co-ordinates*, and that the tangential *equation* of the first degree is a Cartesian *equation of condition*. We shall presently see that tangential equations are *always* Cartesian conditions, and in fact signify that a right line passes through two consecutive points on a given curve: that is, through two points infinitely near to each other: and so, is a tangent to the curve.

246. The truth of this will appear from the following geometric interpretation of tangentials, which will bring the system into relation to curves of all orders, and show that we can represent *any* curve in this notation.

Let *AB*, *CD*, *EF* be any three lines, each passing through two consecutive points of the curve *LM*, and therefore touching it, say at *P*, *Q*, *R*. It is plain from the diagram, that the perimeter of the polygon formed by such 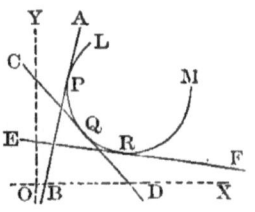 tangents will approach the curve more and more closely as the number of the tangents is increased. Hence, when the number becomes infinite, the perimeter will *coincide* with the curve.

Suppose then that λ, μ, ν being *continuous variables* we connect them by an equation expressing the condition that every line represented by $\lambda x + \mu y + \nu = 0$ shall touch *LM*. We shall then have an infinite system of tangents to *LM*; that is, we shall have the curve itself. Hence, the equation of condition in λ, μ, ν may be taken as the symbol of the curve, and is called its *tangential* equation for obvious reasons.

247. To form the tangential equation to any curve, we therefore have only to find the condition in λ, μ, ν which must be fulfilled in order that the line

$$\lambda x + \mu y + \nu = 0, \text{ or } \lambda \alpha + \mu \beta + \nu \gamma = 0,$$

may touch the curve. This may be done by *eliminating one variable between the equation to the line and that of the curve, and then forming the condition that the resulting equation may have equal roots.* For the roots of the resulting equation are the co-ordinates of the points in which the line cuts the curve (Art. 62, Cor. 1); and, if these roots are equal, the points of section become coincident (that is, consecutive), and the line is a tangent.

Instead of this method, it is often preferable to employ another, which will be illustrated a little farther on.

248. The Right Line in Tangentials.—This is represented not by an *equation*, but by *co-ordinates*. The symbol of a *fixed* right line is therefore

$$\left.\begin{array}{l} a'\lambda + b'\mu + c'\nu = 0 \\ a''\lambda + b''\mu + c''\nu = 0 \end{array}\right\}.$$

For, by solving between these simultaneous equations to two points, we can fully determine the ratios $\lambda : \mu : \nu$; that is, we can determine the line which joins the points represented by the given equations.

This result is in harmony with the fact already noticed, that the Right Line plays in the tangential system the same part that the Point does in the Cartesian. In the abridged notation, the simultaneous equations $a = 0$, $\beta = 0$ are the tangential symbol of a right line. For the sake of brevity, we shall generally speak of the line joining the points a and β (that is, the points whose equations are $a = 0$, $\beta = 0$) as the line $a\beta$: just as, in the trilinear system, we call the point in which the lines a and β intersect, the point $a\beta$.

ENVELOPES.

249. We have called the geometric equivalent of a Cartesian or trilinear equation the *locus of a point*. Similarly, the geometric equivalent of a tangential equation is called the *envelope of a right line*. This term needs explanation.

Every tangential equation takes the co-efficients of $\lambda a + \mu\beta + \nu\gamma = 0$ as *variables*. It therefore implies the existence of an infinite series of right lines, the successive members of which have directions differing by less than any assignable quantity, and intersections lying infinitely near to each other. Such a series of consecutive directions, blending at consecutive points, will of course form a *curve*, whose figure will depend on the equation connecting the variable co-efficients λ, μ, ν. A curve thus defined by a right line whose co-ordinates vary continuously, is what we mean by an *envelope*. From the reasoning of Art. 246, it is evident that a right line always *touches* its envelope.

Definition.—The **Envelope** of a right line is the series of consecutive directions to which it is restricted by given conditions of form.

Or, The envelope of a right line is the curve which it always touches.

Remark.—We thus come upon an essential distinction between the tangential system and the Cartesian. In the latter, a curve is conceived to be the aggregate of an infinite number of *positions;* in the former, it is regarded as the complex of an infinite number of *directions*. It appears from this article, that, instead of calling the condition that a right line shall touch a curve the *tangential equation to the curve*, we may say it is the *equation to the envelope of the line*.

250. Tangential equation to the Conic circumscribed about a Triangle.—We might form this by the method of Art. 247, but can proceed more rapidly as follows:—The trilinear equation to the tangent of a conic, referred to the inscribed triangle, (Art. 240, II) is

$$\frac{l\alpha}{\alpha'^2} + \frac{m\beta}{\beta'^2} + \frac{n\gamma}{\gamma'^2} = 0 \qquad (1),$$

the equation to the conic itself being

$$\frac{l}{\alpha} + \frac{m}{\beta} + \frac{n}{\gamma} = 0 \qquad (2).$$

Hence, in order that the line $\lambda a + \mu\beta + \nu\gamma = 0$ may touch the curve, we must have $\lambda a'^2 = \rho l$, $\mu\beta'^2 = \rho m$, $\nu\gamma'^2 = \rho n$; that is,

$$a' = \sqrt{\frac{\rho l}{\lambda}}, \quad \beta' = \sqrt{\frac{\rho m}{\mu}}, \quad \gamma' = \sqrt{\frac{\rho n}{\nu}}.$$

But a', β', γ' are the co-ordinates of the point of contact, and must satisfy the equation to the curve. Therefore, after substituting in (2) the values just found,

$$\sqrt{l\lambda} + \sqrt{m\mu} + \sqrt{n\nu} = 0$$

is the condition that $\lambda a + \mu\beta + \nu\gamma = 0$ shall touch the conic. In other words, it is the tangential equation to the conic.

Remark.—By clearing this equation of radicals, we shall find that it is of the *second degree* in λ, μ, ν.

251. The further investigation of tangentials requires that we now turn aside from our direct path, to examine the method of finding the envelope of a right line when the constants which enter its equation are subject to given conditions.

By the terms of this problem, the equation to the line may be written so as to involve a single indeterminate quantity: for instance, in the form

$$m^2 a - 2km\gamma + k\beta = 0,$$

where m is indeterminate, and k is fixed. For, by means of the given conditions, we can eliminate one of the arbitrary constants which enter the original equation, and thus leave but one.

Now, by the definition of an envelope (Art. 249), the right line

$$m^2 a - 2km\gamma + k\beta = 0 \tag{1},$$

is tangent to its envelope. In this particular instance, then, the envelope is such a curve that only *two* tangents can be drawn to it from a given point. For if we suppose the line (1) to pass through the given point $a'\beta'\gamma'$, we shall have

$$a'.m^2 - 2k\gamma'.m + k\beta' = 0 \qquad (2),$$

from which to find m, the indeterminate in virtue of which (1) represents a *system* of tangents to the envelope; and, since (2) is a *quadratic* in m, only two lines of the system can pass through $a'\beta'\gamma'$. This property of the envelope is designated by calling it a curve of the *Second class*: the class of a curve being determined by the number of tangents that can be drawn to it from any one point. We learn, then, that *if the equation to a right line involves an indeterminate quantity in the nth degree, the envelope is a curve of the nth class.* *

The question still remains, How shall we obtain the *equation* to this envelope? The definition of Art. 249 answers this; for since the envelope is the series of *consecutive* directions of the tangent, we have only to form the condition that the n values of the indeterminate m may be consecutive, and the required equation is found. Now this condition is of course the same as that which requires the equation in m to have equal roots. For example, the equation to the envelope of (1) is $a\beta = k\gamma^2$.

Hence, *To find the envelope of a right line, throw its equation, by means of the given conditions, into a form involving a single indeterminate quantity in the nth degree: the condition that this equation in* m *shall have equal roots, is the equation to the envelope.*

---

* The student must not confound the *class* of a curve with its *order*. A conic belongs to the Second *order*, and also to the Second *class;* but other curves do not in general show this agreement.

Example.—Given the vertical angle and the sum of the sides of a triangle: to find the envelope of the base.

Take the sides for axes; then the equation to the base is

$$\frac{x}{a} + \frac{y}{b} = 1,$$

in which a and b are subject to the condition $a + b = c$. The equation to the base may therefore be written

$$a^2 + (y - x - c)a + cx = 0.$$

Hence, the equation to the envelope is

$$(y - x - c)^2 = 4cx;$$

or,

$$x^2 - 2xy + y^2 - 2cx - 2cy - c^2 = 0.$$

The required envelope is therefore (Art. 191) a parabola; which touches the sides $x = 0$ and $y = 0$, since the equations for determining its intercepts on the axes are

$$x^2 - 2cx + c^2 = 0, \quad y^2 - 2cy + c^2 = 0.$$

We may now resume the direct course of our investigation, and finish this part of it by establishing, in their proper order, the theorems necessary for interpreting tangential equations of the first and second degrees.

252. Tangential equation to any Conic.—We shall here follow the method of Art. 247, and find the condition that $\lambda a + \mu \beta + \nu \gamma = 0$ may touch the conic

$$A a^2 + B \beta^2 + C \gamma^2 + 2F \beta \gamma + 2G \gamma a + 2H a \beta = 0.$$

Eliminating γ between the equation to the line and that of the conic, and collecting the terms of the result with reference to $a : \beta$, we obtain

$$(A\nu^2 - 2G\nu\lambda + C\lambda^2) a^2 + 2(H\nu^2 - G\mu\nu - F\nu\lambda + C\lambda\mu) a\beta$$
$$+ (B\nu^2 - 2F\mu\nu + C\mu^2) \beta^2 = 0.$$

Forming the condition that this complete quadratic in $a : \beta$ may have equal roots, we get the required tangential equation, namely,

$$(H\nu^2 - G\mu\nu - F\nu\lambda + C\lambda\mu)^2 = (A\nu^2 - 2G\nu\lambda + C\lambda^2)(B\nu^2 - 2F\mu\nu + C\mu^2):$$

An. Ge. 23

which, after expansion and division by ν^2, becomes

$$\left. \begin{aligned} (BC-F^2)\lambda^2+(AC-G^2)\mu^2+(AB-H^2)\nu^2 \\ +2(HG-AF)\mu\nu+2(HF-BG)\nu\lambda+2(GF-CH)\lambda\mu \end{aligned} \right\} = 0 \quad (1).$$

Now it is remarkable that the co-efficients of (1) are the *derived polynomials* (see Alg., 411) of the Discriminant Δ, obtained by supposing the variable to be successively A, B, C, F, G, H. They may therefore be aptly denoted by a, b, c, $2f$, $2g$, $2h$, so that (1) shall be written

$$a\lambda^2 + b\mu^2 + c\nu^2 + 2f\mu\nu + 2g\nu\lambda + 2h\lambda\mu = 0 \quad (2).$$

Corollary.—If we represent the Discriminant of (2), namely,

$$abc + 2fgh - af^2 - bg^2 - ch^2$$

by ∂, and substitute for a, b, c, etc., their values from (1), we shall obtain the relation

$$\partial = \Delta^2 :$$

a property which has some important bearings.

253. Theorem.—*The envelope of a right line whose co-ordinates are connected by any relation of the first degree, is a point.*

For (Art. 243) every such relation is the tangential equation to a point through which the line always passes (that is, to which it is always tangent); and (Art. 249, Rem.) the tangential equation is the equation to the envelope.

Corollary.—By means of the relation $a\lambda + b\mu + c\nu = 0$, we can eliminate either $\lambda : \nu$ or $\mu : \nu$ from the equation $\lambda a + \mu \beta + \nu \gamma = 0$, and so cause it to involve only a single indeterminate, of the first degree. Hence, *If the* *equation to a right line involves a single indeterminate quantity in the first degree, the envelope of the line is a point.* [Compare Art. 116.]

This corollary only carries out the principle of Art. 251. For a point may of course be regarded as a curve of the First class.

254. Theorem.—*The envelope of a right line whose co-ordinates are connected by any relation of the second degree, is a conic.*

We are here required to prove that any tangential equation of the form

$$a\lambda^2 + b\mu^2 + c\nu^2 + 2f\mu\nu + 2g\nu\lambda + 2h\lambda\mu = 0,$$

in which a, b, c, f, g, h are any six constants whatever, represents a conic. Eliminating ν between the given equation and $\lambda a + \mu\beta + \nu\gamma = 0$, we find

$$(a\gamma^2 - 2g\gamma a + ca^2)\lambda^2 + 2(h\gamma^2 - g\beta\gamma - f\gamma a + ca\beta)\lambda\mu$$
$$+ (b\gamma^2 - 2f\beta\gamma + c\beta^2)\mu^2 = 0.$$

Hence, the equation to the envelope of the line is

$$(h\gamma^2 - g\beta\gamma - f\gamma a + ca\beta)^2 = (a\gamma^2 - 2g\gamma a + ca^2)(b\gamma^2 - 2f\beta\gamma + c\beta^2);$$

that is, after expanding and dividing through by γ^2,

$$\left.\begin{array}{l}(bc - f^2)a^2 + (ac - g^2)\beta^2 + (ab - h^2)\gamma^2 \\ + 2(hg - af)\beta\gamma + 2(hf - bg)\gamma a + 2(gf - ch)a\beta\end{array}\right\} = 0 \quad (1):$$

or, since the co-efficients are the derived polynomials of δ, supposing the variable to be successively a, b, c, etc., and may therefore be represented by A, B, C, etc.,

$$Aa^2 + B\beta^2 + C\gamma^2 + 2F\beta\gamma + 2G\gamma a + 2Ha\beta = 0 \quad (2):$$

which is the trilinear equation to a conic. Our proposition is therefore established.

Corollary.—By means of the relation given in the hypothesis of the theorem above, we can cause the equation $\lambda\alpha + \mu\beta + \nu\gamma = 0$ to involve but a single indeterminate, say $\lambda : \nu$, in the second degree. Hence, *If the equation to a right line involves an indeterminate quantity in the second degree, the envelope of the line is a conic.*

This conclusion might have been deduced at once from the equation of Art. 251, namely,

$$m^2\alpha - 2mk\gamma + k\beta = 0.$$

For this is a general type for all equations answering the description of the corollary, and the equation to the envelope of the corresponding line is

$$\alpha\beta = k\gamma^2,$$

which, as we have seen in Ex. 1, p. 221, is the *equation to a conic, referred to two tangents and their chord of contact.*

255. Interchange of the Trilinear and Tangential equations to a Conic.—If we compare equation (1) of the preceding article with equation (1) of Art. 252, it becomes apparent that the former is derived from

$$a\lambda^2 + b\mu^2 + c\nu^2 + 2f\mu\nu + 2g\nu\lambda + 2h\lambda\mu = 0$$

by exactly the same series of operations by which the latter is derived from

$$A\alpha^2 + B\beta^2 + C\gamma^2 + 2F\beta\gamma + 2G\gamma\alpha + 2H\alpha\beta = 0.$$

Hence, *To form the tangential equation to a conic when its trilinear equation is given, replace α, β, γ by λ, μ, ν, and the co-efficients* A, B, C, *etc., by the corresponding derived polynomials of* $\lrcorner$; *and to form the trilinear equation from the tangential, replace λ, μ, ν by α, β, γ, and the co-efficients* a, b, c, *etc., by the derived polynomials of* δ.

256. It is obvious that the application of the foregoing principles will often facilitate the investigation of envelopes. We may make the following summary of results:

I. Every tangential equation of the first degree represents a point.

II. Every tangential equation of the second degree represents a conic.

III. A tangential equation of the nth degree represents a curve of the nth class.

RECIPROCAL POLARS.

257. Reciprocal relation between Points and Lines.—We have already noticed the reciprocity of Cartesian and tangential equations, as suggested by the fact that the Point and the Right Line interchange their offices in passing from one system to the other. This remarkable property, however, does not appear in its full significance until we apply to tangentials the same system of abridged notation that converts a Cartesian into a trilinear equation. When this notation *is* applied, it is found that an equation in a, β, γ or u, v, w is susceptible of two interpretations, according as it is read in trilinears or in tangentials; and gives rise to two distinct theorems (one relating to *points*, the other to *lines*), which in view of their derivation may not inaptly be styled *reciprocal* theorems.

This capability of double interpretation is known among mathematicians as the Principle of Duality, and has led to many of the most striking results of the Modern Geometry. A few illustrations will enable the student to conceive of the principle clearly.

Suppose, for brevity, we write $S = 0$, $S' = 0$ as the equations to two conics, either Cartesian or tangential. Then the equation $S + kS' = 0$, being satisfied either by the co-ordinates of *points* which render S and S' simultaneously equal to zero, or by the co-ordinates of *lines* which effect the same result, denotes in trilinears a conic which passes through the four *points* in which the conic S cuts the conic S', and in tangentials a conic which touches the four *lines* that *touch* S and S' in common.

Similarly $a\gamma = k\beta\delta$, being the equation to a conic since it is of the second degree, may be read either in trilinears or tangentials. It is obviously satisfied by either of the four conditions

$$(a = 0, \ \beta = 0), \quad (\beta = 0, \ \gamma = 0), \quad (\gamma = 0, \ \delta = 0), \quad (\delta = 0, \ a = 0).$$

Hence, in trilinears it denotes a conic passing through the four *points* $a\beta$, $\beta\gamma$, $\gamma\delta$, δa; that is, *circumscribed* about the quadrilateral whose *sides* are the four *lines* a, β, γ, δ: while in tangentials it represents a conic touching the four *lines* $a\beta$, $\beta\gamma$, $\gamma\delta$, δa; that is, *inscribed* in the quadrilateral whose *vertices* are the four *points* a, β, γ, δ.

Again, $S + ka\beta = 0$ is a conic passing through the four *points* in which the *lines* a and β cut the conic S, or touching the four *lines* drawn from the *points* a and β to *touch* the conic S. If, then, we have three conics S, $S + ka\beta$, $S + k'a\gamma$, we may either say that all three pass through the two points in which the line a cuts S, or that all three touch the two lines drawn from the point a to touch S.

We can now exemplify the method of obtaining reciprocal theorems. The three conics S, $S + ka\beta$, $S + k'a\gamma$, as we have just shown, all pass through the two points in which the line a cuts them. Moreover, the line β evidently joins the two remaining points in which S cuts $S + ka\beta$; the line γ joins the two remaining points in which S cuts $S + k'a\gamma$; while, for the line joining the two remaining points in which $S + ka\beta$ cuts $S + k'a\gamma$, we get, by eliminating between these equations, $k\beta - k'\gamma = 0$. Now (Art. 108) this last line must pass through the point $\beta\gamma$. Hence, we have the following theorem:

I. *If three conics have two points common to all, the three lines joining the remaining points common to each two, meet in one point.*

Let us now take the same equations in tangentials. The two tangents from a are common to the three conics, the pair from β is common to the first and second, the pair from γ is common to the first and third, while the pair common to the second and third intersect in the point $k\beta - k'\gamma$. But, on the analogy of Art. 108, the latter point is on the line $\beta\gamma$. Hence the *reciprocal* theorem:

II. *If three conics have two tangents common to all, the three points in which the remaining tangents common to each two intersect, lie on one right line.*

By comparing the phraseology of I and II, we see that either may be derived from the other by simply interchanging the words *point* and *tangent*, and *point* and *line*. In fact, if the reader chooses to push his inquiries by consulting other authors upon this subject, he will find that the entire process of *reciprocation*, as it is called, may be reduced to the operation of interchanging the terms *point* and *line*, *chord* and *tangent*, *circumscribed* and *inscribed*, *locus* and *envelope*, etc.

258. Geometric meaning of the Reciprocal Relation.—The process of reciprocation being so mechanical, the student may very naturally ask how we can be certain that reciprocal theorems are any thing more than fanciful trifling with words. As a sufficient answer to this question, we shall now show that if a given theorem is proved of a certain curve, we can always generate a second curve from the first, to which the reciprocal of the given theorem will surely apply. In short, we shall show that the reciprocity which we have illustrated is not merely a property of trilinear and tangential *equations* identical in form, but that the *curves* to which such equations belong are reciprocal.

The truth of this statement will appear in two steps: we shall first explain the meaning and establish the existence of reciprocal curves; and then prove that the tangential equation to a curve is the trilinear equation to its reciprocal.

259. Generation of Reciprocal Curves.—To explain this, and establish its possibility, we shall have to anticipate a single property of conics. The theorem will be proved in its proper place in Part II, but for the present the student must take it upon trust.

Every conic, then, is characterized by the following twofold property :

I. *If different chords to a conic be drawn through the same point, and tangents to the curve be formed at the extremities of each chord, the intersections of all these pairs of tangents will lie on the same right line.*

II. *If different pairs of tangents be drawn to a conic from points lying on the same right line, and chords be formed joining the points of contact belonging to each pair, all these chords of contact will intersect in the same point.*

From this it appears, that, in relation to any conic, there is a certain right line determined by, and therefore corresponding to, any assumed point; and a certain point determined by, and therefore corresponding to, any assumed right line. This interdependence of points and lines is expressed by calling the point the *pole* of the line, and the line the *polar* of the point.

If the student will now draw diagrams, forming the polar of a point according to I, and the pole of a line according to II, he will find that when a point is *within* the conic (a circle will be most convenient for illustration), its polar is *without;* that when the point is *without* the conic, its polar is *within,* and in fact is the chord of contact of the two tangents drawn from the point; that when the point is *on* the conic, its polar is also *on* the curve — in fact, is the tangent at the point. Conversely, if a right line is *without* a conic, its pole is *within;* if the line is *within* (that is, if it forms a chord), its pole is *without,* and is the intersection of the two tangents drawn at the extremities of the chord; if the line is *on* the curve, the pole is the *point of contact.* Thus, in the diagram, P is the pole of LM, and LM the polar of P; L is the pole of $T'T$, and $T'T$ the polar of L; M is the pole of $V'V$, and $V'V$ the polar of M; T is the pole of LT, and LT the polar of T; and so on. The pole is said to *correspond* to its polar, and reciprocally.

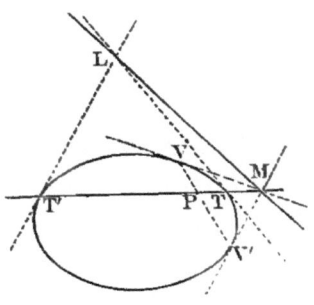

It is obvious that as the polar changes its position, the position of the pole is changed; so that, if the polar determine a curve as its envelope, the pole will determine another as its locus. Suppose, then, we have any curve S, and relate it to the conic Σ by taking the pole with respect to Σ of any tangent to S: the locus of the pole will be a second curve s, which may be called the *polar* curve of S. It is evident that every point of s will *correspond* to (that is, be the pole of) some tangent to S. Therefore, if we take any two points on s, they will at the same time determine a chord of s, and the intersection of two tangents to S: that is, every chord of s is the polar of the intersection of the two tangents to S which are the polars of the extremities of that chord. Hence, supposing the points of s to be consecutive, and the corresponding tangents to S on that account to intersect on the curve, we have: *Every point of S is the pole with respect to Σ of some tangent to s.* That is, as s is the locus of the pole of any tangent to S, so S is the locus of the pole of any tangent to s. Or, in other words, *a curve and its polar with respect to a fixed conic Σ may be generated each from the other in exactly the*

same manner. A curve and its polar are thus seen to be *reciprocal forms.*

Given, then, any curve whatever, by means of a fixed conic Σ we can always generate a second curve, which may properly be called the *reciprocal polar* of the first.

260. The tangential equation to a Curve, the trilinear equation to its Reciprocal.—This theorem is clearly true; for the co-ordinates of any tangent to S may of course be taken as the co-ordinates of its pole, that is, as the co-ordinates of any point on s: hence, the equation to the envelope S may be regarded as the equation to the locus s.

261. Since it thus appears that every curve has its reciprocal, whose equation is identical with the tangential equation to the curve, it follows that all the results obtainable either by mechanical reciprocation or by the double interpretation of equations, are real properties of real curves. Given, then, any equation in a, β, γ, we are not only *justified* in reading it both in trilinears and in tangentials, but *must* so read it, if we wish to exhaust its geometric meaning.

Note.—From the relation now established between the curves corresponding to the two interpretations of an equation in a, β, γ, the method of deriving reciprocal theorems is sometimes called the *Method of Reciprocal Polars*, instead of the *Principle of Duality.* It should be stated, however, that both these terms, as now applied to processes purely analytical, are borrowed from the cognate branch of pure geometry. They both entered the history of Geometry as titles of purely geometric processes, and the larger part of their remarkable results were established by the aid of the diagram alone. From the very process of generating a reciprocal polar, it is evident that the Method of Reciprocals contains in itself the evidence for the truth of all theorems based upon it, and need not invite the aid of analysis.

The Principle of Duality, as a purely geometric method, is due to the French mathematician Gergonne; its first presentation in the analytic form of an equation with double meaning, was made by Plucker, in his *System der analytischen Geometrie*, 1835. The geometric Method of Reciprocal Polars was the invention of Poncelet, who presented an account of its elements in Gergonne's *Annales de Mathématiques*, tom. VIII, 1818; and, afterward, an extended development of its general theory in Crelle's *Journal für die reine und angewandte Mathematik*, Bd. IV, 1829. The latter, however, was previously read in 1824 to the Royal Academy of Sciences at Paris, and led to a dispute between Poncelet and Gergonne as to the prior claims of the Principle of Duality. For the discussion which ensued, the reader is referred to the *Annales*, tom. XVIII.

The conic Σ, upon which the Principle of Duality and the Method of Reciprocal Polars as analytic processes are based, is called the *auxiliary* conic. It may be any fixed conic whatever, but is in practice usually a *circle*; because that curve enjoys certain properties by means of which we can reciprocate theorems concerning magnitude as well as those concerning position. The use of a *parabola* for Σ has been introduced by Chasles, but the applications of which his method is capable are comparatively few.

We shall now demonstrate two or three of the leading properties of reciprocals.

262. Theorem.—*The reciprocal of a right line with respect to Σ is a point; and conversely.*

For the tangential equation of the first degree denotes a point.

263. Theorem.—*The reciprocal of a conic with respect to Σ is a conic.*

For the trilinear equation to a conic is of the second degree, and every equation of the second degree, when interpreted in tangentials, denotes a conic.

264. Theorem.—*The reciprocal of a curve of the* n^{th} *order, is a curve of the* n^{th} *class.*

For the trilinear equation of the nth degree, when interpreted in tangentials, denotes a curve to which n tangents can be drawn from a given point.

265. We add a few exercises upon various subjects treated in this Section, premising that the student must not suppose them to be adequate illustrations of the scope of the tangential method : they are in fact only useful for fixing the leading points of the preceding sketch. The reader who wishes to see the very remarkable results of the principles of Duality and Reciprocal Polars may consult, in addition to the works already mentioned, Chasles's *Traité de Géométrie Supérieure*, Steiner's *Entwickelung der Abhängigkeit geometrischen Gestalten, etc.*, Booth's treatise *On the Application of a New Analytic Method to the Theory of Curves and Curved Surfaces*, Salmon's *Higher Plane Curves* and *Geometry of Three Dimensions*, and Poncelet's *Traité des Propriétés Projectives*.

EXAMPLES.

1. Interpret in tangentials the several equations of the example in Art. 219.

2. Write the equations to the points $(5, 6)$, $(-3, 2)$, $(7, 8, -9)$.

3. What is the tangential symbol of the right line passing through $(2, 3)$ and $(4, 5)$?

4. Form the tangential equation to the circle represented by

$$\frac{\sin A}{a} + \frac{\sin B}{\beta} + \frac{\sin C}{\gamma} = 0.$$

5. Interpret the equation $a\beta = k\gamma^2$ both in trilinears and tangentials.

6. Show that $m^2a - 2mk\gamma + k\beta = 0$ is the tangential equation to any point on the curve $a\beta = k\gamma^2$.

7. Find the equation to the reciprocal of the conic represented by

$$\sqrt{la} + \sqrt{m\beta} + \sqrt{n\gamma} = 0.$$

8. Find the equation to the envelope of the right line whose co-efficients fulfill the condition

$$\frac{l}{\lambda} + \frac{m}{\mu} + \frac{n}{\nu} = 0,$$

and the equation to the envelope of one whose co-efficients satisfy

$$\sqrt{l\lambda} + \sqrt{m\mu} + \sqrt{n\nu} = 0.$$

What is the meaning of the results?

9. Prove that the envelope of the conic represented by

$$\frac{l}{a} + \frac{m}{\beta} + \frac{n}{\lambda} = 0,$$

in which l, m, n are subject to the condition

$$\sqrt{\lambda l} + \sqrt{\mu m} + \sqrt{\nu n} = 0,$$

is the right line $\lambda a + \mu \beta + \nu \gamma = 0.$

10. Prove that the envelope of the conic $\sqrt{la} + \sqrt{m\beta} + \sqrt{n\gamma} = 0$, whose co-efficients satisfy the relation

$$\frac{l}{\lambda} + \frac{m}{\mu} + \frac{n}{\nu} = 0,$$

is the right line $\lambda a + \mu \beta + \nu \gamma = 0.$

11. Find the envelope of a right line, the perpendiculars to which from two given points contain a constant rectangle.

12. The vertex of a given angle moves along a fixed right line while one side passes through a fixed point: to find the envelope of the other side.

13. A triangle is inscribed in a conic, and its two sides pass through fixed points: to find the envelope of its base.

14. Prove that if three conics have two points common to all, their three reciprocals will have two tangents common to all; and conversely.

15. Establish the following reciprocal theorems, and determine the conditions for the cases noticed under them:

If two vertices of a triangle move along fixed right lines while the three sides pass each through a fixed point, the locus of the third vertex is a conic.

But if the points through which the sides pass lie on one right line, the locus will be a right line.

In what other case will the locus be a right line?

If two sides of a triangle pass through fixed points while the three vertices move each along a fixed right line, the envelope of the third side is a conic.

But if the lines on which the vertices move meet in one point, the envelope will be a point. [Compare Ex. 39, p. 125.]

In what other case will the envelope be a point? [Compare Ex. 40, p. 125.]

PLANE CO-ORDINATES.

PART II.

THE PROPERTIES OF CONICS.

266. The investigations of Part I, have taught us the methods of representing geometric forms by analytic symbols; and furnished us, in the resulting equations to curves of the First and Second orders, with the necessary instruments for discussing those curves. We now proceed to apply our instruments to the determination of the properties of the several conics. We shall adhere to the order of Part I, considering first the several varieties in succession, and then, by way of illustrating the method of investigating the common properties of a whole order of curves, determining those of the Conic in General.

CHAPTER FIRST.

THE RIGHT LINE.

267. Under this head, we only purpose developing a few properties, noticeable either on account of their usefulness or their relations to the new or to the higher geometry.

268. Area of a Triangle in terms of its Vertices. — Let the vertices be x_1y_1, x_2y_2, x_3y_3, represented in the diagram by A, B, C. It is obvious that for the required area we have

$$ABC = ALMB + BMNC - CALN;$$

that is,

$$T = \tfrac{1}{2}\left\{ (x_2 - x_1)(y_1 + y_2) + (x_3 - x_2)(y_2 + y_3) - (x_3 - x_1)(y_3 + y_1) \right\};$$

or,

$$2T = y_1(x_2 - x_3) + y_2(x_3 - x_1) + y_3(x_1 - x_2).$$

Remark. — This expression for the double area of the triangle is identical with that which in Art. 112 is equated to zero as the condition that three points may lie on one right line. We thus discover the latter condition to be simply the algebraic statement, that, *when three points lie on one right line, the triangle which they determine vanishes:* which obviously accords with the fact.

269. Area of a Triangle in terms of its inclosing Lines. — Let the three lines be $Ax + By + C = 0$, $A'x + B'y + C' = 0$, $A''x + B''y + C'' = 0$. Their intersections will form the vertices of the triangle; hence, substituting for x_1y_1, x_2y_2, x_3y_3, in the preceding formula, according to Art. 106, we obtain

$$2T = \frac{C\,A' - C'\,A}{A\,B' - A'\,B}\left\{ \frac{B'\,C'' - B''\,C'}{A'\,B'' - A''B'} - \frac{B''C - B\,C''}{A''B - A\,B''} \right\}$$

$$+ \frac{C'\,A'' - C''A'}{A'\,B'' - A''B'}\left\{ \frac{B''C - B\,C''}{A''B - A\,B''} - \frac{B\,C' - B'\,C}{A\,B' - A'\,B} \right\}$$

$$+ \frac{C''A - C\,A''}{A''B - A\,B''}\left\{ \frac{B\,C' - B'\,C}{A\,B' - A'\,B} - \frac{B'\,C'' - B''C'}{A'\,B'' - A''B'} \right\}.$$

Now if we reduce each of the sets of fractions inside the braces to a common denominator, the three new numerators will be respectively

$$B''\{B(C''A'-C'A'')+B'(CA''-C''A)+B''(C'A-CA')\},$$
$$B\ \{B(C''A'-C'A'')+B'(CA''-C''A)+B''(C'A-CA')\},$$
$$B'\{B(C''A'-C'A'')+B'(CA''-C''A)+B''(C'A-CA')\}.$$

Hence, the final expression for the required area may be written

$$2T=\frac{\{B(C''A'-C'A'')+B'(CA''-C''A)+B''(C'A-CA')\}^2\}.}{(AB'-A'B)(A'B''-A''B')(AB''-A''B)}$$

Remark.—The numerator of this expression may be otherwise written

$$\{A''(BC'-B'C)+B''(CA'-C'A)+C''(AB'-A'B)\}^2:$$

so that (Art. 113), if the three lines pass through the same point, $2T$ vanishes. On the other hand, the expression for $2T$ becomes infinite (Art. 96, Cor. 2) whenever any two of the lines are parallel. In both respects, then, the formula accords with the facts.

270. Ratio in which the Distance between Two Points is divided by a Given Line.—Let $m:n$ be the ratio sought, and $x_1y_1,\ x_2y_2$ the given points. The co-ordinates of division (Art. 52) will then be

$$x=\frac{mx_2+nx_1}{m+n},\qquad y=\frac{my_2+ny_1}{m+n}.$$

But these must of course satisfy the equation to the given line : hence,

$$A(mx_2+nx_1)+B(my_2+ny_1)+C(m+n)=0.$$

$$\therefore\ \ \frac{m}{n}=-\frac{Ax_1+By_1+C}{Ax_2+By_2+C}.$$

An. Ge. 24.

Corollary.—If the given line passed through two fixed points x_3y_3 and x_1y_1, we should have (Art. 95, Cor. 1)

$$\frac{m}{n} = -\frac{(y_3 - y_1)\,x_1 - (x_3 - x_4)\,y_1 + x_3y_1 - y_3x_4}{(y_3 - y_4)\,x_2 - (x_3 - x_4)\,y_2 + x_3y_4 - y_3x_4},$$

as the *ratio in which the distance between two fixed points is divided by the line joining two others.*

TRANSVERSALS.

271. Definition.—A **Transversal** of any system of lines is a line which crosses all the members of the system.

Thus, LMN is a transversal of the three sides AB, BC, CA in the triangle ABC.

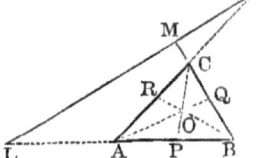

272. Theorem.—*In any triangle, the compound ratio of the segments cut off upon the three sides by any transversal is equal to* -1.

In the above diagram, let the vertices A, B, C be represented by x_1y_1, x_2y_2, x_3y_3, and the transversal LN by $Ax + By + C = 0$. Then (Art. 270)

$$AL : LB = -(Ax_1+By_1+C) : (Ax_2+By_2+C),$$
$$BM : MC = -(Ax_2+By_2+C) : (Ax_3+By_3+C),$$
$$CN : NA = -(Ax_3+By_3+C) : (Ax_1+By_1+C).$$

Multiplying these equations member by member, we obtain the proposition.

273. Theorem.—*In any triangle, the compound ratio of the segments cut off upon the three sides by any three convergents that pass through the vertices is equal to* $+1$.

For, if the point of convergency be O, represented by x_4y_4 while the vertices are denoted as in the preceding article, we shall have (Art. 270, Cor.), after merely re-arranging the terms,

$$\frac{AP}{PB} = \frac{x_1(y_4-y_3)+x_4(y_3-y_1)+x_3(y_1-y_4)}{x_2(y_3-y_4)+x_3(y_4-y_2)+x_4(y_2-y_3)},$$

$$\frac{BQ}{QC} = \frac{x_2(y_4-y_1)+x_1(y_2-y_4)+x_4(y_1-y_2)}{x_1(y_4-y_3)+x_4(y_3-y_1)+x_3(y_1-y_4)},$$

$$\frac{CR}{RA} = \frac{x_2(y_1-y_4)+x_3(y_4-y_2)+x_4(y_2-y_3)}{x_2(y_4-y_1)+x_1(y_2-y_4)+x_4(y_1-y_2)};$$

and the product of these equations is the algebraic expression of the theorem.

We shall next give some illustrations of the uses of abridged notation, as applied to rectilinear figures and to right lines in general.

TRIPLE CONVERGENTS IN A TRIANGLE.

274. Theorem.—*The three bisectors of the internal angles of a triangle meet in one point.*

For their equations are $a-\beta=0$, $\beta-\gamma=0$, $\gamma-a=0$; and (Art. 114) these vanish identically when added together.

275. Theorem.—*The bisectors of any two external angles of a triangle, and the bisector of the remaining internal angle, meet in one point.*

For their equations are $a-\beta=0$, $\beta+\gamma=0$, $\gamma-a=0$, etc. But (Art. 108) $\gamma-a$ is a line passing through the intersection of $a+\beta$ and $\beta+\gamma$.

276. Theorem.—*The three lines which join the vertices of a triangle to the middle points of the opposite sides meet in one point.*

For (Ex. 2, p. 221) their equations are respectively $a\sin A-\beta\sin B=0$, $\beta\sin B-\gamma\sin C=0$, $\gamma\sin C-a\sin A=0$, and therefore vanish identically when added together.

277. Theorem.—*The three perpendiculars let fall from the vertices of a triangle upon the opposite sides meet in one point.*

For (Ex. 3, p. 221) the corresponding equations are

$$a \cos A - \beta \cos B = 0, \beta \cos B - \gamma \cos C = 0, \gamma \cos C - a \cos A = 0.$$

278. Theorem.—*The three perpendiculars erected at the middle points of the sides of a triangle meet in one point.*

For (Ex. 4, p. 221) we have found their equations to be

$$a \sin A - \beta \sin B + \gamma \sin (A - B) = 0,$$
$$\beta \sin B - \gamma \sin C + a \sin (B - C) = 0,$$
$$\gamma \sin C - a \sin A + \beta \sin (C - A) = 0;$$

and, if we multiply these by $\sin^2 C$, $\sin^2 A$, $\sin^2 B$ respectively, we shall cause them to vanish identically. (See Trig., 850, Ex. 2.)

HOMOLOGOUS TRIANGLES.

279. Definitions.—Two triangles, the intersections of whose sides taken two and two lie on one right line, are said to be *homologous.*

The line on which the three intersections lie is called the *axis of homology.* Any two sides that form one of the three intersections are termed *corresponding* sides; and the angles opposite to them, *corresponding* angles.

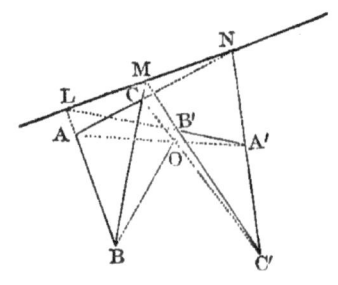

Thus, in the diagram, the triangles ABC, $A'B'C'$ are homologous with respect to the axis LMN. AB, $A'B'$; BC, $B'C''$; CA, $C'A'$ are the corresponding sides; and A, A'; B, B'; C, C'', the corresponding angles.

280. Theorem.—*In any two homologous triangles, the right lines joining the corresponding vertices meet in one point.*

Let a, β, γ be the sides of ABC, and take the latter for the triangle of reference; the equation to the axis LN may then be written $la + m\beta + n\gamma = 0$. Suppose the Cartesian origin to be somewhere between the triangles, and (Art. 108, Cor. 2) the equations to $A'B'$, $B'C'$, $C'A'$, which pass through the intersections of a, β, γ with the axis, will be

$$(l-l')a+m\beta+n\gamma=0, \quad la+(m-m')\beta+n\gamma=0, \quad la+m\beta+(n-n')\gamma=0.$$

Subtracting the second of these from the first, the third from the second, and the first from the third, we get

$$l'a - m'\beta = 0, \quad m'\beta - n'\gamma = 0, \quad n'\gamma - l'a = 0.$$

But these equations (Art. 107) evidently denote the lines BB', CC', AA'; and they vanish identically, when added together.

Remark.—The point in which the lines joining the corresponding vertices meet, is called the *center of homology*.

281. Theorem.—*If the lines joining the corresponding vertices of two triangles meet in one point, the intersections of the corresponding sides lie on one right line.*

This theorem, the converse of the preceding, is obtained at once by merely interpreting the equations of the foregoing article in tangentials. We leave the student to carry out the details.

COMPLETE QUADRILATERALS.

282. Definitions.—A **Complete Quadrilateral** is the figure formed by any four right lines intersecting in six points.

The three remaining right lines by
which the six points of intersection
can be joined two and two, are called
the *diagonals* of the quadrilateral.

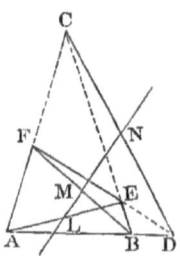

Thus, *ABCDEF* is the complete
quadrilateral of the four lines *AB*,
BE, EF, FA, which meet in the six
points *B, E, F, A, C, D.*   *AE, BF* are
the two *inner* diagonals, and *CD* is the *outer* one, sometimes called the *third*.

283. Theorem.—*In any complete quadrilateral, the
middle points of the three diagonals lie on one right line.*

Let a, β, γ, δ be the equations to the four sides of the
quadrilateral, and let the respective lengths of *BE, EF,
FA, AB* equal a, b, c, d. Then *L, M, N* being the middle
points of the three diagonals, we have the following equations to the lines drawn from the vertices to the middle
points of the bases of the triangles *ABE, EFA, ABF,
FEB:*

$$d\delta - aa = 0 \ (BL), \quad b\beta - c\gamma = 0 \ (FL);$$
$$c\gamma - d\delta = 0 \ (AM), \quad aa - b\beta = 0 \ (EM).$$

Hence, *L* and *M* both lie upon the line

$$aa - b\beta + c\gamma - d\delta = 0 \qquad\qquad (1),$$

as this obviously passes through the intersection of
(*BL, FL*), and of (*AM, EM*). If we now put $Q =$ the
double area of *ABEF*, we shall have

$$aa + b\beta + c\gamma + d\delta = Q,$$

and thence

$$aa - b\beta + c\gamma - d\delta = 2(aa + c\gamma) - Q,$$
$$aa - b\beta + c\gamma - d\delta = -2(b\beta + d\delta) + Q.$$

Therefore (Art. 229) the line (1) is parallel to the two lines $a\alpha + c\gamma$ and $b\beta + d\delta$, and midway between them. It accordingly bisects the distance between $\gamma\alpha$ (which is a point on the first) and $\beta\delta$ (which is a point on the second). That is, N lies on (1): which proves our proposition. *

THE ANHARMONIC RATIO.

284. Definition.—A **Linear Pencil** is a group of four right lines radiating from one point.

Thus, OA, OQ, OB, OP constitute a linear pencil.

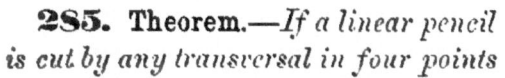

285. Theorem.—*If a linear pencil is cut by any transversal in four points* A, Q, B, P, *the ratio* AP.QB : AQ.PB *is constant.*

For, putting $p =$ the perpendicular from O upon the transversal, we have $p.AP = OA.OP \sin AOP$; $p.QB = OB.OQ \sin QOB$; $p.AQ = OA.OQ \sin AOQ$: and $p.PB = OB.OP \sin POB$: whence

$$p^2.AP.QB = OA.OB.OP.OQ \sin AOP \sin QOB,$$
$$p^2.AQ.PB = OA.OB.OP.OQ \sin AOQ \sin POB.$$

$$\therefore \frac{AP.QB}{AQ.PB} = \frac{\sin AOP \sin QOB}{\sin AOQ \sin POB}.$$

a value which is independent of the position of the transversal.

Remark.—The constant ratio just established is called the *anharmonic* ratio of the pencil. By reasoning similar to that just used, it may be shown that the ratios $AP.QB : AB.QP$ and $AB.QP : AQ.BP$ are also con-

---

stant. To these, accordingly, the term anharmonic is at times applied; but it is generally reserved for the particular ratio to which we have assigned it, and we shall always intend that ratio when we use it hereafter.

Note.—The Anharmonic Ratio has an important place in the Modern Geometry, especially in connection with the doctrine of the Conic. Its existence, however, has been known since the time of the Alexandrian geometer PAPPUS, who gives the property in his *Mathematicæ Collectiones*, Book VII, 129, and who probably belongs to the fourth century. The name *anharmonic* was given by CHASLES. But the bearing of the ratio upon the new geometry had been previously investigated by MÖBIUS, who called it the *Ratio of Double Section* (Doppelschnittsverhältniss).

286. Definition.—An **Harmonic Proportion** subsists between three quantities, when the first is to the third as the difference between the first and second is to the difference between the second and third.

Thus, if the pencil in the above diagram cut the transversal so as to make $AP:AQ :: AP - AB : AB - AQ$, the whole line AP would be in harmonic proportion with its segments AB and AQ.

Corollary.—A line is divided *harmonically*, when it is cut into three segments such that the whole is to either extreme as the other extreme is to the mean. For the proportion given above may obviously be written

$$AP : AQ :: BP : BQ.$$

287. Definitions.—An **Harmonic Pencil** is one which cuts its transversals harmonically.

I. From the final equation of Art. 285, it is evident that when the anharmonic ratio of a pencil is numerically equal to 1, the pencil is harmonic.

II. The four points in which a line is cut by an harmonic pencil, are called *harmonic points*.

III. Linear pencils are in general termed *anharmonic*, because they do not in general cut off harmonic segments from their transversals.

288. Theorem.—*The anharmonic of the pencil formed by the four lines* a, β, $a + k\beta$, $a + k'\beta$ *is equal to* k : k'.

Let OA be the position of a, OB of β, OP of $a + k\beta$, and OQ of $a + k'\beta$. Then, k being the ratio of the perpendiculars from OP upon OA, OB, and k' the ratio of those from OQ on the same lines, we shall 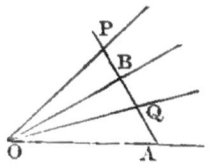 have $k = \sin AOP : \sin POB$, and $k' = \sin AOQ : \sin QOB$. Hence,

$$\frac{k}{k'} = \frac{\sin AOP \sin QOB}{\sin AOQ \sin POB} :$$

which (Art. 285) proves the proposition.

Corollary.—If $k' = -k$, the anharmonic of the pencil becomes numerically equal to 1. Hence (Art. 287, I) the important property: *The four lines* a, β, $a + k\beta$, $a - k\beta$ *form an harmonic pencil.*

Also, by the analogy of tangentials: *The four points* a, β, $a + k\beta$, $a - k\beta$ *are harmonic.*

289. Theorem.—*The anharmonic of any four lines* $a + k\beta$, $a + l\beta$, $a + m\beta$, $a + n\beta$ *is equal to*

$$\frac{(n - k)\,(m - l)}{(n - m)\,(l - k)} .$$

Let OA, OB represent the lines a and β, and OK, OL, OM, ON the four lines $a + k\beta$, $a + l\beta$, $a + m\beta$, $a + n\beta$. Then, if rs be any transversal, the anharmonic of the four given lines will be $au . eo : ae . ou :$ or, what amounts to the same thing, it will be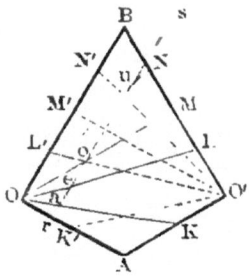

$(ru-ra)\ (ro-re) : (re-ra)\ (ru-ro)$. Now, taking the lengths of the perpendiculars from $a,\ e,\ o,\ u$ upon a, both trigonometrically and from the equations to the given lines, we obtain

$$ru = -n\beta \text{ cosec } Ors, \quad ra = -k\beta \text{ cosec } Ors,$$
$$ro = -m\beta \text{ cosec } Ors, \quad re = -l\beta \text{ cosec } Ors.$$

Substituting these values in the ratio last written, and recollecting that, since the anharmonic is independent of the position of the transversal, we may take rs parallel to the line β, and thus render the perpendicular β a constant, we find after reductions

$$R = \frac{(n-k)\ (m-l)}{(n-m)\ (l-k)}.$$

Remark.—The student can easily convince himself that the harmonic and anharmonic properties above obtained are true for lines whose equations are of the more general form $L=0$, $M=0$, $L+kM=0$, etc.

290. **Theorem.**—*If there be two systems of right lines, each radiating from a fixed point, and if the several members of the one be similarly situated with those of the other in regard to any two lines of the respective groups, then will the anharmonic of any pencil in the first system be equal to that of the pencil formed by the four corresponding lines of the second.*

For, in such a case, the two reference-lines of the first group being $L=0$, $L'=0$, and those of the second $M=0$, $M'=0$, the corresponding pencils will be $(L+kL', L+lL', L+mL', L+nL')$ and $(M+kM', M+lM', M+mM', M+nM')$. Hence, the theorem follows directly from the result of Art. 289.

291. Definition.—Systems of lines whose corresponding pencils have equal anharmonics are called *homographic systems.*

Thus, in the diagram of Art. 289, the two systems (OA, OK, OL, OM, ON, OB) and ($O'A$, $O'K'$, $O'L'$, $O'M'$, $O'N'$, $O'B$) are intended to represent a particular case of homographics.

292. In the examples which follow, some are best adapted for solution by the old notation, and others by the abridged. We have room for only a few of the manifold properties of the Right Line.

EXAMPLES.

1. If from the vertices of any triangle any three convergents be drawn, and the points in which these meet the opposite sides be joined two and two by three right lines, the points in which the latter cut the sides again will lie on one right line.

2. Any side of a triangle is divided harmonically by one of the three convergents mentioned in Ex. 1, and the line joining the feet of the other two.

3. In the figure drawn for the two preceding examples, determine by tangentials all the *points* that are harmonic.

4. In any triangle, the two sides, the line drawn from the vertex to the middle of the base, and the parallel to the base through the vertex, form an harmonic pencil.

5. The intersection of the three perpendiculars to the sides of a triangle, the intersection of the three lines drawn from the vertices to the middle points of the sides, and the center of the circumscribed circle, lie on one right line.

6. APB, CQD are two parallels, and $AP:PB::DQ:QC$: to prove that the three right lines AD, PQ, BC are convergent.

7. From three points A, B, D, in a right line $ABCD$, three convergents are drawn to a point P; and through C is drawn a right line parallel to AP, meeting PB in E and PD in F: to prove that

$$AD.BC : AB.CD :: EC : CF.$$

8. The six bisectors of the angles of any triangle intersect in only four points besides the vertices.

9. If through the vertices of any triangle lines be drawn parallel to the opposite sides, the right lines which join their intersections to the three given vertices will meet in one point. [Use both notations in succession.]

10. If through the vertices of any triangle there be drawn any three convergents whatever, to prove that these three lines and the three sides of the triangle may be respectively represented by the equations

$$v - w = 0, \quad w - u = 0, \quad u - v = 0,$$
$$v + w = \lambda, \quad w + u = \lambda, \quad u + v = \lambda.$$

11. If $OAA'A''$, $OBB'B''$ are two right lines harmonically divided, the former in A and A', the latter in B and B', the lines AB, $A'B'$, $A''B''$ either meet in one point or are parallel.

12. If on the three sides of a triangle, taken in turn as diagonals, there be constructed parallelograms whose sides are parallel to two fixed right lines, the three remaining diagonals of the parallelograms will meet in one point.

13. The three external bisectors of the angles of any triangle meet the opposite sides in three points which lie on one right line.

14. If three right lines drawn from the vertices of any triangle meet in one point, their respective parallels drawn through the middle of the opposite sides also meet in one point.

15. In every quadrilateral, the three lines which join the middle points of the opposite sides and the middle points of the diagonals, meet in one point.

16. If the four inner angles A,B,E,F of a complete quadrilateral (see diagram, Art. 219) are bisected by four right lines, the diagonals of the quadrilateral formed by these bisectors will pass through the two outer vertices of the complete one, namely, one through C and the other through D.

17. Let the two inner diagonals of any complete quadrilateral (same diagram) intersect in O: the diagonals of the two quadrilaterals into which either CO or DO divides $ABEF$, intersect in two points which lie on one right line with O.

18. In any complete quadrilateral, any two opposite sides form an harmonic pencil with the outer diagonal and the line joining their intersection to that of the two inner diagonals.

19. Also, the two inner diagonals are harmonically conjugate to the two lines which join their intersection to the two outer vertices.

20. Also, two adjacent sides are harmonically conjugate to their inner diagonal and the line joining their intersection to that of the outer diagonal and the remaining inner one; etc.

CHAPTER SECOND.

THE CIRCLE.

293. Before attempting the discussion of the three Conics strictly so called, it will be advantageous to illustrate the analytic method by applying it to that case of the Ellipse with whose properties the reader is already familiar from his studies in pure geometry: we mean, of course, the Circle. As we proceed in this application, we shall be enabled to define those elements of curves in general, which constitute at once the leading objects and principal aids of geometric analysis.

THE AXIS OF X.

294. The rectangular equation to the Circle referred to its center (Art. 136) is

$$x^2 + y^2 = r^2.$$

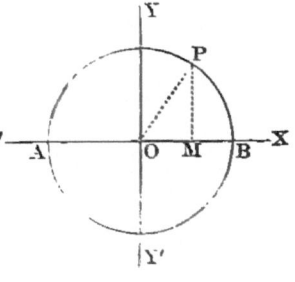

If we solve this for y, we obtain $y = \sqrt{(r+x)(r-x)}$. But, from the diagram, $r + x = AM$, and $r - x = MB$: and we have the well-known property of Geom., 325.

The ordinate to any diameter of a circle is a mean proportional between the corresponding segments.

295. If we eliminate between the equation to a circle $x^2 + y^2 = r^2$ and that of any right line $y = mx + b$ by substituting for y in the former from the latter, we shall obtain, as determining the abscissas of intersection between a right line and a circle, the quadratic

$$(1 + m^2)\, x^2 + 2mb.x + b^2 - r^2 = 0.$$

Now the roots of this quadratic are *real and unequal, equal*, or *imaginary*, according as $(1 + m^2)\, r^2$ is greater than, equal to, or less than b^2. Hence, when the first of these conditions occurs, the right line will meet the circle in two real and *different* points; when the second, in two *coincident* points; when the third, in two *imaginary* points. Adhering, then, to the distinction between these three classes of points, we may assert, with full generality,

Every right line meets a circle in two points, real, coincident, or imaginary.

296. It is so important that the distinction just alluded to shall be exactly understood in our future investigations, that we consider it worth while to illustrate it somewhat more at length.

I. The conception of two real points, situated at a finite distance from each other, is of course already clear to the student. We therefore merely add, that such points are sometimes called *discrete*, or *discontinuous* points.

II. The conception of *coincident*, or, as they are more significantly called, *consecutive* points, is peculiar to the analytic method. The most general definition of consecutive points is, that they are *points whose distance from each other is infinitely small*. It may aid in rendering this definition clear, to think of two points which are drawing closer and closer together, which tend to meet *but not to pass* each other, and whose mutual approach is never for an instant interrupted. The distance between two such points is evidently *less than any assignable quantity;* for however small a distance we may assign as the true one, the points will have drawn nearer together in the very instant in which we assign it : so that their distance eludes all attempts at finite statement, and can only be represented by the phrase *infinitely small.*

The geometric meaning of this analytic conception varies with its different applications. Thus, it may signify exactly the same thing as the *single* point which, in the language of pure geometry, is common to a curve and its tangent. For since the distance between consecutive points is *infinitely* small ; that is, so small that we can not assign a value too small for it ; we may assign the value 0, and take the points as absolutely coincident. It is in this aspect, mainly, that we shall use the conception in our future inquiries. Hence, as from the infinite series of continuous values which the distance between two consecutive points must have, we thus select the one corresponding to the moment of coincidence, we have preferred to designate the conception by the equivalent and for us more pertinent phrase *coincident points.*

The student should be sure that he always thinks of the distance between coincident points as a true *infinitesimal.* The error into which the beginner almost always falls is, to think of a *very* small, instead of an *infinitely* small, distance. He thus confounds with two *consecutive* points, two *discrete* ones extremely close together, between which there is of course a *finite* distance. The consequence is, that he finds in such a distance, however small, an infinite number of points lying between his supposed consecutives, and fancies that all the arguments based on the conception of consecutiveness are fallacious. Whereas, if he excludes from his thoughts, as he should, all points separated by any *finite* distance however small, he will have points absolutely *consecutive*, in the only sense that mature reflection attaches to that term.

III. The phrase *imaginary points* is also peculiar to analytic investigation. It really means, when translated into the language of pure geometry, that *the corresponding points not only do not exist, but are impossible*. But, as we have mentioned once before, the expression, together with the accessories which serve to carry out its use, is found to be of real value in developing certain remote analogies in the properties of curves. We shall therefore retain it, only cautioning the student not to be misled by a false interpretation of it.

297. Definition.—A **Chord** of any curve is any right line that meets it in two points.

298. Equation to a Circular Chord.—Let the equation to the given circle be $x^2 + y^2 = r^2$. Since the chord passes through two points of the curve, its equation (Art. 95) will be of the form

$$\frac{y - y'}{x - x'} = \frac{y'' - y'}{x'' - x'} :$$

in which $x'y'$, $x''y''$, since they both lie upon the circle, are subject to the condition

$$x'^2 + y'^2 = r^2 = x''^2 + y''^2.$$

Hence, $x'^2 - x''^2 = y''^2 - y'^2$; and we obtain

$$\frac{y'' - y'}{x'' - x'} = -\frac{x' + x''}{y' + y''}.$$

Therefore the required equation to a chord is

$$\frac{y - y'}{x - x'} = -\frac{x' + x''}{y' + y''} :$$

in which $x'y'$, $x''y''$ are the points in which the chord cuts the circle.

Corollary.—By a course of analysis exactly similar to that just used, we find the *equation to any chord of the circle* $(x - g)^2 + (y - f)^2 = r^2$, namely,

$$\frac{y - y'}{x - x'} = -\frac{x' + x'' - 2g}{y' + y'' - 2f} :$$

in which $x'y'$, $x''y''$ are the intersections of the circle with its chord, and gf is its center.

DIAMETERS.

299. Definition.—A **Diameter** of any curve is the locus of the middle points of parallel chords.

300. Equation to a Circular Diameter.—To find this, we must form the equation to the locus of the middle points of parallel chords in a circle. Let x, y be the co-ordinates of any middle point: the formula for the length of the chord from xy to the point $x'y'$ of the curve (Art. 102) gives us either

$$x' = x + cl \quad \text{or} \quad y' = y + sl.$$

But since $x'y'$ is a point on the circle $x^2 + y^2 = r^2$, we have

$$(x + cl)^2 + (y + sl)^2 = r^2 :$$

or, remembering that $s^2 + c^2 = 1$, we get, for determining the distance l of the point xy from the circle, the quadratic

$$l^2 + 2(cx + sy)l + (x^2 + y^2 - r^2) = 0.$$

Now, as xy is the *middle* point of a chord, the two values of l in this quadratic must be numerically equal with contrary signs. Hence, (Alg., 234, Prop. 3d,) the co-efficient of l must vanish, and we get

$$cx + sy = 0.$$

But, in the present inquiry, s and c are the sine and cosine of the angle which a chord through xy makes with the axis of x; and as this angle is the same for all parallel chords, the equation

$$cx + sy = 0$$

is a constant relation between the co-ordinates of the middle points of a series of parallel chords. That is, it is the required equation to any diameter of the circle.

301. If θ be the inclination of a series of parallel chords to the axis of x, the equation just obtained may be written $y = -x \cot \theta$. Hence, (Arts. 63; 96, Cor. 3,) we have the familiar property,

Every diameter of a circle passes through the center, and is perpendicular to the chords which it bisects.

Corollary.—From this we immediately obtain the important principle: *If a diameter bisects chords parallel to a second, the second bisects those parallel to the first.*

302. Definition.—By **Conjugate Diameters** of a curve, we mean two diameters so related that each bisects chords parallel to the other.

303. We have seen (Art. 301) that the equation to any diameter of a circle may be written $y = -x \cot \theta$, in which expression θ is the inclination of the chords which the diameter bisects. Hence, by the preceding definition, the equation to its conjugate will be

$$y = x \tan \theta \qquad (1).$$

But, putting $\theta' =$ the inclination of the chords bisected by this conjugate, its equation will take the form

$$y = - x \cot \theta' \qquad (2).$$

Therefore, as the *condition that two diameters of a circle may be conjugate*, we have, since (1) and (2) are only different forms of the same equation,

$$\tan \theta \tan \theta' = - 1 \qquad (3):$$

in which θ and θ' may be taken as the inclinations of the two *diameters* (since these are each perpendicular to the chords which they bisect), and we learn (Art. 96, Cor. 3) that

The conjugate diameters of any circle are all at right angles to each other.

THE TANGENT.

304. Definition.—A **Tangent** of any curve is a chord which meets it in two coincident points.

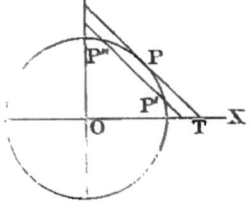

In applying this definition, the student must keep in mind the principles of Art. 296, II. The annexed diagram will aid him in apprehending the definition correctly. Let $P'P''$ be any chord passing through the two distinct points P' and P'', and let PT be a tangent parallel to $P'P''$. Suppose

$P'P''$ to move parallel to itself until it coincides with PT. It is evident that as $P'P''$ advances toward PT, the points P' and P'' will move along the curve toward each other, and that when $P'P''$ at length coincides with PT, they will become coincident in P, which is called the *point of contact.* We shall often allude to the position of the two coincident points through which a tangent passes, by the term *contact* alone.

305. Equation to a Tangent of a Circle.—Let the given circle be represented by $x^2 + y^2 = r^2$. Then, to obtain the required equation, we have only to suppose, in the equation to any chord (Art. 298), that the co-ordinates x' and x'', y' and y'', become identical. Making this supposition, reducing, and recollecting that $x'^2 + y'^2 = r^2$, we obtain

$$x'x + y'y = r^2:$$

in which $x'y'$ is the point of contact.

Corollary.—To obtain the equation to a tangent of the circle $(x - g)^2 + (y - f)^2 = r^2$, which differs from $x^2 + y^2 = r^2$ only in having the origin removed to the point $(-g, -f)$, we simply transform the expression just found to parallel axes passing through the last-named point, by replacing x and y, x' and y', by $x - g$, $y - f$, $x' - g$, $y' - f$. We thus get

$$(x' - g)(x - g) + (y' - f)(y - f) = r^2.$$

We may also obtain this less directly, by applying the condition for coincidence to the equation in the corollary to Art. 298.

306. Condition that a right line shall touch a Circle.—In order that the line $y = mx + b$ may touch the circle $x^2 + y^2 = r^2$, it must intersect the latter in two coincident points; that is, the co-ordinates of its two intersections with the circle must become identical. Hence, the required condition will be found by eliminating between $y = mx + b$ and $x^2 + y^2 = r^2$, and forming the condition that the resulting equation may have equal roots.

The resultant of this elimination is the quadratic

$$(1 + m^2) x^2 + 2mb.x + (b^2 - r^2) = 0;$$

and (see third equation of Art. 127, *et seq.*) the condition that this may have equal roots is

$$m^2b^2 = (1 + m^2) (b^2 - r^2) \quad \therefore \quad b = r \sqrt{1 + m^2}.$$

Corollary.—Hence, every line whose equation is of the form

$$y = mx + r \sqrt{1 + m^2},$$

touches the circle $x^2 + y^2 = r^2$. This equation belongs to a group of analogous expressions for the tangents of the several Conics; and, on account of its great usefulness, especially in problems where the point of contact is not involved, is called the *Magical Equation to the Tangent*.

307. Auxiliary Angle.—In problems that concern the intersection of lines with a circle, it is often advantageous to express the co-ordinates of the point on the curve in terms of the angle which the radius drawn to such point makes with the axis of x. By so doing, we obtain formulæ involving only one variable.

Thus, if $\theta' =$ the angle $P'OM$, it is evident that we shall have, for the co-ordinates of any point P',

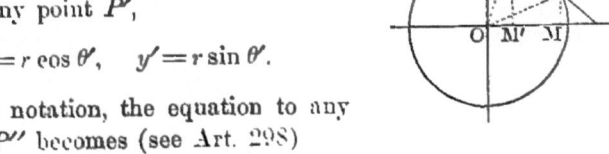

$$x' = r \cos \theta', \quad y' = r \sin \theta'.$$

In this notation, the equation to any chord $P'P''$ becomes (see Art. 298)

$$x \cos \tfrac{1}{2} (\theta'' + \theta') + y \sin \tfrac{1}{2} (\theta'' + \theta') = r \cos \tfrac{1}{2} (\theta'' - \theta');$$

in which $\theta' =$ the angle $P'OM$, and $\theta'' =$ the angle $P''OM'$. Hence, the corresponding equation to any tangent is

$$x \cos \theta' + y \sin \theta' = r.$$

This may also be obtained by substituting directly in the equation of Art. 305; and it expresses (Art. 80, Cor.) the well-known property, that the tangent to a circle is perpendicular to the radius at contact.

308. To draw a Tangent to a Circle from a Fixed Point.—This problem, so far as it belongs to analytic investigation, requires us to find the point of contact corresponding to the tangent which passes through the arbitrary point $x'y'$.

Let $x''y''$ be the unknown point of contact. The equation to the tangent is then (Art. 305)

$$x''x + y''y = r^2.$$

But, since this tangent is drawn through $x'y'$, we have

$$x''x' + y''y' = r^2 \qquad (1).$$

Moreover, since $x''y''$ is upon the circle,

$$x''^2 + y''^2 = r^2 \qquad (2).$$

Solving for x'' and y'' between (1) and (2), we get the co-ordinates of the point of contact, namely,

$$x'' = \frac{r^2 x' \pm r y' \sqrt{x'^2 + y'^2 - r^2}}{x'^2 + y'^2}, \quad y'' = \frac{r^2 y' \mp r x' \sqrt{x'^2 + y'^2 - r^2}}{x'^2 + y'^2}.$$

Corollary.—Hence, through any fixed point there can be drawn *two* tangents to a given circle, real, coincident, or imaginary. *Real,* when $x'^2 + y'^2 > r^2$; that is, when $x'y'$ is *outside* the circle. *Imaginary,* when $x'^2 + y'^2 < r^2$; that is, when $x'y'$ is *within* the circle. *Coincident,* when $x'^2 + y'^2 = r^2$; that is, when $x'y'$ is *on* the circle.

309. Length of the Tangent from any point to a Circle.—Let xy be any point in the plane of a circle whose center is the point gf. Then (Art. 51, I, Cor. 1) for the square of the distance between xy and gf, we have

$$\delta^2 = (x-g)^2 + (y-f)^2.$$

Putting t = the required length of the tangent, we get (since the tangent is perpendicular to the radius at contact) $t^2 = \delta^2 - r^2$. That is,

$$t^2 = (x-g)^2 + (y-f)^2 - r^2.$$

From this we learn that *if the co-ordinates* x *and* y *of any point be substituted in the equation to any circle, the result will be the square of the length of the tangent drawn from that point to the circle.*

Remark.—We have shown (Art. 142) that the general equation

$$A(x^2 + y^2) + 2Gx + 2Fy + C = 0$$

is equivalent to $(x-g)^2 + (y-f)^2 = r^2$, provided we take out A as a common factor. Hence, the property just proved applies to every equation denoting a circle, provided it be reduced to the form in question by the proper division. If, then, we write S as a convenient abbreviation for the left member of the equation to a circle, in which the common co-efficient of x^2 and y^2 is unity, we get

$$t^2 = S.$$

Corollary.—The square of the length of the tangent from the origin of the circle.

$$A(x^2 + y^2) + 2Gx + 2Fy + C = 0,$$

is equal to $C : A$; that is, to the quotient of the absolute term by the common co-efficient of x^2 and y^2.

310. Definition.—The **Subtangent** of a curve, with respect to any axis of x, is the portion of that axis intercepted between the foot[*] of the tangent and that of the ordinate of contact.

Thus, if OT is considered the axis of x, MT is the subtangent corresponding to PT of the inner curve, and to $P'T$ of the outer.

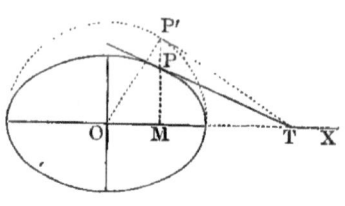

311. Subtangent of the Circle.—To obtain the length of this, we find the intercept OT, cut off from the axis of x by the tangent, and subtract from it the abscissa of contact OM. The equation to the tangent being

$$x'x + y'y = r^2,$$

to find the intercept on the axis of x, we make $y = 0$, and take the corresponding value of x. Thus,

$$x = \frac{r^2}{x'} = OT.$$

Hence, for the subtangent MT, we have

$$\text{subtan} = \frac{r^2 - x'^2}{x'}.$$

That is, *Any subtangent of a circle is a fourth proportional to the abscissa of contact and the two segments into which the ordinate of contact divides the corresponding diameter.*

THE NORMAL.

312. Definition.—The **Normal** of a curve is the right line perpendicular to a tangent at the point of contact.

---

[*] The point in which a line meets the axis of x is termed the *foot* of the line. Similarly, the point where a line meets *any* other is sometimes called its foot.

313. Equation to a Normal of a Circle.—Since the normal is perpendicular to the line $x'x + y'y = r^2$, and passes through the point of contact $x'y'$, its equation (Art. 103, Cor. 1) is

$$y - y' = \frac{y'}{x'} (x - x');$$

or, after reduction,

$$y'x - x'y = 0.$$

The form of this expression (Art. 95, Cor. 2) shows that *every normal of a circle passes through the center :* — a property which we might have gathered at once from the definition.

314. Definition.—The portion of the normal included between the point of contact and the axis on which the corresponding subtangent is measured, is called the *length* of the normal. For example, $P'O$ in the diagram of Art. 310.

From the result of Art. 313, it follows that *the length of the normal in any circle is constant, and equal to the radius.*

315. Definition.—The **Subnormal** of a curve, with respect to any axis of x, is the portion of that axis intercepted between the foot of the normal and that of the corresponding ordinate of contact.

Thus, in the diagram of Art. 310, OM is the subnormal corresponding to the point P'.

Hence, for the Circle, we have

$$\text{subnor} = x'.$$

That is, *Any subnormal of a circle is equal to the corresponding abscissa of contact.*

SUPPLEMENTAL CHORDS.

316. Definition.—By **Supplemental Chords** of a circle, we mean two chords passing respectively through the extremities of a diameter, and intersecting on the curve.

An. Ge. 26.

Thus, AP, BP are supplemental chords of the circle whose center is O and whose radius is OA.

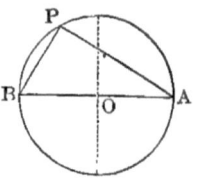

317. Condition that Chords of a Circle be Supplemental.—Take for the axis of x the diameter through whose extremities the chords pass, and for the axis of y a second diameter perpendicular to the first. Let $\varphi =$ the inclination of one chord, say of AP, and $\varphi' =$ that of the other, say of BP. Then, as the two chords pass through the opposite extremities of the same diameter, their equations (Art. 101, Cor. 1) will be

$$y = (x - r)\tan\varphi, \quad y = (x + r)\tan\varphi'.$$

Hence, at their point of intersection (Art. 62, Rem.) we shall have the condition

$$y^2 = (x^2 - r^2)\tan\varphi\tan\varphi';$$

or, since they intersect upon the circle $x^2 + y^2 = r^2$,

$$\tan\varphi\tan\varphi' = -1.$$

Corollary.—From the form of this condition (Art. 96, Cor. 3) we infer the property: *Any two supplemental chords of a circle are at right angles to each other.*

This is only another way of stating the familiar principle, that every angle inscribed in a semicircle is a right angle.

POLE AND POLAR.

318. The terms *pole* and *polar* are used, as already mentioned (p. 238), to call up a very remarkable relation between points and right lines, which depends upon a property common to the whole order of Conics. We shall now endeavor to develop the conception of the pole and polar with respect to the Circle, in the order according to which they naturally appear in analysis.

319. Chord of Contact belonging to Two Tangents which pass through a Fixed Point.—By the *chord of contact* here mentioned, we mean the right line joining the two points of contact corresponding to the pair of tangents which (Art. 308, Cor.) we have seen can be drawn to a circle from any external point. Let $x'y'$ be the fixed external point, and x_1y_1, x_2y_2 the two corresponding points of contact. The equations to the two tangents (Art. 305) will then be

$$x_1 x + y_1 y = r^2, \quad x_2 x + y_2 y = r^2.$$

Now, since both the tangents pass through $x'y'$, we have the two conditions

$$x_1 x' + y_1 y' = r^2, \quad x_2 x' + y_2 y' = r^2.$$

Hence, the co-ordinates of both points of contact satisfy the equation

$$x' x + y' y = r^2:$$

which is therefore the *equation to the chord of contact.*

320. Locus of the intersection of Tangents at the extremities of a Chord passing through a Fixed Point.—Let $x'y'$ be the fixed point through which the chord passes, and x_1y_1 the point in which the two tangents drawn at its extremities intersect. The equation to the chord (Art. 319) will be $x_1 x + y_1 y = r^2$: and, as $x'y'$ is a point on the chord, we shall have the condition

$$x_1 x' + y_1 y' = r^2,$$

no matter what be the direction of the chord. Hence, supposing the chord to be movable, and the intersection of the two corresponding tangents to be a variable point, the co-ordinates of the latter will always satisfy the equation

$$x'x + y'y = r^2:$$

which is therefore the equation to the locus required, and shows (Art. 85) that this locus is a *right line.*

321. Relation of the Tangent to this Locus and to the Chord of Contact.—The equation to the chord of contact of the two tangents drawn from any point outside of a circle, is

$$x'x + y'y = r^2 \qquad (1),$$

in which $x'y'$ is the *point from which the tangents are drawn.* The equation to the locus of the intersection of two tangents drawn at the extremities of any chord passing through a fixed point, is

$$x'x + y'y = r^2 \qquad (2),$$

in which $x'y'$ is the *point through which the movable chord is drawn.* The equation to any tangent of the same circle to which the chord (1) and the locus (2) refer, is

$$x'x + y'y = r^2 \qquad (3),$$

in which $x'y'$ is the *point of contact.* What, then, is the significance of this remarkable identity in the equations to these three lines? It certainly means that *there is some law of form common to the tangent, the chord of contact, and the locus mentioned.* For (1), (2), (3) assert that the chord of contact is connected with the point from which the two corresponding tangents are drawn, and that the right line forming the locus of the intersection of two tangents at the extremities of a chord passing through a fixed point is connected with that point, in exactly the same way that the tangent is connected with its point of contact. Now a right line in the plane of a circle must be either a tangent, a chord of contact, or a locus corresponding to (2): hence we learn that the Circle possesses the remarkable property of imparting to any right line in its plane the power of determining a point: and reciprocally.

This property is known as the *principle of polar reciprocity;* or, as it is sometimes called, the *principle of reciprocal polarity.* It is fully expressed in the following twofold theorem :

I. *If from a fixed point chords be drawn to any circle, and tangents to the curve be formed at the extremities of each chord, the intersections of the several pairs of tangents will lie on one right line.*

II. *If from different points lying on one right line pairs of tangents be drawn to any circle, their several chords of contact will meet in one point.*

The truth of (I) is evident from the equation of Art. 320 : that of (II) appears as follows : — Let $Ax + By + C = 0$ be any right line. The chord of contact corresponding to any point $x'y'$ of this line (Art. 319) is $x'x + y'y = r^2$. Now the co-efficients of this equation are connected by the linear relation

$$Ax' + By' + C = 0.$$

and the chord passes through a fixed point by Art. 117.

We shall find, as we go on, that the property just proved of the Circle is common to all conics. The reciprocal relation between a point and a right line is expressed by calling the point the *pole* of the line, and the line the *polar* of the point.

322. Definitions.—The **Polar** of any point, with respect to a circle, is the right line which forms the locus of the intersection of the two tangents drawn at the extremities of any chord passing through the point. Thus, $LL'L''$ is the polar of P.

The **Pole** of any right line, with respect to a circle, is the point in which all the chords of contact corresponding to different points on the line intersect each other. Thus, P is the pole of $LL'L''$.

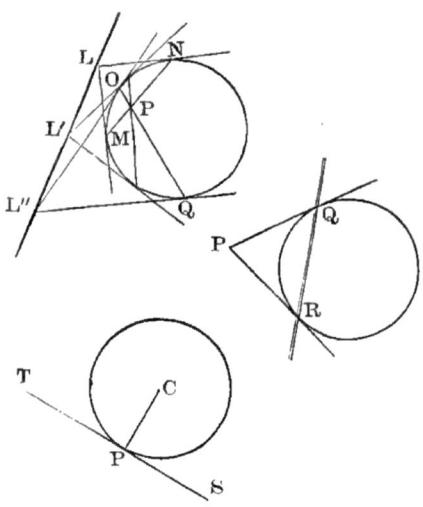

These definitions enable us to construct the polar when the point is given, or the pole when we have the line. Thus, if P be the point, we draw any two chords through it, as MPN, OPQ, and the corresponding pairs of tangents, ML, NL; OL'', QL''. The line which joins the points L and L'', in which the respective pairs of tangents intersect, is the polar of P. On the other hand, if LL'' be the given line, we draw from any two of its points, as L and L'', two pairs of tangents, LM, LN; $L''O, L''Q$, and the two corresponding chords of contact, MN, OQ: the point P in which the latter meet, is the required pole.

We have given the construction in the form above because it answers in all cases. It is evident, however, from the results of Art. 321, that when P is *without* the circle, its polar is the corresponding chord of contact QR; and that when it is *on* the curve, its polar is the corresponding tangent TS. In these cases, then, our drawing may be modified in accordance with these facts.

323. From all that has now been shown, it follows that the equation to the polar of any point $x'y'$, with respect to the circle $x^2 + y^2 = r^2$, is

$$x'x + y'y = r^2.$$

Now the equation to the line which joins $x'y'$ to the center of the same circle (Art. 95, Cor. 2) is

$$y'x - x'y = 0.$$

Hence, (Art. 99,) *The polar of any point is perpendicular to the line which joins that point to the center of the corresponding circle.*

Corollary.—This property affords a method of constructing the polar, simpler than that explained in the preceding article. For (Art. 92, Cor. 2) the distance of the polar from the center of its circle is

$$p = \frac{r^2}{\sqrt{(x'^2 + y'^2)}},$$

in which (Art. 51, I, Cor. 2) $\sqrt{x'^2 + y'^2}$ is the distance of the *pole* from the center; hence, *To construct the polar, join the pole to the center of the circle, and from the latter as origin lay off upon the resulting line a distance forming a third proportional to its whole length and the radius: the perpendicular to the first line, drawn through the point thus reached, will be the polar required.*

In the next four articles, we will present a few striking properties of polars.

324. The condition that a point $x''y''$ shall lie upon the polar of $x'y'$ is of course

$$x'x'' + y'y'' = r^2.$$

Now, obviously, this is also the condition that $x'y'$ shall lie upon the polar of $x''y''$. Therefore, *If a point lie upon the polar of a second, the second will lie upon the polar of the first.*

Corollary.—Hence, *The intersection of two right lines is the pole of the line which joins their poles.*

Remark.—We shall find hereafter that these properties are common to all conics.

325. The distance of $x'y'$ from the polar of $x''y''$ (Art. 105, Cor. 2) is

$$P' = \frac{x''x' + y''y' - r^2}{\sqrt{(x''^2 + y''^2)}} ;$$

and the distance of $x''y''$ from the polar of $x'y'$ is

$$P'' = \frac{x'x'' + y'y'' - r^2}{\sqrt{(x'^2 + y'^2)}} .$$

Hence, *The distances of two points from each other's polars are proportional to their distances from the center of the corresponding circle.*

326. Definitions.—Two triangles so situated with respect to any conic that the sides of the one are polars to the vertices of the other, are called *conjugate* triangles.

Thus, in the diagram, ABC and abc are conjugate triangles with respect to the circle RQS.

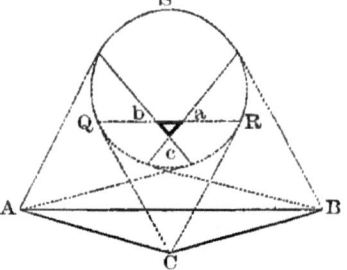

The *corresponding sides* of two conjugate triangles are those sides of the second which are *opposite to the poles of the sides of the first, and reciprocally.* The *corresponding angles* lie *opposite to the polars of the several vertices.* Thus, AB and ab are corresponding sides; A and a, corresponding angles; etc.

A triangle whose sides are the polars of its own vertices is called *self-conjugate.* To draw a self-conjugate triangle, take any point P, and form its polar; on the latter, take any point T, and form *its* polar: this new polar (Art. 324) will pass through P, and will of course intersect the polar of P in some point Z; join PT, and PTZ will be the required triangle. For TZ is the polar of P, and ZP of T, by construction; while PT is the polar of Z by the corollary to Art. 324.

327. Theorem.—*The three lines which join the corresponding vertices of two conjugate triangles meet in one point.*

Let the vertices of one triangle be $x'y'$, $x''y''$, $x'''y'''$; the sides of the other will then be

$$x'x + y'y = r^2, \quad x''x + y''y = r^2, \quad x'''x + y'''y = r^2.$$

For brevity, write these equations $P'=0$, $P''=0$, $P'''=0$; and let P_1'', P_1''' denote the results of substituting $x'y'$ in P'' and P'''; P_2''', P_2', the results of substituting $x''y''$ in P''' and P'; and P_3', P_3'', the results of substituting $x'''y'''$ in P' and P''. For the three lines joining the corresponding vertices (see diagram, Art. 326) we shall then have (Art. 108, Cor. 1)

$$P_1'''.P'' - P_1''.P''' = 0 \qquad (Aa),$$
$$P_2'.P''' - P_2'''.P' = 0 \qquad (Bb),$$
$$P_3''.P' - P_3'.P'' = 0 \qquad (Cc).$$

Now, writing the abbreviations in full, we get $P_1'''=P_3'$; $P_1''=P_2'$; $P_2'''=P_3''$. Hence, the three equations just written vanish identically when added, and the proposition is proved.

Corollary.—By Art. 281, it follows that *the intersections of the corresponding sides of two conjugate triangles lie on one right line.* That is, conjugate triangles are *homologous.*

SYSTEMS OF CIRCLES.

I. SYSTEM WITH COMMON RADICAL AXIS.

328. Any two circles lying in the same plane give rise to a very remarkable line, which is called their *radical axis.* Its existence and its fundamental property will appear from the following analysis:

An. Ge. 27.

Let $S = 0$, $S' = 0$ be the equations to two circles, so written that in each the common co-efficient of x^2 and y^2 is unity. Then will the equation $S + kS' = 0$ in general denote a circle passing through the points in which S and S' intersect; for in it x^2 and y^2 will have the common co-efficient $(1 + k)$, and obviously it will vanish when S and S' vanish simultaneously. To this theorem, however, there is one exception, namely, when $k = -1$. The resulting equation is then

$$S - S' = 0,$$

and, being necessarily of the first degree, denotes a *right line*.

Moreover, this equation is satisfied by an infinite series of continuous values, whether any can be found to satisfy S and S' simultaneously or not. That is, *The line* S — S' *is real even when the two common points of the circles, through which it passes, are imaginary.* When the circles intersect in *real* points, the line is of course their *common chord;* and it might still be called by that name even when the points of intersection are imaginary, if we chose to extend the usage in regard to imaginary points and lines which has been so frequently employed. To avoid this apparent straining of language, however, the name *radical axis* has been generally adopted.

The equation $S = S'$ asserts (Art. 309, Rem.) that the tangents to S and S', drawn from any point in its locus, are equal. Hence, *The radical axis of two circles is a right line, from any point of which if tangents be drawn to both of them, the two tangents will be of equal length.*

329. Writing $S — S'$ in full, and reducing, the equation to the radical axis becomes (Art. 134)

$$(g'-g)x + (f'-f)y = \tfrac{1}{2}\{(r^2-r'^2) - (g^2+f^2) + (g'^2+f'^2)\}.$$

Now the equation to the line joining the centers of the two circles (Art. 95, Cor. 1) may be written

$$(f'-f)x-(g'-g)y=f'g-g'f.$$

Therefore, (Art. 99, Cor.,) *The radical axis of two circles is perpendicular to the line which joins their centers.*

Corollary.—Hence, *To construct the radical axis of two circles, find its intercept on the axis of* x *by making* y $= 0$ *in the equation* S $-$ S$' = 0$, *and through the extremity of the intercept draw a perpendicular to the line of the centers.*

Remark.—This construction is applicable in all cases; but, when the circles intersect in real points, the axis is obtained at once by drawing the common chord.

330. If S, S', S'' be any three circles, the equations to the three radical axes to which the group gives rise will be

$$S-S'=0, \quad S'-S''=0, \quad S''-S=0:$$

which evidently vanish identically when added. Hence, *The three radical axes belonging to any three circles meet in one point, called the* RADICAL CENTER.

Corollary.—We may therefore construct the radical axis as follows: *Find the radical center of the two given circles with respect to any third, and through it draw a perpendicular to the line of their centers.* The annexed diagram will illustrate the details of the process. In it, c and c' are the centers of the two given circles, C the radical center, and CQC' the radical axis.

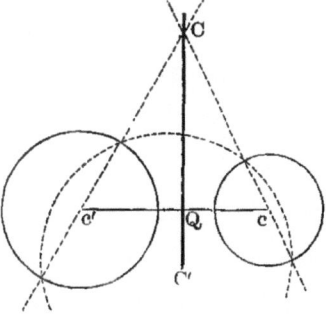

331. Two special cases of the radical axis deserve notice. First: We have seen that a point may be regarded as an infinitely small circle. Hence, a point and a circle have a radical axis; that is, given any point and any circle, we can always find a right line, from any point of which if we draw a tangent to the circle and a line to the given point, the two will be of equal length. The axis is of course perpendicular to the line drawn from the given point to the center of the circle. It lies *without* the circle, whether the point be within or without; for, as the radical axis always passes through the points common to its two circles, if it cut the given circle, the given point would form two consecutive points of that curve. From this it appears, that, when the given point is *on* the given circle, the axis is the tangent at the point.

Second: If *both* circles to which a radical axis belongs become points, we have a line every point of which is equally distant from two given ones. Hence, the radical axis of two *points* is the perpendicular bisecting the distance between them.

We now proceed to the properties of the entire system of circles formed about a common radical axis.

332. Definition.—By a **System of Circles with a Common Radical Axis,** we mean a system so situated with respect to a fixed right line, that, if a tangent be drawn to each circle from any point in the line, all these tangents will be of equal length.

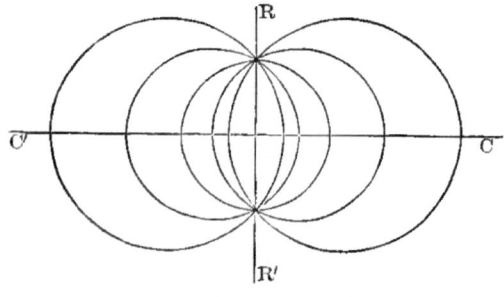

The simplest case of such a system is that of the infinite series of circles which can be passed through two given points. The diagram illustrates this case. Another is that of a series of circles touching each other at a common point. The appearance of the system when the two common points are imaginary, will be presented farther on, in connection with the method of constructing the system.

It follows directly from Art. 329, that *all the centers of the system lie on one right line at right angles to their radical axis.*

333. **Equation to any member of the System.**—The equation to any circle whose center lies on the axis of x, at a distance g from the origin, may be written (Art. 134)

$$x^2 + y^2 - 2gx = r^2 - g^2;$$

so that the circle will cut the axis of y in real or imaginary points according as $r^2 - g^2$ is positive or negative, the quantity $\sqrt{r^2 - g^2}$ representing half the intercept on the axis of y. Hence, if in the system of circles with a common radical axis, the common line of centers be taken for the axis of x, and the common radical axis for the axis of y: by putting $k =$ the arbitrary distance of the center from the origin, and $\delta^2 = constant = r^2 - k^2$, we may write the equation to any member of the system

$$x^2 + y^2 - 2kx = \delta^2;$$

and the corresponding system will cut the radical axis in *real* or in *imaginary* points according as δ^2 is positive or negative.

Corollary.—Hence, *To trace the system from the equation, assume different centers corresponding to arbitrary values of* k, *and from them, with radii in each case equal to* $\sqrt{k^2 \pm \delta^2}$, *describe circles.*

334. **The Orthogonal Circle.**—From the definition of the system (Art. 332), it follows that the locus of the point of contact of the tangent drawn from any point in the common radical axis to any member of the system, is a circle. Since, then, the corresponding tangents of the system are all radii of this circle, tangents to this circle at the points where it cuts the several members of the system will be perpendicular to the respective tangents of the system. In other words, the circle in question cuts every member of the system at right angles,* and may therefore be called the *orthogonal circle* of the system.

---

* The angle between two curves is the angle contained by their respective tangents at the point of intersection.

Since the center of such a circle is *any* point on the radical axis, there is an infinite series of orthogonal circles for every system with a common axis. But for the special purpose to which we are about to apply it, any one of these may be selected, of which we shall speak as *the* orthogonal circle.

335. Construction of the System.—We can now construct the system geometrically, in all cases. The only case that needs illustration, however, is that of the system passing through two imaginary common points. Since (Art. 334) the tangents of the orthogonal circle are all radii of the circles forming the system, we may draw any number of these circles as follows:—Lay down any right line *MN*, and any perpendicular to it *RQ*. On the latter,

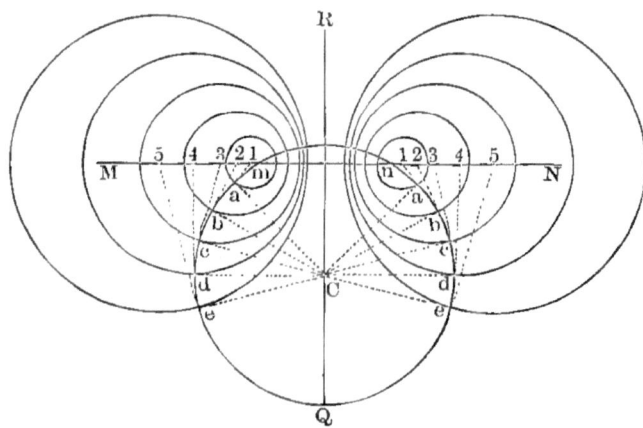

take any point C' as a center, and, with any radius CQ, describe a circle cutting *MN* in the points m and n. At any points a, b, c, d, e of this circle, draw tangents to meet *MN* in 1, 2, 3, 4, 5; and from these points as centers, with radii in each case equal to the corresponding tangent, describe circles. It is evident that the radii Ca, Cb, Cc, etc., of the fundamental circle, will all be tangents to the respective circles last drawn. Hence RQ is the common radical axis of all these circles, and the circle C'-Qmn is orthogonal to them.

336. Properties of the System: Limiting Points.—From the nature of the foregoing construction and the resulting diagram, we obtain the following properties :

I. The circles of a system having two imaginary common points, in which the orthogonal circle cuts the line of centers in two *real* points *m* and *n*, exist in pairs: to every circle on the right of the radical axis, corresponds an equal one on the left, with an equally distant center.

II. In a system of this character, the center of no circle can lie nearer to the radical axis than *m* or *n*. For the radius of the variable member of the system continually diminishes as the tangent of the orthogonal circle advances from *Q* toward *m* and *n*, and at *m* and *n* it vanishes.

III. But the center of a member of the system may be as remote from the radical axis as we please. For the tangents of the orthogonal circle at *Q* meet the line of centers at infinity.

IV. Hence, the points *m* and *n*, and the axis *RQ* form the inferior and superior limits of the system; in short, are the corresponding *limiting members* of it: *m* and *n* being equal infinitesimal circles at equal distances from the axis, and the axis itself being the resultant of two coincident circles having equal infinite radii.

V. The circle *C-Qmn* is drawn in the diagram to cut *MN* in real points; but if the student will draw a new diagram, in which *C-Qmn* fails to cut *MN*, he will find that the circles of the resulting system all cut each other in two real points on the line *RQ*. Hence, a system of circles with a common radical axis intersect each other in two real or imaginary points, according as the limiting points *m* and *n* are imaginary or real.

VI. The limiting points *m* and *n* are by construction equally distant from every point in *RQ*. Moreover, every orthogonal circle is described from some point in *RQ*, and cuts *every* member of the system at right angles. Hence *every* orthogonal circle passes through the limiting points. That is, *The orthogonals of any system of circles with a common radical axis, form a complemental system, whose radical axis is the line joining the centers of the conjugate system.*

VII. Hence, if a system of circles intersect in two real points, the conjugate system of orthogonal circles will intersect in two imaginary ones; and reciprocally.

337. The Limiting Points by Analysis.—If a system of circles cut its radical axis in two imaginary points, the equation to any member of the system (Art. 333) is

$$x^2 + y^2 - 2kx = -\delta^2 \qquad (1),$$

in which δ^2 is constant for the whole system, while k varies for each different member. Now $-\delta^2 = r^2 - k^2$ $\therefore$ $r = \sqrt{k^2 - \delta^2}$: therefore r vanishes when $k = \pm \delta$, and becomes imaginary when $k < \delta$ or $> -\delta$. Hence, the two points $(y = 0, x = \delta)$ and $(y = 0, x = -\delta)$ are the infinitesimal circles which we have called the *limiting points* of the system; for we have just shown that they have the property of Art. 336, II, and they are represented (Art. 61, Rem.) by the equation

$$(x \mp \delta)^2 + y^2 = 0; \quad \text{that is,} \quad x^2 + y^2 \mp 2\delta x = -\delta^2,$$

which conforms to the type of (1).

To exhibit the singular nature of these limiting points, we will now develop one more property of the system to which they belong. The equation to any of its members may be thrown into the form

$$(x - k)^2 + y^2 = r^2.$$

Hence, (Art. 305, Cor.,*) the polar of any point $x'y'$, with respect to any member of the system, will be represented by

$$x'x + y'y + \delta^2 - k(x + x') = 0 \qquad (2).$$

Now (Art. 108) the line denoted by (2) passes through the intersection of the two lines $x'x + y'y + \delta^2 = 0$ and $x + x' = 0$, whatever be the value of k. Therefore, *If the polars of a given point be taken with respect to the whole system of circles having a common radical axis, they will all meet in one point.*

Suppose, then, that $x'y'$ be either of the limiting points. The polar will then become

$$x = \mp \delta \qquad (3).$$

Hence, *The polar of either limiting point is a line drawn through the other at right angles to the line of centers, and is therefore absolutely fixed for the whole system.*

II. TWO CIRCLES WITH A COMMON TANGENT.

338. The problem of constructing a common tangent to two given circles (which properly belongs to Deter-

---

* The student will remember that the equations to the tangent and polar of the Circle are identical in form.

minate Geometry, and which we solved under that head on pp. 18—21) leads to some important results when treated by the methods of Indeterminate Geometry. A few of these, we shall now present.

339. The problem, as coming within the sphere of pure analysis, consists in finding co-ordinates of contact such that the corresponding tangent may touch both circles. Suppose, then, that the equations to the two circles are

$$(x - g)^2 + (y - f)^2 = r^2 \qquad (S),$$
$$(x - g')^2 + (y - f')^2 = r'^2 \qquad (S').$$

The equation to a tangent of S will then be (Art. 305, Cor.)

$$(x' - g)(x - g) + (y' - f)(y - f) = r^2;$$

and our problem is, so to determine $x'y'$ that this line may also touch S'.

In settling what condition $x'y'$ must satisfy in order that this result may take place, it will be convenient to employ the auxiliary angle mentioned in Art. 307. Let $\theta =$ the inclination of the radius through $x'y'$: then will $x' - g = r \cos \theta$, and $y' - f = r \sin \theta$. The equation to the tangent of S may therefore be written

$$x \cos \theta + y \sin \theta = g \cos \theta + f \sin \theta + r \qquad (1),$$

and, similarly, the equation to a tangent of S',

$$x \cos \theta' + y \sin \theta' = g' \cos \theta' + f' \sin \theta' + r' \qquad (2).$$

Now (1) will represent the *same* line as (2), if the mutual ratios of its co-efficients are the same as those of (2); that is, if simultaneously

$$\tan \theta = \tan \theta',$$
$$(g \cos \theta + f \sin \theta + r) \cos \theta' = (g' \cos \theta' + f' \sin \theta' + r') \cos \theta.$$

The first of these conditions is satisfied either by $\theta' = \theta$, or $\theta' = \pi + \theta$. Combining the two, then, on both suppositions, we obtain

$$(g' - g) \cos \theta + (f' - f) \sin \theta = r - r',$$
$$(g' - g) \cos \theta + (f' - f) \sin \theta = r + r';$$

or, after replacing $\cos\theta$ and $\sin\theta$ by their values, $\dfrac{x'-g}{r}$ and $\dfrac{y'-f}{r}$,

$$(y'-g)(x'-g) + (f'-f)(y'-f) = r(r-r') \qquad (\text{A}),$$
$$(g'-g)(x'-g) + (f'-f)(y'-f) = r(r+r') \qquad (\text{B});$$

and we learn that if $x'y'$ satisfies either (A) or (B), the tangent of S will touch S' Since (A) arose from the supposition $\theta'=\theta$, that is, from the supposition that the radii of contact in the two circles were parallel, and lay in the *same* direction, a moment's inspection of the diagram will show that, when (A) is satisfied, the com-mon tangent is *direct*, as $MN;$ while as (B) arose from $\theta' = \pi + \theta$, that is, from sup-

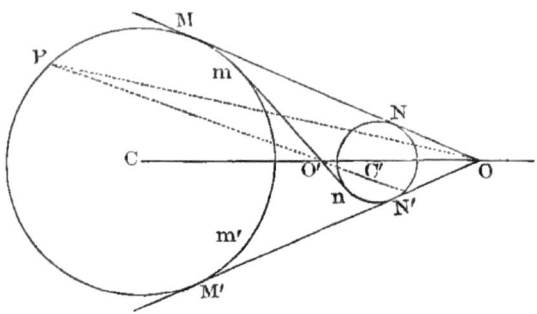

posing the radii of contact parallel, but lying in *opposite* directions, the corresponding common tangent must be *transverse*, as mn.

To find, then, the required points on S, at which if a tangent be drawn, it will also touch S', we merely eliminate between (S) and either (A) or (B). Now the result of this elimination will be a quadratic; hence, there are in all *four* tangents common to S and S': two direct, and two transverse.

340.   The Chords of Contact.—Since the points of contact, M and M', as we have just seen, *both* satisfy the condition (A), it follows that

$$(g'-g)(x-g) + (f'-f)(y-f) = r(r-r') \qquad (1),$$

is the *equation to the chord of contact for the direct tangents.* Similarly,

$$(g'-g)(x-g) + (f'-f)(y-f) = r(r+r') \qquad (2),$$

is the *equation to* mm′, *the chord of contact for the transverse tangents.*

Corollary.—If the origin be transferred to the center of S, g and f will vanish, and the chords of contact will be represented by

$$g'x + f'y = r(r \mp r').$$

If the axes be now revolved until the line of centers becomes the axis of x, f' will vanish, and the equation assume the form $x = constant$. Hence, *The chords of contact corresponding to common tangents of two circles are perpendicular to the line of their centers.*

341. Segments formed on the Line of Centers.—The points O and O', in which the two direct and two transverse common tangents have their respective intersections, are the poles of (1) and (2) in Art. 340. In these, if we multiply by r, and then divide by $r - r'$ and $r + r'$ respectively, the co-efficients of x and y (Art. 305, Cor.) will be the co-ordinates of O and O', diminished by the co-ordinates of the center of S. We thus get

$$x' - g = \frac{r(g'-g)}{r \mp r'} \quad \therefore \quad x' = \frac{rg' \mp r'g}{r \mp r'},$$

$$y' - f = \frac{r(f'-f)}{r \mp r'} \quad \therefore \quad y' = \frac{rf' \mp r'f}{r \mp r'}.$$

Now (Art. 95) these values satisfy the equation to the line of centers, and show (Art. 52) that $x'y'$ divides the distance between gf and $g'f'$ in the ratio of r and r'. Hence, *Common tangents of two circles intersect on the line of their centers, and divide the distance between those centers in the ratio of the radii.*

342. Centers of Similitude.—If the common tangent be made the initial line, and either O or O' be taken for the pole, the polar equation to the circle S (Art. 138) may be written

$$\rho_1^2 - 2\frac{r\cos(\theta - a)}{\sin a}\rho_1 + \frac{r^2\cos^2 a}{\sin^2 a} = 0,$$

by merely substituting for d its value $r : \sin a$. Hence, for the circle S,

$$\rho_1 = \frac{r\{\cos(\theta - a) \pm \sqrt{\cos^2(\theta - a) - \cos^2 a}\}}{\sin a}.$$

Similarly, for the circle S', we get

$$\rho_2 = \frac{r'\{\cos(\theta - a) \pm \sqrt{\cos^2(\theta - a) - \cos^2 a}\}}{\sin a}$$

Therefore, $\rho_1 : \rho_2 :: r : r'$.

Now these vectors of S and S' are the segments formed by the two circles on any right line drawn through O or O'. Hence, *All right lines drawn through the intersection of the common tangents of two circles are cut similarly by the circles, namely, in the ratio of the radii.*

Remark.—On account of this property, the points in which the common tangents intersect are called *centers of similitude.*

343. **Axis of Similitude.**—This name is given to a certain right line whose relation to three circles we will now develop.

Let gf, $g'f'$, $g''f''$ be the centers of any three circles, and r, r', r'' their radii. The co-ordinates of the external center of similitude for the first and second (Art. 341) will then be

$$x = \frac{rg' - r'g}{r - r'}, \qquad y = \frac{rf' - r'f}{r - r'} \qquad (1);$$

those of the corresponding center for the second and third,

$$x' = \frac{r'g'' - r''g'}{r' - r''}, \qquad y' = \frac{r'f'' - r''f'}{r' - r''} \qquad (2);$$

and those of the corresponding center for the third and first,

$$x'' = \frac{r''g - rg''}{r'' - r}, \qquad y'' = \frac{r''f - rf''}{r'' - r} \qquad (3).$$

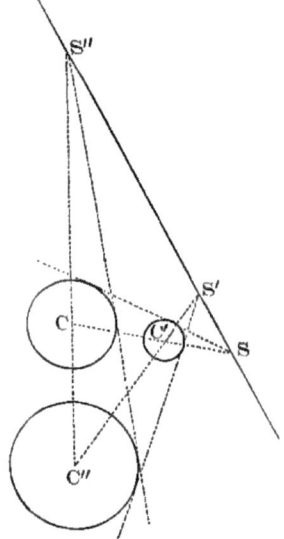

Now, if we make the necessary substitutions and reductions, we shall find that (1), (2), (3) satisfy the condition of Art. 112. Hence, *Any three homologous centers of similitude belonging to three circles lie on one right line, called the* AXIS OF SIMILITUDE.

Corollary.—If two circles touch each other, one of their centers of similitude becomes the point of contact. Hence, *If in a group of three circles the third touches the other two, the line joining the points of contact passes through a center of similitude of the two.*

Remark.—The homologous centers of similitude are either all three external, as in the diagram, or else two internal and the third external. Corresponding to the latter case, there will of course be three different axes of similitude; making in all *four* such axes for every group of three circles.

THE CIRCLE IN THE ABRIDGED NOTATION.

344. We have room for only a few examples of the uses to which this notation can be advantageously applied in the case of the Circle. The illustrations given will afford the beginner some further insight into the method, and the reader who desires fuller information must consult the larger works to which we have already referred in connection with this subject.

345. Since a, β, γ are the perpendiculars dropped from any point P to the three sides of a triangle, it is evident that the function

$$\beta\gamma \sin A + \gamma a \sin B + a\beta \sin C$$

denotes (Trig., 874) the double area of the triangle formed by joining the feet of those perpendiculars; for the angle A, included between the *sides* β and γ, will be either the supplement of the angle between the *perpendiculars* β and γ, or else equal to it: and so, also, of the angles B and C. Now (Art. 236) if the point P be on the circumference of the circumscribed circle, we have

$$\beta\gamma \sin A + \gamma a \sin B + a\beta \sin C = 0;$$

that is, the triangle contained between the feet of the perpendiculars from P vanishes, and we obtain the following theorem: *The feet of the perpendiculars dropped from any point in a circle upon the sides of an inscribed triangle lie on one right line.*

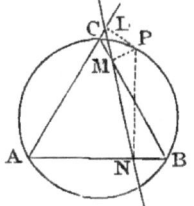

346. The equation to the circle circumscribed about a triangle may, by factoring, be written

$$\gamma (a \sin B + \beta \sin A) + a\beta \sin C = 0:$$

which shows that the line $a \sin B + \beta \sin A$ meets the circle on the line a, and also on the line β; since, if $a \sin B + \beta \sin A = 0$ in the above equation, we get $a\beta = 0$, a condition satisfied by either $a = 0$ or $\beta = 0$. But the only point in which either a or β meets the circle is their intersection: hence $a \sin B + \beta \sin A$ is the *tangent* of the circle at $a\beta$. Now (Ex. 8, p. 222) $a \sin A + \beta \sin B$ is the parallel to the base of the triangle, passing through its vertex; and (Ex. 6, p. 222) this parallel, and therefore the base, has the same inclination to a or β as the tangent $a \sin B + \beta \sin A$ has to β or a. Hence, the tangent of the circumscribed circle, at the vertex of a triangle, makes the same angle with either side as the base does with the other; or we have the well-known theorem: *The angle contained by a tangent and chord of any circle is equal to that inscribed under the intercepted arc.*

347. The equation obtained in the preceding article denotes the tangent at the vertex C of a triangle inscribed in a circle; and analogous equations may at once be written for the tangents at the other two vertices A and B. The equation to the tangent at *any* point of a circle circumscribed about a given triangle (compare Arts. 236; 240, II) is

$$\frac{a \sin A}{a'^2} + \frac{\beta \sin B}{\beta'^2} + \frac{\gamma \sin C}{\gamma'^2} = 0.$$

348. The equations to the tangents at the vertices of an inscribed triangle (Art. 346) may be written

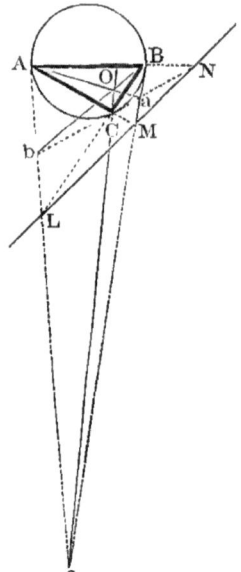

$$\frac{a}{\sin A} + \frac{\beta}{\sin B} = 0 \qquad (1),$$

$$\frac{\beta}{\sin B} + \frac{\gamma}{\sin C} = 0 \qquad (2),$$

$$\frac{\gamma}{\sin C} + \frac{a}{\sin A} = 0 \qquad (3).$$

Now (Art. 108) the line

$$\frac{a}{\sin A} + \frac{\beta}{\sin B} + \frac{\gamma}{\sin C} = 0$$

passes through the intersection of (1) with γ, of (2) with a, and of (3) with β. Hence, *The tangents at the vertices of an inscribed triangle cut the opposite sides in points which lie on one right line.*

349. Subtracting (2) from (3) above, (3) from (1), and (1) from (2), we get

$$\frac{a}{\sin A} - \frac{\beta}{\sin B} = 0, \qquad \frac{\beta}{\sin B} - \frac{\gamma}{\sin C} = 0, \qquad \frac{\gamma}{\sin C} - \frac{a}{\sin A} = 0:$$

which (Art. 108) are the equations to the three lines which join the intersections of the tangents at the vertices to the intersections of the sides. Hence, (Art. 114,) *The lines which join the vertices of a triangle to those of the triangle formed by drawing to its circumscribed circle tangents at its vertices, meet in one point.*

Remark.—The theorems of the last two articles, which are illustrated in the diagram, are evidently a particular case of homology (Art. 327) due to a pair of conjugate triangles.

350. Radical Axis in Trilinears.—The equations to any two circles, in the abridged notation, (Arts. 236, 237) are

$$\frac{\sin A}{\alpha} + \frac{\sin B}{\beta} + \frac{\sin C}{\gamma} + M\,(l\alpha + m\beta + n\gamma) = 0,$$

$$\frac{\sin A}{\alpha} + \frac{\sin B}{\beta} + \frac{\sin C}{\gamma} + M\,(l'\alpha + m'\beta + n'\gamma) = 0.$$

Hence, their radical axis, $S - S'$, is denoted by

$$l\alpha + m\beta + n\gamma = l'\alpha + m'\beta + n'\gamma.$$

Corollary.—The radical axis of any circle and the circle circumscribed about the triangle of reference, is represented by

$$l\alpha + m\beta + n\gamma = 0.$$

EXAMPLES ON THE CIRCLE.

1. Find the intersections of the line $4x + 3y = 35c$ with the circle
$$(x - c)^2 + (y - 2c)^2 = 25c^2.$$
Also, the tangents from the origin to the circle $x^2 + y^2 - 6x - 2y + 8 = 0$.

2. Show that the equation to any chord of a circle may be written
$$(x - x')\,(x - x'') + (y - y')\,(y - y'') = x^2 + y^2 - r^2,$$
the origin being at the center, and $x'y'$, $x''y''$ being the extremities of the chord.

3. Find the polar of $(4, 5)$ with respect to $x^2 + y^2 - 3x - 4y = 8$, and the pole of $2x + 3y = 6$ with respect to $(x-1)^2 + (y-2)^2 = 12$.

4. Prove that the condition upon which $Ax + By + C = 0$ will touch the circle $(x - g)^2 + (y - f')^2 = r^2$ is

$$\frac{Ag + Bf + C}{\sqrt{(A^2 + B^2)}} = r.$$

5. Find the length of the chord common to the two circles

$$(x - a)^2 + (y - \beta)^2 = r^2, \qquad (x - \beta)^2 + (y - a)^2 = r^2.$$

Also, the equations to the right lines which touch $x^2 + y^2 = r^2$ at the two points whose common abscissa is 1.

6. Find the equation to the circle of which $y = 2x + 3$ is a tangent, the center being taken for the origin.

7. Prove that the bisectors of all angles inscribed in the same segment of a circle pass through a fixed point on the curve.

8. Given the hypotenuse of a right triangle: the locus of the center of the inscribed circle is the quadrant of which the given hypotenuse is the chord.

9. Given two sides and the included angle of a triangle: to find the equation to the circumscribed circle.

10. The locus of a point from which if lines be drawn to the vertices of a triangle, their perpendiculars through the vertices will meet in one point, is the circle circumscribed about the triangle.

11. If any chord be drawn through a fixed point on the diameter of a circle, and its extremities joined to either end of the diameter, the joining lines will cut from the tangent at the other end, portions whose rectangle is constant. [See Art. 137.]

12. The locus of the intersection of tangents drawn to any circle at the extremities of a constant chord is a concentric circle. [See Art. 307.]

13. If a chord of constant length be inscribed in a given circle, it will always touch a concentric circle.

14. If through a fixed point O any chord of a circle be drawn, and OP be taken an *harmonic mean* between its segments OQ, OQ', the locus of P will be the polar of O.

15. If through any point O of a circle, any three chords be drawn, and on each, as a diameter, a circle be described, the three circles which thus meet in O will meet in three other points, lying on one right line.

16. If several circles pass through two fixed points, their radical axes with a fixed circle will pass through a fixed point.

[This example may be best solved by means of the Abridged Notation, but can be done very neatly without it.]

17. Form the equation to the system of circles which cuts at right angles any system with a common radical axis, and prove, by means of it, that every member of the former system passes through the limiting points of the latter.

18. If PQ be the diameter of a circle, the polar of P with respect to any circle that cuts the first at right angles, will pass through Q.

19. The square of the tangent drawn to any circle from any point on another is in a constant ratio to the perpendicular drawn from that point to their radical axis.

20. If a movable circle cut two fixed ones at constant angles, it will cut at constant angles all circles having the same radical axis as these two.

[First prove that the angle ϕ at which two circles cut each other, is determined by the formula

$$D^2 = R^2 + r^2 - 2Rr \cos \phi,$$

in which R, r are the radii of the circles, and D the distance between their centers.]

21. Find the equations to the common tangents of the two circles

$$x^2 + y^2 - 4x - 2y + 4 = 0, \qquad x^2 + y^2 + 4x + 2y - 4 = 0.$$

What is the equation to their radical axis?

22. If a movable circle cut three fixed ones, the intersections of the three radical axes will move along three fixed right lines which meet in one point.

23. The radical axis of any two circles that do not intersect, bisects the distances between the two points of contact corresponding to each of the four common tangents.

24. If through a center of similitude belonging to any two circles, we draw any two right lines meeting the first circle in the points R and R', S and S' respectively, and the second in r and r', s and s': then will the chords RS and rs, $R'S'$ and $r's'$, be parallel; while RS and $r's'$, $R'S'$ and rs, will each intersect on the radical axis.

An. Ge. 28.

25. Find the trilinear equation to the circle passing through the middle points of the sides of any triangle, and prove that this circle passes through the feet of the three perpendiculars of the triangle, and bisects the distances from the vertices to the point in which the three perpendiculars meet. [This circle is celebrated in the history of geometry, and, on account of passing through the points just mentioned, is called the *Nine Points Circle.*]

Find, also, the radical axis of this and the circumscribed circle.

CHAPTER THIRD.

THE ELLIPSE.

I. The Curve referred to its Axes.

351. We may most conveniently begin the discussion of the Ellipse by means of the equation which we obtained in Art. 147, namely,

$$\frac{x^2}{a^2} + \frac{y^2}{b^2} = 1.$$

At a later point in our investigations, we shall refer the curve to lines which have a relation to it more generic than that of the two known as *the axes* (Art. 146), which give rise to the equation just written.

THE AXES.

352. If in the above equation we make $y = 0$, we shall obtain, as the intercept of the curve upon the transverse axis,

$$x = \pm a \qquad (1);$$

and, making $x = 0$, we get, for the intercept upon the conjugate axis,

$$y = \pm b \qquad (2).$$

Comparing (1) and (2), we see that the curve cuts both axes in two points, and that in each case these two points are equally distant from the focal center, which (Art. 147) was taken for the origin. Hence we have

Theorem I.—*The focal center of any ellipse bisects the transverse axis, and also the conjugate.*

Corollary.—We must therefore from this time forward interpret the constants a and b in the equation

$$\frac{x^2}{a^2} + \frac{y^2}{b^2} = 1$$

as respectively denoting half the transverse axis and half the conjugate axis.

Remark.—This theorem follows, of course, directly from that of Art. 149. We have purposely developed it by a separate analysis, however, in order that the student may see the consistency of the analytic method.

353. If in the equation of Art. 147, which may be written

$$y = \pm \frac{b}{a} \sqrt{a^2 - x^2},$$

we suppose $x > a$ or $< -a$, the corresponding values of y are imaginary; so that no point of the curve is farther from the origin, either to the right or to the left, than the extremities of the transverse axis. Now (Art. 147), for the distance from the origin to either focus, we have

$$c^2 = a^2 - b^2.$$

Hence, c can not be greater than a, though it may approach infinitely near to the value of a, as b diminishes toward zero. Therefore,

Theorem II.—*The foci of any ellipse fall* within *the curve*.

354. Moreover, $a - c$ measures the distance of either focus from the adjacent vertex; while the distance of either from the remote vertex $= a + c$. We accordingly get

Theorem III.—*The vertices of the curve are equally distant from the foci.*

355. From Art. 352, the length of the transverse axis $= 2a$. But (Art. 147) $2a =$ the constant sum of the focal radii of any point on the curve. That is,

Theorem IV.—*The sum of the focal radii of any point on an ellipse is equal to the length of its transverse axis.*

Corollary.—This property gives rise to the following construction of the curve by points:
Divide the transverse axis at any point M between the foci F' and F. From F' as a center, with a radius equal to the segment MA', strike two small arcs, one above the axis, and the other below it. Then from F, with the remaining segment MA 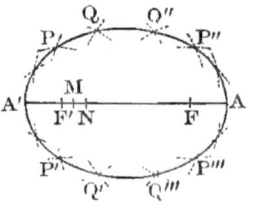 as radius, strike two more arcs, intersecting the two former in P and P': these points will be upon the required ellipse; for $F'P + FP = AA' = F'P' + FP'$. By using the radius MA' from F, and MA from F', two more points, P'' and P''', may be found; so that every division of the transverse axis will determine four points

of the curve. Thus, the point of division N will give rise to the four points Q, Q', Q'', Q'''. When enough points have been found to mark the outline of the curve distinctly, it may be drawn through them; if necessary, with the help of a curve-ruler. It is evident that this construction implies that the transverse axis and the foci are given.

356. The abbreviation $b^2 = a^2 - c^2$ adopted (Art. 147) for the Ellipse, gives us

$$b = \sqrt{(a + c)(a - c)}.$$

Hence, attributing to a, b, c the meanings now known to belong to them, we have

Theorem V.—*The conjugate semi-axis of any ellipse is a geometric mean between the segments formed upon the transverse axis by either focus.*

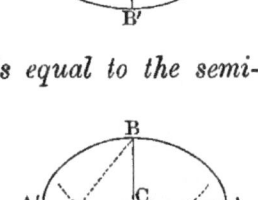

Corollary.—Transposing in the abbreviation above, we have $b^2 + c^2 = a^2$. But, from the diagram, $b^2 + c^2 = F'B^2 = FB^2$. Therefore, *The distance from either focus of an ellipse to the vertex of the conjugate axis* is equal to the semi-transverse. We have, then, the following construction for the foci, when the two axes are given:— From B, the vertex of the conjugate axis, with a radius equal to the semi-transverse, describe an arc cutting the transverse axis $A'A$ in F' and F: the two points of intersection will be the foci sought.

357. Let $x'y'$, $x''y''$ be any two points of an ellipse. Then, from the equation of Art. 147,

$$y'^2 = \frac{b^2}{a^2}(a^2 - x'^2), \quad y''^2 = \frac{b^2}{a^2}(a^2 - x''^2) \quad (1);$$

$$x'^2 = \frac{a^2}{b^2}(b^2 - y'^2), \quad x''^2 = \frac{a^2}{b^2}(b^2 - y''^2) \quad (2).$$

Dividing the first equation of (1) by the second, we get

$$y'^2 : y''^2 :: (a + x')(a - x') : (a + x'')(a - x'').$$

By a like operation in (2), we obtain

$$x'^2 : x''^2 :: (b + y')(b - y') : (b + y'')(b - y'').$$

Now $a + x'$, $a - x'$ are evidently the segments formed by y' upon the transverse axis, and $a + x''$, $a - x''$ are those formed by y''. Similarly, $b + y'$, $b - y'$ are the segments formed by x' upon the conjugate axis, and $b + y''$, $b - y''$ those formed by x''. Hence,

Theorem VI.—*The squares on the ordinates drawn to either axis of an ellipse are proportional to the rectangles under the corresponding segments of that axis.*

Corollary.—If in the first expression of (1) we make $x' = \pm c$, we get

$$y'^2 = \frac{b^2}{a^2}(a^2 - c^2).$$

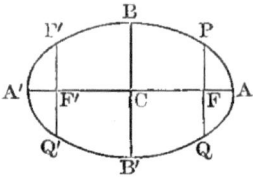

But $a^2 - c^2 = b^2$. Hence, after reductions,

$$y' = FP \text{ or } F'P' = \frac{b^2}{a}.$$

Now, either of the double ordinates that pass through the foci, PQ or $P'Q'$, is called the *latus rectum* of the ellipse to which it belongs. Hence,

$$\text{latus rectum} = \frac{2b^2}{a} = \frac{(2b)^2}{2a}.$$

That is, *The latus rectum of any ellipse is a third proportional to the transverse axis and the conjugate.*

358. The equation to the Ellipse (Art. 147) may be thrown into either of the forms

$$\frac{y^2}{(a+x)(a-x)} = \frac{b^2}{a^2}, \qquad \frac{x^2}{(b+y)(b-y)} = \frac{a^2}{b^2}.$$

Hence, since a ratio is not altered when both its terms are multiplied by the same number,

Theorem VII.—*The squares on the axes of any ellipse are to each other as the rectangle under any two segments of either is to the square on the ordinate which forms the segments.*

Note.—It may be worth while to observe, in passing, that, in this theorem and the one of Art. 357, the word *ordinate* has been used in a wider sense than we originally assigned to it. We shall frequently have occasion to employ it in this larger meaning, of *a line drawn to either axis of co-ordinates parallel to the other.*

359. The equation to the Ellipse being put into the form

$$y^2 = \frac{b^2}{a^2}(a^2 - x^2) \qquad (1),$$

the equation to the circle described on the transverse axis as a diameter (Art. 161) will be

$$y^2 = a^2 - x^2 \qquad (2).$$

Hence, supposing the x of (1) and (2) to become identical, we get

$$y_e : y_c :: b : a.$$

That is to say,

Theorem VIII.—*The ordinate of any ellipse is to the corresponding ordinate of the circumscribed circle, as the conjugate semi-axis is to the semi-transverse.*

Corollary 1.—By similar reasoning, we should find that *the abscissa of an ellipse is to the corresponding abscissa of the* inscribed *circle, as the transverse semi-axis is to the semi-conjugate.* When the axes are given, we may therefore construct the curve by either of the following methods:

First: Describe circles upon the given axes $A'A$, $B'B$. At any point M of the transverse axis, erect a perpendicular, and join the point Q, in which it meets the outer circle, with the common center C. Through R, in which QC cuts the inner circle, draw RP at right angles to the con-

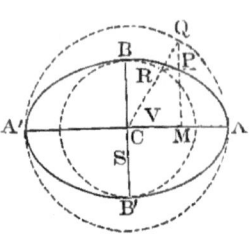

jugate axis: the point P, in which RP cuts MQ, will be upon the required ellipse. For, by similar triangles, $MP : MQ :: CR : CQ :: b : a$.

Second: Suppose PS to be a ruler whose length $= CA$. From the end P, lay off upon it $PV = CB$. Set the end S against the conjugate axis, say at some point below C, and rest the point V upon the transverse axis, say to the right of C. Move the ruler so that S and V may slide along the axes: the extremity P will describe an ellipse. For, if MPQ be drawn through P perpendicular to CA, and QC be joined, the latter will be equal and thence parallel to the ruler PS, and we shall have the proportion $MP : MQ :: PV : PS :: b : a$.

The second of these methods constitutes the principle of the so-called *elliptic compasses* — an instrument used for describing ellipses, and consisting of two bars, $A'A$, $B'B$, fixed at right angles to each other, along which a third, PS, slides freely upon two points, S and V, whose distance apart is constant. [See Ex. 6, p. 167.]

Corollary 2.—We are now enabled to interpret the abbreviation

$$\frac{a^2 - b^2}{a^2} = e^2,$$

adopted in Art. 150. In the first place, since $a^2 - b^2 = c^2$, we learn that c is *the ratio which the distance from the center to either focus of an ellipse bears to its transverse semi-axis.* But (Art. 161) a circle is an ellipse in which $c = 0$, or in which, therefore, $a = b$. In any circle, then, the ratio e is equal to zero. Hence, if we compare ellipses having a common transverse axis $= 2a$ with their common circumscribed circle, it is evident (since e will increase as the conjugate semi-axis b diminishes from the maximum value a toward zero) that e may be taken to measure the deviation of any of these ellipses from the circumscribed circle. For this reason, the ratio e is called the *eccentricity* of the ellipse to which it belongs.

The eccentricity of any ellipse evidently lies between the limits 0 and 1. In fact the name *ellipse* (derived from the Greek ἐλλείπειν, *to fall short*) may be taken as signifying, that, in this curve, the eccentricity is *less than unity*.

Since e increases as b diminishes, it is evident that the greater the eccentricity, the flatter will be the corresponding ellipse.

360. The distance of any point on an ellipse from either focus, may be expressed in terms of the abscissa of the point. For, putting ρ to denote any such focal distance, we have (Art. 147)

$$\rho^2 = (c \pm x)^2 + y^2.$$

But, from the equation to the Ellipse,

$$y^2 = \frac{b^2}{a^2} (a^2 - x^2).$$

An. Ge. 29.

Substituting and reducing, and remembering (Art. 151) that $b^2 + c^2 = a^2$, $a^2 - b^2 = a^2e^2$, and $c = ae$, we get

$$\rho = a \pm ex,$$

in which the upper sign corresponds to the left-hand focus, and the lower sign to the right-hand one. Hence, since ρ and x are of the first degree,

Theorem IX.—*The focal radius of any point on an ellipse is a linear function of the corresponding abscissa.*

Remark.—The expression just obtained is accordingly known as the *Linear Equation to the Ellipse.*

361. The form of the Ellipse is already familiar to the student, from the method of generating it given in Art. 145. Its appearance shows, or at least suggests, that it is an oblong, closed curve, continuous in extent, and symmetric to both of its axes. But it may be interesting, at this point, to show how we might have discovered each of these peculiarities of form from the equation itself, without the aid of any drawing.

I. The curve is *oblong.* For, no matter where we take it between its two limiting cases, the Point and the Circle, in its equation

$$y^2 = \frac{b^2}{a^2} (a^2 - x^2)$$

b (since it is equal to $\sqrt{a^2 - c^2}$) must be less than a; that is, the conjugate axis must be less than the transverse.

II. It is *closed,* i. e., *limited in the directions of both axes.* For, if we suppose $x > a$ or $< -a$, the corresponding values of y are imaginary; and, if we suppose $y > b$ or $< -b$, the corresponding values of x are imaginary.

III. It is *continuous in extent.* For, between the limits $x = -a$ and $x = a$, all the values of y are real.

IV. It is *symmetric to both axes.* For, corresponding to every value of x between the limits $-a$ and a, the two values of y are numerically equal with opposite signs; and the same is true of the values of x corresponding to any value of y between $-b$ and b.

DIAMETERS.

362. Equation to any Diameter.—We are required to find the equation to the locus of the middle points of any system of parallel chords in an ellipse. Let xy be the variable point of this locus, θ' the common inclination of the bisected chords, and $x'y'$ the point in which any chord of the system cuts the curve. Then (Art. 101, Cor. 3)

$$x' = x - l \cos \theta', \quad y' = y - l \sin \theta'.$$

But $x'y'$ is a point on the curve; hence (Art. 147)

$$\frac{(x - l \cos \theta')^2}{a^2} + \frac{(y - l \sin \theta')^2}{b^2} = 1.$$

That is, to determine the distance l between xy and $x'y'$, we get

$$(a^2 \sin^2 \theta' + b^2 \cos^2 \theta') l^2 - 2 (a^2 y \sin \theta' + b^2 x \cos \theta') l$$
$$+ (b^2 x^2 + a^2 y^2 - a^2 b^2) = 0.$$

Now, xy being the *middle* point of any chord, the two values of l must be numerically equal with opposite signs. Therefore (Alg., 234, Prop. 3d) the co-efficient of l vanishes, and we obtain, as the required equation,

$$y = - \frac{b^2}{a^2} x \cot \theta'.$$

Corollary.—Let $\theta =$ the inclination of any diameter to the transverse axis. The co-efficient of x in the equation just found is equal (Art. 78, Cor. 1) to $\tan \theta$. Hence, as the *condition connecting the inclination of any diameter with that of the chords which it bisects,*

$$\tan \theta \, \tan \theta' = - \frac{b^2}{a^2}.$$

363. The equation to any diameter conforms to the type $y = mx$. Therefore (Art. 78, Cor. 5) we have

Theorem X.—*Every diameter of an ellipse is a right line passing through the center.*

Corollary.—Since θ' in the foregoing condition is arbitrary, θ is also arbitrary. Hence, the converse of this theorem is true; that is, *Every right line that passes through the center of an ellipse is a diameter.*

364. If we eliminate between the equation to the Ellipse and that of any diameter, the roots of the resulting equation will be

$$x = \pm \frac{a^2 \sin \theta'}{\sqrt{(a^2 \sin^2\theta' + b^2 \cos^2\theta')}} :$$

which, it is evident, are necessarily real. Hence,

Theorem XI.—*Every diameter of an ellipse cuts the curve in two real points.*

365. Length of any Diameter.—This is of course double the radius vector given by the central polar equation (Art. 150), namely,

$$\rho^2 = \frac{b^2}{1 - e^2 \cos^2 \theta} .$$

Hence, given the inclination θ, the length of the corresponding diameter can at once be found.

366. From the preceding formula, it is evident that the diameter is longest when $\theta = 0$, and shortest when $\theta = 90°$. That is,

Theorem XII.—*In every ellipse, the transverse axis is the maximum, and the conjugate axis the minimum diameter.*

Remark.—For this reason, the transverse axis is called the *axis major*, and the conjugate, the *axis minor*.

367. Moreover, since θ enters the foregoing formula by the square of its cosine, the value of ρ is the same for θ and $\pi - \theta$. Hence,

Theorem XIII.—*Diameters which make supplemental angles with the axis major of an ellipse are equal.*

Corollary.—The converse of this is also given by the formula; so that, having the curve, we can always construct the axes as follows:—Draw any two pairs of parallel chords, and, by means of them, two diameters DC, $D'C$: their intersection C (Art. 363) will be the

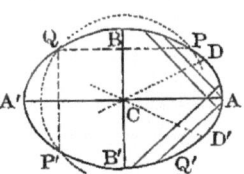

center. From C describe any circle cutting the curve in four points P, Q, P', Q'. The diameters PP', QQ' will then be equal, and, by the converse of the theorem above, the two bisectors of the angles between them will be the axes. These bisectors may be drawn most readily by forming the chords PQ, QP' which subtend the angles: they will be perpendicular to each other (Art. 317, Cor.), and their parallels through C will be the bisecting axes required.

368. Let θ and θ' be the inclinations of any two diameters to the axis major. Then the condition that the first shall bisect chords parallel to the second (Art. 362, Cor.) is

$$\tan \theta \tan \theta' = -\frac{b^2}{a^2}.$$

But this is also the condition upon which the second would bisect chords parallel to the first. Hence,

Theorem XIV.—*If one diameter of an ellipse bisects chords parallel to a second, the second bisects chords parallel to the first.*

369. Two diameters of an ellipse which are thus related, are called *conjugate* diameters, as in the case of the Circle. The re-appearance of this relation in connection with the Ellipse, gives occasion to define what is meant by *ordinates to a diameter.*

Definition.—The **Ordinates to a Diameter** are the right lines drawn from the curve to such diameter, parallel to its conjugate; or, they are the halves of the chords which the diameter bisects.

Corollary.—Hence, To construct a pair of conjugate diameters, draw any diameter $D'D$, and any two chords MN, PQ parallel to it. Join the middle points of the latter by the line $S'S$, which will be the required conjugate. When the center C is not given, the construction is effected by drawing $S'S$ through the middle of $D'D$, parallel to the two chords (double ordinates to $D'D$) by the aid of which this first diameter must in such a case be determined.

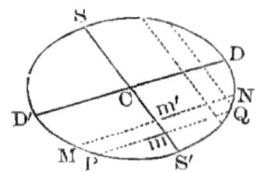

370. Equation of Condition for Conjugate Diameters.—This, as we have already seen (Art. 368), is

$$\tan \theta \tan \theta' = -\frac{b^2}{a^2}.$$

371. This condition, since it shows that the tangents of inclination belonging to any two conjugate diameters have opposite signs, indicates that one of two conjugates makes an *acute* angle with the axis major, and the other an *obtuse.* Now the axis minor makes a *right* angle with the axis major; hence,

Theorem XV.—*Conjugate diameters of an ellipse lie on* opposite *sides of the axis minor.*

372. Equation to a Diameter conjugate to a Fixed Point.—For brevity, we shall say that a diameter is *conjugate to a fixed point*, when it is conjugate to the diameter drawn through such point. If, now, $x'y'$ be any fixed point, the diameter drawn through it (Art. 95, Cor. 2) will be

$$y'x - x'y = 0 \qquad (1).$$

The equation to the conjugate will be of the form

$$y = x \tan \theta' \qquad (2),$$

in which (Art. 370) $\tan \theta'$ is determined by the condition

$$\tan \theta \tan \theta' = -\frac{b^2}{a^2}.$$

But (1), $\tan \theta = y' : x'$. Hence, the equation sought is

$$\frac{x'x}{a^2} + \frac{y'y}{b^2} = 0.$$

Corollary.—The equation to the diameter conjugate to that which passes through the point $(a, 0)$ is evidently $x = 0$. But the diameter through $(a, 0)$ is the axis major, while $x = 0$ denotes the axis minor. Hence, *The axes of an ellipse constitute a case of conjugate diameters.*

It is from this fact, that the axis minor derives its name of the *conjugate axis.*

373. Problem.—*Given the co-ordinates of the extremity of a diameter, to find those of the extremity of its conjugate.*

Let $x'y'$ be the extremity of the given diameter. The required co-ordinates, found by eliminating between the equation of Art. 372 and that of the Ellipse, are

$$x_c = \pm \frac{ay'}{b}, \qquad y_c = \mp \frac{bx'}{a}.$$

374. The expressions just obtained, transformed into the proportions

$$x_c : y' :: a : b, \quad x' : y_c :: a : b,$$

give us

Theorem XVI.—*The abscissa of the extremity of any diameter is to the ordinate of the extremity of its conjugate, as the axis major is to the axis minor.*

375. Squaring the second expression of Art. 373, adding y'^2, and remembering that $x'y'$ satisfies the equation to the Ellipse, we find

$$y'^2 + y_c^2 = b^2.$$

Hence, in ordinary language, we have

Theorem XVII.—*The sum of the squares on the ordinates of the extremities of conjugate diameters is constant, and equal to the square on the semi-axis minor.*

Remark.—The student may prove the analogous property:—*The sum of the squares on the abscissas of the extremities of conjugate diameters is constant, and equal to the square on the semi-axis major.*

376. Problem.—*To find the length of a diameter in terms of the abscissa of the extremity of its conjugate.*

Let a' be half the length required. Then, $x'y'$ being the extremity of a', we have (Art. 51, I, Cor. 2) $a'^2 = x'^2 + y'^2$. Substituting for x' and y' from Art. 373, we get

$$a'^2 = \frac{a^2}{b^2} y_c^2 + \frac{b^2}{a^2} x_c^2.$$

But x_c and y_c satisfy the equation to the Ellipse; hence,

$$a'^2 = (a^2 - x_c^2) + \frac{b^2}{a^2} x_c^2 = a^2 - \frac{a^2 - b^2}{a^2} x_c^2.$$

Therefore (Art. 151), for determining the required length, we have

$$a'^2 = a^2 - e^2 x_c^2.$$

By a precisely similar analysis, we should find

$$b'^2 = a^2 - e^2 x'^2.$$

377. Comparing the expressions last found with the formula of Art. 360, we obtain

Theorem XVIII.—*The square on any semi-diameter of an ellipse is equal to the rectangle under the focal radii drawn to the extremity of its conjugate.*

378. The result of Art. 376 leads to a noticeable property of the Ellipse, which we may as well develop in passing.

Let $x'y'$ be the extremity D of any diameter DD'. The equation to the conjugate diameter $S'S$ (Art. 372) will be

$$\frac{x'x}{a^2} + \frac{y'y}{b^2} = 0 \qquad (1).$$

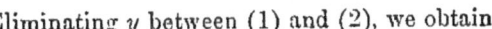

The equation to DF, which joins D to the focus, (Arts. 95, 151) may be written

$$y(x' - ae) = y'(x - ae) \qquad (2).$$

Eliminating y between (1) and (2), we obtain

$$(a^2 y'^2 + b^2 x'^2 - b^2\, ae\, x')\, x = a^3 e\, y'^2;$$

or, since x' and y' satisfy the equation to the Ellipse,

$$(a^2 b^2 - b^2\, ae\, x')\, x = b^2\, ae\, (a^2 - x'^2).$$

Hence, for the co-ordinates of M, we have

$$x = \frac{e(a^2 - x'^2)}{a - ex'}, \qquad y = -\frac{e\, x' y'}{a - ex'}.$$

The length of DM (Art. 51, I, Cor. 1) will therefore be found from

$$\delta^2 = \left\{ \frac{e(a^2 - x'^2)}{a - ex'} - x' \right\}^2 + \left\{ \frac{e\, x' y'}{a - ex'} + y' \right\}^2$$

Reducing the last expression, we obtain

$$\delta^2 = \frac{a^2(a^2 e^2 - 2ae\, x' + x'^2 + y'^2)}{(a - ex')^2}.$$

Now $x'^2 + y'^2 = a'^2 =$ (Art. 376) $a^2 - e^2 x_c^2 =$ (Art. 375, Rem.) $a^2 - e^2(a^2 - x'^2)$. Hence,

$$\delta = DM = a :$$

a relation which may be expressed by

Theorem XIX.—*The distance from the extremity of any diameter to its conjugate, measured upon the corresponding focal radius, is constant, and equal to the semi-axis major.*

379. Let b' denote the length of the semi-diameter conjugate to that whose extremity is $x'y'$. Then (Art. 51, I, Cor. 2) we shall have

$$b'^2 = x_c^2 + y_c^2 = (\text{Art. 147}) \ x_c^2 + \frac{b^2}{a^2} (a^2 - x_c^2).$$

Reducing, and applying the abbreviation of Art. 150, we get

$$b'^2 = b^2 + e^2 x_c^2.$$

Now (Art. 376) $a'^2 = a^2 - e^2 x_c^2$. Adding this expression to the preceding, we find

$$a'^2 + b'^2 = a^2 + b^2,$$

giving us the following important property:

Theorem XX.—*The sum of the squares on any two conjugate diameters of an ellipse is constant, and equal to the sum of the squares on the axes.*

380. Angle between two Conjugates.—Let φ denote the angle required. We shall then have $\varphi = \theta' - \theta$; whence (Trig., 845, III)

$$\sin \varphi = \sin \theta' \cos \theta - \cos \theta' \sin \theta.$$

But, putting a', b' for the lengths of the semi-diameters, and $x'y'$, $x_c y_c$ for their extremities, we have

$$\sin \theta' = y_c : b', \quad \cos \theta' = x_c : b';$$
$$\cos \theta = x' : a', \quad \sin \theta = y' : a'.$$

Hence,
$$\sin \varphi = \frac{x' y_c - y' x_c}{a' b'};$$

or, substituting for x_c and y_c from Art. 373, and reducing,

$$\sin \varphi = \frac{b^2 x'^2 + a^2 y'^2}{ab.a'b'}.$$

Now $x'y'$ being on the curve, (Art. 147) $b^2x'^2 + a^2y'^2 = a^2b^2$. Therefore,

$$\sin \varphi = \frac{ab}{a'b'}.$$

381. The expression just found, by a single transformation gives the relation

$$a'b' \sin \varphi = ab.$$

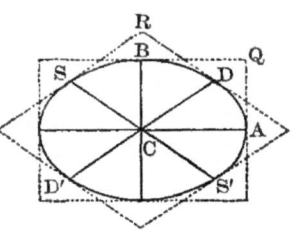

Now it is evident from the diagram, in which $CD = a'$, and $CS = b'$, that the first member of this equation denotes the parallelogram $CDRS$; and the second, the rectangle $CAQB$. Hence,

Theorem XXI.—*The parallelogram under any two conjugate diameters is constant, and equal to the rectangle under the axes.*

Remark.—We have drawn the parallelogram and rectangle in question as *circumscribed.* Future investigations will justify the figure. The property last obtained may be otherwise stated: *The triangle formed by joining the extremities of any two conjugate diameters is constant, and equal to that included between the semi-axes.*

Corollary 1.—If we suppose $\varphi = 90°$, then $\sin\varphi = 1$; and we obtain

$$a'b' = ab.$$

But (Art. 379),

$$a'^2 + b'^2 = a^2 + b^2.$$

Combining these equations, we get, after the proper reductions,

$$a' + b' = a + b, \quad a' - b' = a - b.$$

That is,

$$a' = a, \quad b' = b.$$

In other words, *In any ellipse there is but one pair of conjugate diameters at right angles to each other, namely, the axes.*

Corollary 2.—From the formula (Art. 380), φ is obviously greatest [*] when $a'b'$ is greatest. But since $a'^2 + b'^2$ is constant, the rectangle $a'b'$ has a constant diagonal, and is therefore greatest when $a' = b'$. Hence, *The inclination of two conjugate diameters is greatest when the diameters are of equal lengths.*

382. The diameters corresponding to the condition $a' = b'$ may be appropriately termed the *equi-conjugate diameters* of the ellipse to which they belong. Now (Art. 367, Cor.) for the case of equi-conjugates,

$$\tan \theta' = - \tan \theta \,;$$

hence (Art. 370), for the inclinations of the equi-conjugates to the axis major,

$$\tan \theta = \pm \frac{b}{a} \,.$$

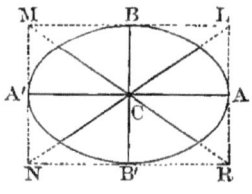

If we form the rectangle of the axes, $LMNR$, it is evident that the first of the two values just found, corresponds to the angle ACL; and the second, to the angle ACM. Hence,

Theorem XXII.—*The equi-conjugates of an ellipse are the diagonals of the rectangle contained under its axes.*

Corollary.—It follows directly from this, that *an ellipse can have but one pair of equi-conjugates.* In this case, (Art. 379,) $2a'^2 = a^2 + b^2$; so that (Art. 380)

$$\sin \varphi = \frac{2ab}{a^2 + b^2} \,.$$

383. We shall find hereafter that the two lines just brought to our notice have a striking significance with

---

[*] The angle ϕ is supposed to be that angle between two conjugates which is *not acute.*

respect to the analogy between the Ellipse and the Hyperbola. They in fact foreshadow the two remarkable lines known as the *asymptotes* of the latter curve, which, though still the diagonals of the rectangle formed upon its axes, meet it only at infinity.

THE TANGENT.

384. Equation to any Chord.—Let $x'y'$, $x''y''$ be the extremities of any chord in an ellipse : then (Art. 147)

$$b^2 x'^2 + a^2 y'^2 = b^2 x''^2 + a^2 y''^2.$$

Hence, after transposing and factoring,

$$\frac{y'' - y'}{x'' - x'} = - \frac{b^2}{a^2} \cdot \frac{x' + x''}{y' + y''} \qquad (1).$$

Now the equation to the chord (Art. 95) must be of the form

$$\frac{y - y'}{x - x'} = \frac{y'' - y'}{x'' - x'}.$$

Substituting from (1) for the second member, the required equation is

$$\frac{y - y'}{x - x'} = - \frac{b^2}{a^2} \cdot \frac{x' + x''}{y' + y''}.$$

385. Equation to the Tangent.—Suppose the two points in which the chord cuts the curve to become coincident: then, in the preceding expression, $x'' = x'$, $y'' = y'$; and the required equation to the tangent, in terms of the point of contact $x'y'$, is

$$\frac{y - y'}{x - x'} = - \frac{b^2}{a^2} \cdot \frac{x'}{y'};$$

or, by clearing of fractions and remembering that x' and y' satisfy the equation to the curve,

$$\frac{x'x}{a^2} + \frac{y'y}{b^2} = 1.$$

386. Condition that a Right Line shall touch an Ellipse.—Eliminating y between the line $y = mx + n$ and the ellipse

$$y^2 = \frac{b^2}{a^2}(a^2 - x^2),$$

we obtain, as the equation determining the intersections of the line and the curve, the quadratic

$$(m^2a^2 + b^2)x^2 + 2ma^2nx + a^2(n^2 - b^2) = 0.$$

The condition that this may have equal roots is

$$(ma^2n)^2 = a^2(n^2 - b^2)(m^2a^2 + b^2).$$

Hence, after the necessary reductions, the required condition of tangency is

$$n = \sqrt{m^2a^2 + b^2}.$$

Corollary.—Every right line, therefore, whose equation is of the form

$$y = mx + \sqrt{m^2a^2 + b^2}$$

is a tangent to the ellipse whose semi-axes are a and b. This expression, like the corresponding one belonging to the Circle (Art. 306, Cor.), affords singularly rapid solutions of problems which do not involve the point of contact; and for this reason is called the *Magical Equation to the Tangent.*

387. The Eccentric Angle.—If the ordinate of any point P on an ellipse be produced to meet the circumscribed circle in Q, and Q be joined to the center C, the angle QCM is called the *eccentric angle* of the point P. We introduce it here, because it serves the important purpose of *expressing the position of any point on an ellipse in terms of a single variable:* a purpose sometimes especially 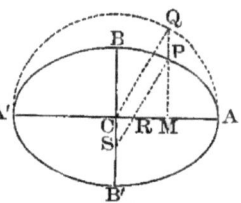 useful in connection with the equation to a chord or a tangent. The eccentric angle is usually denoted by φ.

It is evident from the diagram that $CM = CQ \cos QCM$, and $MP = PR \sin PRM = CQ\,(MP : MQ) \sin QCM =$ (Art. 359) $CB \sin QCM$. That is, if $x'y'$ be any point P of an ellipse,

$$x' = a \cos \phi, \quad y' = b \sin \phi.$$

By means of this relation, we can always express a point on an ellipse in terms of the single variable ϕ. Thus, the equation to the tangent at $x'y'$ becomes

$$\frac{x}{a} \cos \phi + \frac{y}{b} \sin \phi = 1.$$

388. Problem.—*If a tangent to an ellipse passes through a fixed point, to find the co-ordinates of contact.*

Let $x'y'$ be the required point of contact, and $x''y''$ the given point. Then, since $x''y''$ must satisfy the equation to the tangent, and $x'y'$ the equation to the curve, we shall have the two conditions

$$\frac{x'x''}{a^2} + \frac{y'y''}{b^2} = 1, \quad \frac{x'^2}{a^2} + \frac{y'^2}{b^2} = 1.$$

Solving these for x' and y', we find

$$x' = \frac{a^2b^2x'' \pm a^2y''\sqrt{b^2x''^2 + a^2y''^2 - a^2b^2}}{b^2x''^2 + a^2y''^2},$$

$$y' = \frac{a^2b^2y'' \mp b^2x''\sqrt{b^2x''^2 + a^2y''^2 - a^2b^2}}{b^2x''^2 + a^2y''^2}.$$

Corollary.—From these values it appears, that from any given point *two* tangents can be drawn to an ellipse: *real* when $b^2x''^2 + a^2y''^2 > a^2b^2$, that is, when the point is *without* the curve; *coincident* when $b^2x''^2 + a^2y''^2 = a^2b^2$, that is, when the point is *on* the curve; *imaginary* when $b^2x''^2 + a^2y''^2 < a^2b^2$, that is, when the point is *within* the curve.

389. The equation to the tangent at $x'y'$ (Art. 385) is

$$\frac{x'x}{a^2} + \frac{y'y}{b^2} = 1 \qquad (1);$$

and the equation to the diameter conjugate to that whose extremity is $x'y'$ (Art. 372) is

$$\frac{x'x}{a^2} + \frac{y'y}{b^2} = 0 \qquad (2).$$

Now (Art. 98, Cor.) the lines (1) and (2) are parallel. That is,

Theorem XXIII.—*The tangent at the extremity of any diameter of an ellipse is parallel to the conjugate diameter.*

Corollary.—If we replace x' and y' by $-x'$ and $-y'$, equations (1) and (2) still satisfy the condition of parallelism. Hence, *Tangents at the extremities of a diameter are parallel to each other.*

Remark.—If the student will form the equation to the parallel of (2) passing through $x'y'$, he will find that it is (1). In other words, the converse of our theorem is also true, and we can always construct a tangent at any point P, by drawing the diameter PD and its conjugate, and making LPM parallel to the latter. In this way we can form the circumscribed parallelogram

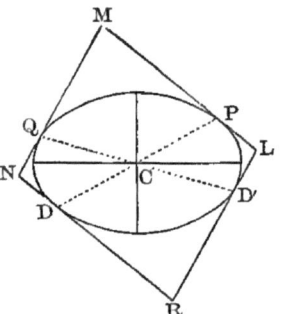

corresponding to any two diameters PD, QD'; and we here find the promised justification of the statement (Art. 381, Rem.), that the parallelogram under two conjugates is *circumscribed*, since its sides must be parallel to the conjugates, and therefore be tangents to the curve at their extremities.

390. Let PT be a tangent to an ellipse at any point P, and let FP, $F'P$ be the focal radii of contact. The equations to these lines may be written (Arts. 385, 95)

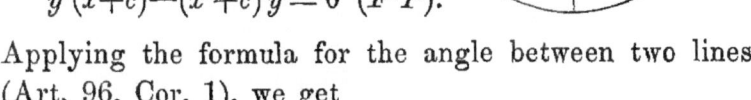

$$b^2 x'x + a^2 y'y = a^2 b^2 \quad (PT),$$
$$y'(x-c)-(x'-c)\,y = 0 \quad (FP),$$
$$y'(x+c)-(x'+c)\,y = 0 \quad (F'P).$$

Applying the formula for the angle between two lines (Art. 96, Cor. 1), we get

$$\tan FPT = \frac{a^2 y'^2 + b^2 x'(x'-c)}{b^2 x'y' - a^2 y'(x'-c)} = \frac{b^2(a^2-cx')}{cy'(a^2-cx')} = \frac{b^2}{cy'},$$

$$\tan F'PT = \frac{a^2 y'^2 + b^2 x'(x'+c)}{b^2 x'y' - a^2 y'(x'+c)} = -\frac{b^2(a^2+cx')}{cy'(a^2+cx')} = -\frac{b^2}{cy'}.$$

Hence, $FPT = 180° - F'PT = QPT$; and we obtain

Theorem XXIV.—*The tangent of an ellipse bisects the external angle between the focal radii drawn to the point of contact.*

Corollary 1.—We therefore have the following solution of the problem: *To construct a tangent to an ellipse at a given point.* Through the given point P, draw the focal radii FP, $F'P$, and produce the latter until $PQ = FP$. Join QF, and draw SPT perpendicular to it: SPT will be the required tangent. For the construction gives us $FPT = QPT$, according to the theorem that the perpendicular from the vertex to the base of an isosceles triangle bisects the vertical angle.

Corollary 2.—Since in Optics the angle of reflection is equal to the angle of incidence, while $FPT = QPT = SPF'$, all rays emanating from F and striking the curve will be reflected to F'; and reciprocally. Hence it is, that the two points F, F' are called the *foci*, or *burning points*, of the curve.

An. Ge. 30.

391. If we make $y = 0$ in the equation to the tangent, namely, in

$$b^2 x' x + a^2 y' y = a^2 b^2,$$

we shall obtain, as the intercept of the tangent on the axis major,

$$x = CT = \frac{a^2}{x'}.$$

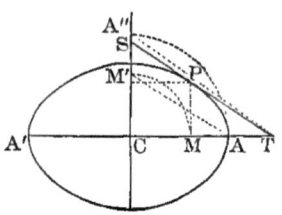

This intercept being thus a third proportional to the abscissa of contact and the semi-axis major, we have the following constructions:

I. *To draw a tangent at any point* P *of the curve.* On the axis major, take CT a third proportional (Art. 10, I, 2d) to the abscissa of contact CM and the semi-axis CA. Join PT, which will be the tangent required.

II. *To draw a tangent from any point* T *of the axis major.* Take CM a third proportional to CT and CA, and at M draw the ordinate MP: its extremity P will be the point of contact. Join TP, which will be the tangent sought.

392. The Subtangent.—The portion MT of the axis major, included between the foot of the tangent and the foot of the ordinate of contact, is called the *subtangent of the curve*, to distinguish it from the subtangent formed on any other diameter. For its length, we have $MT = CT - CM$; that is (Art. 391),

$$\text{subtan} = \frac{a^2 - x'^2}{x'} = \frac{(a + x')(a - x')}{x'}.$$

Now $a + x' = A'M$, and $a - x' = MA$; so that we get

Theorem XXV.—*The subtangent of an ellipse is a fourth proportional to the abscissa of contact and the two segments into which the ordinate of contact divides the axis major.*

Corollary.—It appears from the formula just found, that the subtangent is independent of b. Hence, *All ellipses described upon a common axis major will have a common subtangent for any given abscissa of contact.* We thus get a construction of the tangent by means of the circumscribed circle. For, circumscribe the circle AQA'; and at Q, where the

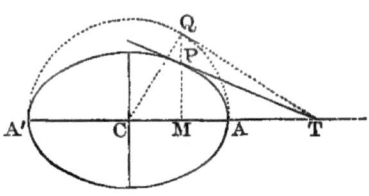

prolonged ordinate of any point P of the ellipse meets the circle, draw the tangent QT: then, by what has just been shown, T will be the foot of the tangent at P, which may be drawn by joining PT.

If T, the foot of the tangent, were given instead of the point of contact P, we should draw TQ tangent to the circumscribed circle, and, from the point of contact Q, let fall the ordinate QM. The point P in which the latter would cut the ellipse, would be the required point of contact; and, on joining this with the given point T, we should have the tangent sought.

393. Perpendicular from the Center to any Tangent.—The length of this is of course the length of the perpendicular from the origin upon the line

$$b^2 x' x + a^2 y' y = a^2 b^2.$$

Hence, (Art. 92, Cor. 2,) to determine it we have

$$p = \frac{a^2 b^2}{\sqrt{(b^4 x'^2 + a^4 y'^2)}} = \frac{ab}{\sqrt{(a^2 - e^2 x'^2)}}.$$

But (Art. 376) $a^2 - e^2x'^2 = b'^2$. Hence, finally,

$$p = \frac{ab}{b'}.$$

Expressing this relation in ordinary language, and observing the principle of Art. 389, we obtain

Theorem XXVI.—*The central perpendicular upon any tangent of an ellipse is a fourth proportional to the parallel semi-diameter and the semi-axes.*

394. **Central Perpendicular in terms of its inclination to the Axis Major.**—For the length of the perpendicular from the origin upon the tangent whose equation (Art. 386, Cor.) is

$$y = mx + \sqrt{m^2a^2 + b^2},$$

we have (Art. 92, Cor. 2)

$$p = \frac{\sqrt{(m^2a^2 + b^2)}}{\sqrt{(1 + m^2)}}.$$

Now let $\theta =$ the inclination of the perpendicular: then will $m = -\cot\theta$, and we get

$$p = \sqrt{a^2\cos^2\theta + b^2\sin^2\theta}.$$

395. The following investigation will illustrate the usefulness of the expression last obtained, and of the equation to the tangent from which it is derived.

Let it be required to find the locus of the intersection of tangents to an ellipse which cut at right angles. The inclinations of the two tangents will be θ and $90° + \theta$; and we shall have, for their central perpendiculars,

$$p^2 = a^2\cos^2\theta + b^2\sin^2\theta, \quad p'^2 = a^2\sin^2\theta + b^2\cos^2\theta.$$

$$\therefore p^2 + p'^2 = a^2 + b^2.$$

Now, if xy be the intersection of the tangents, the square of its distance from the center will be $x^2 + y^2 = p^2 + p'^2$. Hence, from what has just been proved, the co-ordinates of intersection are connected by the constant relation

$$x^2 + y^2 = a^2 + b^2:$$

which is the equation to a circle concentric with the ellipse, and circumscribed about the rectangle of the axes.* That is,

Theorem XXVII.—*The locus of the intersection of tangents which cut each other at right angles, is the circle circumscribed about the rectangle formed on the axes.*

396. Perpendiculars from the Foci to any Tangent.—The co-ordinates of the *right-hand* focus are $x = ae$, $y = 0$: hence, for the length of the perpendicular from F upon the tangent $b^2x'x + a^2y'y = a^2b^2$, we have (Art. 105, Cor. 2)

$$p = \frac{b^2x'ae - a^2b^2}{\sqrt{(b^4x'^2 + a^4y'^2)}} = \frac{ab^2(ex' - a)}{ab\sqrt{(a^2 - e^2x'^2)}} = \frac{b(ex' - a)}{\sqrt{(a^2 - e^2x'^2)}} \cdot$$

Now (Art. 360) $a - ex' = \rho$, the right-hand focal radius of contact; and (Art. 376) $a^2 - e^2x'^2 = b'^2$, the square of the semi-diameter conjugate to the point of contact. Hence, for the right-hand focus,

$$p = \frac{b\rho}{b'} \cdot$$

Similarly, for the left-hand focus, we should find

$$p' = \frac{b\rho'}{b'} \cdot$$

Corollary.—By Art. 377, $b'^2 = \rho\rho'$: hence, after squaring the expressions just obtained,

$$p^2 = \frac{b^2\rho}{\rho'}, \quad p'^2 = \frac{b^2\rho'}{\rho} :$$

formulæ which, in certain cases, are more useful than the preceding.

---

* This locus may be obtained even more readily, as follows: — The equations to the two tangents (Arts. 386, Cor.; 99, Cor.) may be written

$$y - mx = \sqrt{m^2a^2 + b^2}, \quad my + x = \sqrt{m^2b^2 + a^2}.$$

Squaring and adding these expressions, we eliminate m, and get

$$x^2 + y^2 = a^2 + b^2.$$

397. Comparing the two results of Art. 396, we get

$$p : p' :: \rho : \rho',$$

a relation expressed in ordinary terms by

Theorem XXVIII.—*The focal perpendiculars upon any tangent of an ellipse are proportional to the adjacent focal radii of contact.*

398. Multiplying together the values of p and p', and observing Art. 377, we obtain

$$pp' = b^2,$$

which is the algebraic expression of

Theorem XXIX.—*The rectangle under the focal perpendiculars upon any tangent is constant, and equal to the square on the semi-axis minor.*

399. The equation to any tangent of an ellipse (Art. 386, Cor.) being

$$y - mx = \sqrt{m^2 a^2 + b^2},$$

that of the focal perpendicular, which passes through $(\sqrt{a^2 - b^2}, 0)$, may be written (Art. 103, Cor. 2)

$$my + x = \sqrt{a^2 - b^2}.$$

Squaring these equations, and adding them together, we obtain

$$x^2 + y^2 = a^2$$

as the equation to the locus of the point in which the focal perpendicular meets the tangent. Hence, (Art. 136,)

Theorem XXX.—*The locus of the foot of the focal perpendicular upon any tangent of an ellipse, is the circle circumscribed about the curve.*

Corollary.—From this property, we obtain the following method of constructing a tangent to any ellipse,—

a method which deserves special attention, because it is applicable alike to all the Conics, and holds good whether the point through which the tangent is drawn be without the curve or upon it.

To draw a tangent to an ellipse through any given point.—Join the given point P with either focus F, and upon PF as a diameter describe a semicircle. Then through Q, where this semicircle cuts the circumscribed circle, draw PQ: 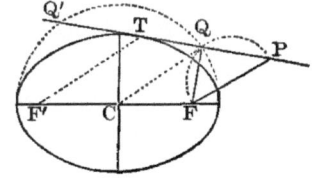 it will touch the ellipse at some point T. For the angle FQP is inscribed in a semicircle, and Q is therefore the foot of the focal perpendicular upon PQ.

In case, as in the second diagram annexed, the point P is on the ellipse, the circle described on PF will be found to *touch* 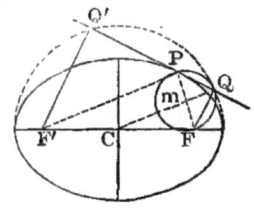 the circumscribed circle at Q (see Ex. 8, p. 359). The construction still holds, however; for the point of contact Q must lie on the line joining the centers of the auxiliary and circumscribed circles, and may therefore be found at once by joining C with the middle point of PF, and producing the line thus formed until it meets the circumscribed circle.

Remark.—It is obvious that the ordinary method of drawing a tangent to a circle through a given point (Geom., 230), is only a particular case of the method here described: the case, namely, where the two foci of the ellipse become coincident at C, when of course the ellipse becomes identical with the circumscribed circle. We have seen (Art. 388, Cor.) that *two* tangents can, in general, be drawn to an ellipse from a given point P; and the construction evidently corroborates this, since the auxiliary circle PQF must in general cut the circumscribed circle in two points.

400. From what has been shown in the preceding article, it follows that every chord drawn through the focus of an ellipse to meet the circumscribed circle is a focal perpendicular to some tangent of the ellipse. Now it is evident, that, a *cirele* being given, any point within it may be considered as the focus of an *inscribed* ellipse. We have, then,

Theorem XXXI.—*If from any point* within *a circle a chord be drawn, and a perpendicular to it at its extremity, the perpendicular will be tangent to the inscribed ellipse of which the point is a foeus.*

Corollary.—This is of course equivalent to saying that the inscribed ellipse is the *envelope* of the perpendicular. Advantage may be taken of this principle, to construct an ellipse approximately by means of right lines; for it is evident that by taking the perpendiculars sufficiently near together, we can approach the line of the curve as closely as we please. The diagram presents an example of this method.

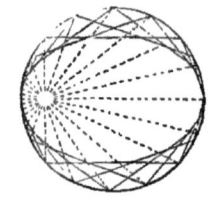

401. If the student will now form, by the method of Art. 108, Cor. 1, the equation to the diameter CQ, that is, the equation to the line joining the origin to the intersection of the tangent

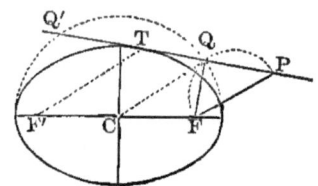

$$b^2x'x + a^2y'y = a^2b^2$$

with its focal perpendicular (Art. 103, Cor. 2)

$$a^2y'x - b^2x'y = a^2ey',$$

he will find that it may easily be reduced to the form

$$y'x - (x' + c)\,y = 0 \qquad (CQ).$$

Now the equation to the focal radius of contact $F'T$, which passes through the focus $(-c, 0)$ and the point of contact $x'y'$, (Art. 95) may be written

$$y'x - (x' + c)y + cy' = 0 \qquad (F'T).$$

But (Art. 98, Cor.) these equations show that CQ and $F'T$ are parallel; and, by like reasoning, the same may be proved of CQ' and FT. Hence,

Theorem XXXII.—*The diameters which pass through the feet of the focal perpendiculars upon any tangent of an ellipse, are parallel to the alternate focal radii of contact.*

Corollary.—The equations to CQ and $F'T$ evidently involve the converse theorem, *Diameters parallel to the focal radii of contact meet the tangent at the feet of its focal perpendiculars.* Hence, if in drawing a tangent through a given point P it becomes desirable, after obtaining (Art. 399, Cor.) the foot Q of the focal perpendicular, to find the point of contact, we can do so by merely drawing $F'T$ parallel to CQ.

By combining this property with Arts. 389, 399, we learn that *the distance between the foot of the perpendicular drawn from either focus to a tangent, and the foot of the perpendicular drawn from the remaining focus to the parallel tangent, is constant, and equal to the length of the axis major.*

THE NORMAL.

402. Equation to the Normal.—The expression for the perpendicular drawn through the point of contact $x'y'$ to the tangent

$$\frac{x'x}{a^2} + \frac{y'y}{b^2} = 1,$$

An. Ge. 31.

according to Art. 103, Cor. 2, is

$$\frac{y'}{b^2}(x - x') = \frac{x'}{a^2}(y - y').$$

Clearing of fractions, dividing through by $x'y'$, and putting (Art. 151) c^2 for $a^2 - b^2$, we may write the equation sought

$$\frac{a^2 x}{x'} - \frac{b^2 y}{y'} = c^2.$$

403. If we seek the angle φ made by the normal with the left-hand focal radius of contact $F'T$, whose equation is $y'x - (x' + c)y + cy' = 0$, we get (Art. 96, Cor. 1)

$$\tan \varphi = \frac{b^2 - \dfrac{a^2(x' + c)}{x'}}{\dfrac{b^2(x' + c)}{y'} + \dfrac{a^2 y'}{x'}} = -\frac{cy'}{b^2}.$$

Similarly, for the angle φ' made with the normal by the right-hand focal radius of contact FT, we get

$$\tan \varphi' = \frac{\dfrac{a^2(x' - c)}{x'} - b^2}{\dfrac{a^2 y'}{x'} + \dfrac{b^2(x' - c)}{y'}} = -\frac{cy'}{b^2}.$$

Hence $\varphi' = \varphi$; or, the normal makes equal angles with the two focal radii drawn to the point of contact, and we have

Theorem XXXIII.—*The normal of an ellipse bisects the internal angle between the focal radii of contact.*

Corollary 1.—This property enables us to construct a normal at a given point on the curve. For let P be the given point, and draw the corresponding focal radii FP, $F'P$. On $F'P$ lay off PQ equal to FP, join QF, and draw PN perpendicular to the latter: PN (Geom., 271) will bisect the angle $F'PF$, and will therefore be the normal required.

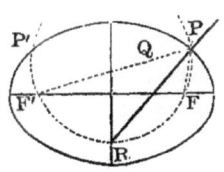

Corollary 2.—We can also draw a normal through any point on the axis minor. Let R be such a point. Pass a circle through the given point and the foci: it will cut the ellipse in P and P'. Join R with either of these points, as P: then will RP be a normal. For the arc $F'RF$ will be bisected in R, and the inscribed angles $F'PR$, FPR will therefore be equal; that is, RP will bisect the angle $F'PF$.

404. Intercept of the Normal.—If in the equation to the normal (Art. 402) we make $y = 0$, we find, as the length of the intercept on the axis major,

$$x = CN = \frac{c^2}{a^2} x' = e^2 x'.$$

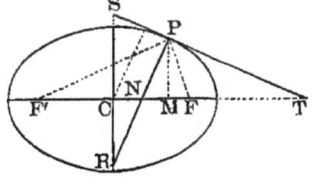

Corollary.—By means of this value, we can construct a normal either through a given point on the axis major or at a given point on the curve. For, in the former case, we have CN given, to find $x' = CM$; and, in the latter, CM is given, to find $x = CN$.

405. By Art. 151, we have $F'C = ae = CF$. From the preceding article, therefore,

$$F'N = e\,(a + ex'), \quad FN = e\,(a - ex').$$

Hence (Art. 360), $F'N : FN = F'P : FP$; that is,

Theorem XXXIV.—*The normal of an ellipse cuts the distance between the foci in segments proportional to the adjacent focal radii of contact.*

406. Length of the Subnormal.—The portion NM of the axis major, in-cluded between the foot of the normal and that of the ordinate of contact, is called the *subnormal of the curve,* to distinguish it from that formed on any other diameter.

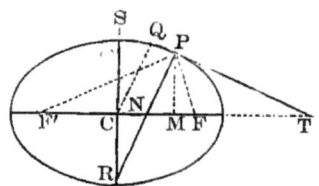

For its length, we have $NM = CM - CN = x' - e^2 x' = (1-e^2)x'$. That is (Art. 151),

$$\text{subnor} = \frac{b^2}{a^2}\, x'.$$

407. Comparing the results of Arts. 404 and 406, $CN : NM = c^2 : b^2$. Hence, as $c^2 = a^2 - b^2$, we get

Theorem XXXV.—*The normal of an ellipse cuts the abscissa of contact in the constant ratio* $(a^2 - b^2) : b^2$.

408. Length of the Normal.—By this is meant the portion of the normal intercepted between the point of contact and either axis. We have, then,

$$PN^2 = PM^2 + NM^2 = y'^2 + \frac{b^4}{a^4} x'^2 = \frac{b^2}{a^2}\left(a^2 - e^2 x'^2\right).$$

Hence, since (Art. 376) $a^2 - e^2 x'^2 = b'^2$,

$$PN = \frac{bb'}{a}.$$

By similar reasoning, the details of which are left for the student to supply,

$$PR = \frac{ab'}{b}.$$

409. From the foregoing, it follows immediately that $PN.PR = b'^2$. In other words, we have obtained

Theorem XXXVI.—*The rectangle under the segments formed by the two axes upon the normal is equal to the square on the semi-diameter conjugate to the point of contact.*

Corollary.—We have proved (Art. 377) that $b'^2 = \rho\rho'$. Hence, $PN.PR = \rho\rho'$; and we get the additional property: *The rectangle under the segments of the normal is equal to the rectangle under the focal radii of contact.*

410. It has been shown (Art. 393) that, for the length of the central perpendicular upon any tangent, we have

$$CQ = \frac{ab}{b'}.$$

Therefore, $CQ.PR = a^2$, and $CQ.PN = b^2$. That is,

Theorem XXXVII.—*The rectangle under the normal and the central perpendicular upon the corresponding tangent is constant, and equal to the square on the semi-axis other than the one to which the normal is measured.*

SUPPLEMENTAL AND FOCAL CHORDS.

411. Definition.—By **Supplemental Chords** of an ellipse, are meant two chords passing through the opposite extremities of any diameter, and intersecting on the curve.

Thus, DP, $D'P$ are supplemental with respect to the diameter $D'D$; and AQ, $A'Q$, with respect to the axis major.

412. Condition that Chords of an Ellipse be Supplemental.—Let φ be the inclination of any chord DP, and ς' that of the supplemental chord $D'P$. Then, since (Art. 149) every diameter is bisected by the center C, if the co-ordinates of D be x', y', those of D' will be $-x'$, $-y'$; and the equations to DP, $D'P$ (Art. 101, Cor. 1) may be written

$$y - y' = (x - x') \tan \varphi, \quad y + y' = (x + x') \tan \varphi'.$$

Hence, at the intersection P, we shall have the condition

$$y^2 - y'^2 = (x^2 - x'^2) \tan \varphi \tan \varphi',$$

in which xy, $x'y'$, being both upon the curve, are so connected (Art. 147) that

$$\frac{y^2 - y'^2}{x^2 - x'^2} = -\frac{b^2}{a^2}.$$

The supplemental chords are therefore subject to the constant condition

$$\tan \varphi \tan \varphi' = -\frac{b^2}{a^2}.$$

413. If θ and θ' are the respective inclinations of two diameters drawn parallel to a pair of supplemental chords, then $\theta = \varphi$ and $\theta' = \varphi'$; and, from the preceding condition, we have

$$\tan \theta \tan \theta' = -\frac{b^2}{a^2}.$$

But this (Art. 370) is the condition that the diameters corresponding to θ and θ' shall be conjugate. Hence,

Theorem XXXVIII.—*Diameters of an ellipse which are parallel to supplemental chords are conjugate.*

Remark.—This theorem may be otherwise stated: *If a diameter be parallel to one of two supplemental chords, its conjugate will be parallel to the other.* It therefore gives rise to the following constructions.

Corollary 1.—*To construct a pair of conjugate diameters at a given inclination.* On any diameter, describe (Geom., 231) an arc containing the given angle, and join either of the remaining points in which the circle cuts the ellipse with the extremities of the diameter : the diameters drawn parallel to the supplemental chords thus formed will be the conjugates required.

Caution.—It should be borne in mind, in connection with this problem, that the inclination of two conjugates in an ellipse is subject to a restriction, and is not *any angle we please*, but only (Arts. 381, Cor. 1; 382, Cor. 2) *any angle between the limits* 90° *and* $\sin^{-1} 2ab : (a^2 + b^2)$.

Corollary 2.—*To construct a tangent parallel to a given right line.* Let LM represent the given line. The point of contact of the required tangent (Art. 389) is the extremity of the diameter conjugate to that drawn parallel to LM. Draw, then, the chord AQ parallel to LM from the extremity of the axis major, and the diameter DP parallel to the supplemental chord QA':

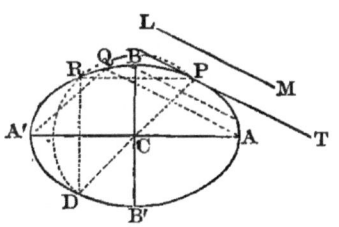

the line PT drawn through the extremity P of this diameter, parallel to the given line LM, will be the tangent sought.

Corollary 3.—*To construct the axes in the empty curve.* Draw any two parallel chords, bisect them, and form the corresponding diameter, say DP. On the latter, describe a semicircle cutting the ellipse in $R.$ Join DR, RP :

they will be supplemental chords of the circle, and therefore (Art. 317, Cor.) at right angles. They will also be
supplemental chords of the ellipse : hence $A'A$, $B'B$,
drawn parallel to them through C', will be the rectangular
conjugates of the curve ; that is, its axes.

414. Definition.—A **Focal Chord** of an ellipse, and in
fact of any conic, is simply a chord drawn through a focus.

The focal chords possess some special properties,
several of which, in the form corresponding to the
Ellipse, will be given in the examples at the close of
this Chapter. One of them, however, has an important
bearing upon the properties of a certain element of the
curve, and we shall therefore develop it here.

415. The equation to any focal chord, having the
inclination θ to the axis major, may be written (Art.
101, Cor. 3)

$$\frac{y}{\sin\theta} = \frac{x - ae}{\cos\theta} = l,$$

l being the distance from the focus to any point on the
chord. At the intersections of the chord with the curve,
we shall therefore have (Art. 147)

$$(a^2\sin^2\theta + b^2\cos^2\theta)\, l^2 + 2b^2 ae\cos\theta.l = b^4.$$

The roots of this quadratic are readily found to be

$$l' = \frac{b^2(1 + e\cos\theta)}{a(1 - e^2\cos^2\theta)}, \quad l'' = -\frac{b^2(1 - e\cos\theta)}{a(1 - e^2\cos^2\theta)}.$$

But these roots are the values of the two opposite segments into which the focus divides the chord. Neglecting,
then, the sign of l'', we have, for the length of the whole
chord,

$$\text{cho} = \frac{2}{a}\cdot\frac{b^2}{1 - e^2\cos^2\theta}.$$

Now (Art. 365) the second factor in this expression is the square on the semi-diameter whose inclination is θ. Hence, if we put $a' =$ the semi-diameter parallel to the chord, we get

$$\text{cho} = \frac{2a'^2}{a} = \frac{(2a')^2}{2a} :$$

a property which we may express by

Theorem XXXIX.—*Any focal chord of an ellipse is a third proportional to the axis major and the diameter parallel to the chord.*

Remark.—This result is exemplified in the value found (Art. 357, Cor.) for the *latus rectum*, which is the focal chord parallel to the axis minor.

II. The Curve referred to any two Conjugates.

DIAMETRAL PROPERTIES.

416. We are now ready to consider the Ellipse from a point of view somewhat higher than the one we have hitherto occupied, and shall presently discover that many of the properties we have developed are only particular cases of theorems more generic. Heretofore, we have referred the curve to its axes: let us now refer it to *any* two conjugate diameters.

417. Equation to the Ellipse, referred to any two Conjugate Diameters.—To obtain this, we must transform the equation of Art. 147,

$$\frac{x^2}{a^2} + \frac{y^2}{b^2} = 1,$$

from rectangular axes to oblique ones having the respective inclinations θ and θ' to the axis major. Replacing (Art. 56, Cor. 1) x and y by $x\cos\theta + y\cos\theta'$ and $x\sin\theta + y\sin\theta'$, we obtain, after obvious reductions,

$$(a^2\sin^2\theta + b^2\cos^2\theta)\, x^2 + (a^2\sin^2\theta' + b^2\cos^2\theta')\, y^2$$
$$+ 2\,(a^2\sin\theta\sin\theta' + b^2\cos\theta\cos\theta')\, xy = a^2 b^2.$$

But, as the new reference-axes are conjugate diameters, we have, by a single transformation of the condition in Art. 370,

$$a^2\sin\theta\sin\theta' + b^2\cos\theta\cos\theta' = 0.$$

The transformed equation is therefore in fact

$$(a^2\sin^2\theta + b^2\cos^2\theta)\, x^2 + (a^2\sin^2\theta' + b^2\cos^2\theta')\, y^2 = a^2 b^2.$$

In this, the co-efficients are still functions of the semi-axes; but if we seek the values of the semi-conjugates a' and b' by finding (Art. 73) the intercepts of the curve upon the new axes of reference, we readily obtain

$$a^2\sin^2\theta + b^2\cos^2\theta = \frac{a^2 b^2}{a'^2}, \quad a^2\sin^2\theta' + b^2\cos^2\theta' = \frac{a^2 b^2}{b'^2}.$$

Hence, the required equation, in its final form, is

$$\frac{x^2}{a'^2} + \frac{y^2}{b'^2} = 1.$$

418. Comparing this equation with that of Art. **147,** and remembering (Art. 372, Cor.) that the axes are a case of conjugates, it becomes evident that the equation hitherto used, namely,

$$\frac{x^2}{a^2} + \frac{y^2}{b^2} = 1,$$

is only the particular form assumed by the general one we have now obtained, when the reference-conjugates have the specific lengths 2a, 2b, and are at right angles to each other. Moreover, from the identity of form in the two equations, we see at once that the transformations applied to

$$\frac{x^2}{a^2} + \frac{y^2}{b^2} = 1$$

are equally applicable to

$$\frac{x^2}{a'^2} + \frac{y^2}{b'^2} = 1,$$

and that the theorems derived from the former may therefore be immediately extended to any conjugate diameters, provided they do not involve the inclination of the reference-axes. Thus, we learn that the theorems of Arts. 357, 358 are particular cases of the following:

Theorem XL.—*The squares on the ordinates to any diameter of an ellipse are proportional to the rectangles under the corresponding segments of the diameter.*

Theorem XLI.—*The square on any diameter of an ellipse is to the square on its conjugate, as the rectangle under any two segments of the diameter is to the square on the corresponding ordinate.*

419. The equation of Art. 417 may of course be written

$$y^2 = \frac{b'^2}{a'^2}(a'^2 - x^2).$$

The equation to a circle described upon the diameter 2a' is

$$y^2 = a'^2 - x^2.$$

Supposing, then, that we consider those ordinates of the two curves that correspond to a common abscissa, we get

$$y_e : y_c = b' : a',$$

and therefore have the following extension of the relation (Art. 359) between the Ellipse and the circumscribed circle:

Theorem XLII.—*The ordinate to any diameter of an ellipse is to the corresponding ordinate of the circle described on that diameter, as the conjugate of the diameter is to the diameter itself.*

Corollary 1.—Hence, given two conjugate diameters in position and magnitude, we may construct the curve by points, as follows: — On each of the given diameters DL, $D'L'$, describe a circle. At M, any point on the diameter selected for the axis of x, set up a rectangular ordinate of the

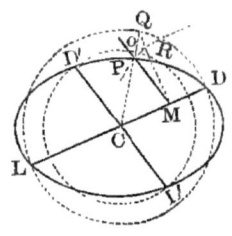

corresponding circle, meeting the curve in Q, and draw MP parallel to the conjugate diameter $D'L'$. Join Q, the extremity of the circular ordinate, to the center C, and through O, where QC cuts the inner circle, draw OR parallel to DL: then will RM measure the distance of O from DL. From M as a center, with the radius MR, describe an arc cutting MP in P: then will P be a point of the required ellipse. For $MP = MR =$ the perpendicular from O upon CD. Hence, $MP : MQ = CO : CQ = b' : a'$.

Corollary 2.—The use of the equation $x^2 + y^2 = a'^2$ in the preceding investigation, to denote a circle described upon a diameter of an ellipse, suggests a point of considerable importance. The equation denotes such a circle, only on the supposition that the axes of reference are rectangular; and, in the construction just explained, the ordinate MQ was drawn in accordance with this principle.

The equation, however, is susceptible of a more general interpretation. It evidently arises from the general equation

$$\frac{x^2}{a'^2} + \frac{y^2}{b'^2} = 1,$$

whenever $a' = b'$. In other words, the equation

$$x^2 + y^2 = a'^2,$$

when the reference-axes are *oblique*, denotes *an ellipse referred to its equi-conjugate diameters.*

420. By throwing the equation of Art. 417 into the form

$$y^2 = \frac{b'^2}{a'^2}\,(a'^2 - x^2),$$

and subjecting it to an analysis precisely like that of Art. 361, we can determine the figure of the Ellipse with respect to any two conjugates. It will thus appear that the curve is oblong, closed, continuous in extent, and symmetric not only to the axis major and axis minor, but to any diameter whatever.

CONJUGATE PROPERTIES OF THE TANGENT.

421. Equation to the Tangent, referred to any two Conjugate Diameters.—From the relation established (Art. 418) between the equations

$$\frac{x^2}{a^2} + \frac{y^2}{b^2} = 1 \quad \text{and} \quad \frac{x^2}{a'^2} + \frac{y^2}{b'^2} = 1,$$

it follows that the application to the latter of the method used in Arts. 384, 385 must result in

$$\frac{x'x}{a'^2} + \frac{y'y}{b'^2} = 1.$$

422. Intercept of the Tangent on any Diameter.—Making $y = 0$ in the equation just found, we get, for the intercept in question,

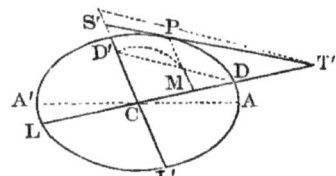

$$x = CT' = \frac{a'^2}{x'}.$$

We have, then, as the extension of Art. 391,

Theorem XLIII.—*The intercept cut off by a tangent upon any diameter of an ellipse is a third proportional to the abscissa of contact and the semi-diameter.*

Corollary.—*To construct a tangent from any given point.* From the given point T', draw the diameter $T'C$, and form its conjugate CD'. On CT' take CM a third proportional to the intercept CT' and the semi-diameter CD, and draw MP parallel to CD': then will MP be the ordinate of contact. Join $T'P$, which will be the tangent required.

423. We may conveniently group at this point a few properties of tangents and their intercepts, which will serve to illustrate the advantages of using conjugate diameters as axes of reference.

I. Let $D'T$, $L'T'$ be any two fixed parallel tangents of an ellipse, intersected by any variable tangent $T'T$. $D'L'$ joining the points of contact of the parallel tangents, and DL drawn parallel to them,

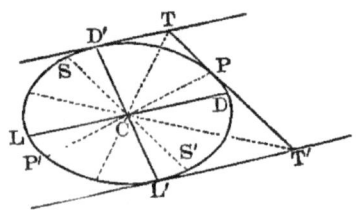

(Art. 389) will be conjugate diameters. Taking these for reference-axes, the equation to the variable tangent $T'T$ is

$$\frac{x'x}{a'^2} + \frac{y'y}{b'^2} = 1.$$

In this, making y successively equal to $+ b'$ and $- b'$, we obtain

$$x = D'T = \frac{a'^2}{x'}\left(1 - \frac{y'}{b'}\right), \quad x = L'T' = \frac{a'^2}{x'}\left(1 + \frac{y'}{b'}\right)$$

Hence, multiplying together the two values of x, and substituting for y'^2 its value from the equation to the Ellipse, we get

$$D'T . L'T' = a'^2.$$

Interpreting this relation in ordinary language, we have

Theorem XLIV.—*The rectangle under the intercepts cut off upon two fixed parallel tangents by any variable tangent of an ellipse is constant, and equal to the square on the semi-diameter parallel to the two tangents.*

II. If we take for axes of reference the diameter CP drawn to the point of contact of any variable tangent, and the conjugate diameter SS', the equations to any two fixed parallel tangents $D'T$, $L'T'$ will be

$$\frac{x'x}{a'^2} + \frac{y'y}{b'^2} = 1, \quad \frac{x'x}{a'^2} + \frac{y'y}{b'^2} = -1.$$

Making $x = a'$ in each of these, and remembering (Art. 389) that the axis of y (SS') is parallel to the variable tangent $T'T$, we get

$$y = PT = \frac{b'^2}{y'}\left(1 - \frac{x'}{a'}\right), \quad y = PT' = -\frac{b'^2}{y'}\left(1 + \frac{x'}{a'}\right).$$

Hence, after substituting for y'^2 from the equation to the curve,

$$PT . PT' = b'^2,$$

the sign of the second factor being disregarded, as we are only concerned with the *area* of the rectangle. We have, then,

Theorem XLV.—*The rectangle under the intercepts cut off upon any variable tangent of an ellipse by two fixed parallel tangents is variable, being equal to the square on the semi-diameter parallel to the tangent.*

III. Using the same axes of reference as in II, the equations to any two conjugate diameters, for instance CT and CT', (Arts. 372, 418) may be written

$$\frac{x}{x'} - \frac{y}{y'} = 0, \quad \frac{x'x}{a'^2} + \frac{y'y}{b'^2} = 0.$$

Making $x = a'$ in each, we obtain

$$y = PT = \frac{a'y'}{x'}, \quad y = PT' = -\frac{b'^2 x'}{a'y'}.$$

Hence, neglecting the sign of the second intercept,

$$PT . PT' = b'^2,$$

and we thus arrive at

Theorem XLVI.—*The rectangle under the intercepts cut off upon any variable tangent of an ellipse by two conjugate diameters is equal to the square on the semi-diameter parallel to the tangent.*

Remark.—It is evident that, by a single change in the interpretation of the symbols, we might have stated the theorem thus: *The rectangle under the intercepts cut off upon a fixed tangent by any two conjugates is constant, and equal to the square on the parallel semi-diameter.*

Corollary 1.—It is obvious from the equations, that none but conjugate diameters will cut off such intercepts as will form the rectangle mentioned. Hence, the converse of the theorem is also true; and, combining it with the result of II, we get: *Diameters drawn through the intersections of any tangent with two parallel tangents are conjugate.*

Corollary 2.—The theorem of III also furnishes us with the following neat solution of the problem,

Given two conjugate diameters of an ellipse in position and magnitude, to construct the axes. Let CD, CD' be the given conjugate semi-diameters. Through the extremity of either, as D, draw AB parallel to the other: it will be a tangent of the corresponding ellipse (Art. 389). Produce CD to P, so that $CD.DP = CD'^2$. Bisect CP by the perpendicular MO, and from O, where MO meets AB, describe a circle through C and P. Join the points A and B, in which this circle cuts AB, with the center C of the ellipse: CA, CB will be

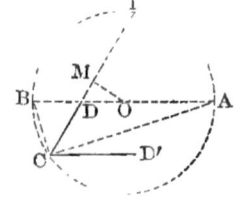

the axes sought. For (Geom., 331) $AD.DB = CD.DP = CD'^2$, and therefore (Th. XLVI) CA, CB are conjugate diameters. Moreover, since ACB is inscribed in a semicircle, they are the rectangular conjugates of the curve; or, in other words, the axes.

424. Subtangent to any Diameter.—This being the portion of any diameter intercepted between the foot T' of a tangent, and the foot M of the ordinate of contact, we have, for its length, $MT' = CT' - CM$. Hence, (Art. 422,)

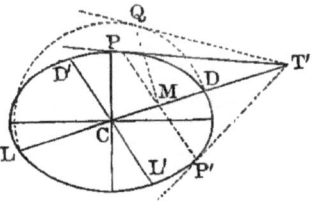

$$\text{subtan}' = \frac{a'^2 - x'^2}{x'}.$$

That is, since $a' + x' = LM$, and $a' - x' = MD$, *The subtangent to any diameter of an ellipse is a fourth proportional to the abscissa of contact and the corresponding segments of the diameter.*

Corollary.—This value is independent of the length of the conjugate semi-diameter b'; and, if we compare it with that of the subtangent in the circle $x^2 + y^2 = a'^2$ described upon the diameter which serves as the axis of x, we find (Art. 311) that the two are equal. Hence the following construction :

To draw a tangent to an ellipse from any given point. Let T' be the given point. Through it draw the diameter $T'DL$; upon DL describe a circle, and form its tangent TQ passing through the given point. Let fall QM perpendicular to DL, and through its foot M draw PP' a double ordinate to the diameter DL: the points P, P' in which this meets the curve will be the points of contact of the two possible tangents from T', either of which may be obtained by joining T' to the proper point of contact.

An. Ge. 32.

Remark.—It is obvious that the same principle may be used to construct a tangent at any point P of the curve, by simply drawing any diameter and its conjugate, forming the corresponding ordinate of the ellipse from P, and erecting at its foot the ordinate MQ of the circle described on the diameter first drawn : the tangent to this circle at Q will determine the foot T'' of the required tangent to the ellipse. This method is very convenient when the curve only, or an arc of it, is given ; but, when the axes are known, the construction described in the corollary to Art. 392 is preferable.

425. If we multiply the value of the subtangent by the abscissa of contact x', we get x' subtan$' = a'^2 - x'^2$. Comparing this with the square on the ordinate of contact, as given by the equation to the curve, namely,

$$y'^2 = \frac{b'^2}{a'^2}(a'^2 - x'^2),$$

we obtain x' subtan$' : y'^2 = a'^2 : b'^2$, a relation expressed by

Theorem XLVII.—*The rectangle under the subtangent and the abscissa of contact is to the square on the ordinate of contact, as the square on the corresponding diameter is to the square on its conjugate.*

426. The equations to the tangents at the extremities of any chord of an ellipse, by taking for axes the diameter parallel to the chord, and its conjugate, may be written

$$\frac{x'x}{a'^2} + \frac{y'y}{b'^2} = 1, \quad \frac{x'x}{a'^2} - \frac{y'y}{b'^2} = 1.$$

Eliminating between these, we find that the co-ordinates of the point in which the tangents intersect are

$$x = \frac{a'^2}{x'}, \quad y = 0.$$

Hence, comparing Arts. 369; 49, Cor. 1, we have

Theorem XLVIII.—*Tangents at the extremities of any chord of an ellipse meet on the diameter which bisects that chord.*

PARAMETERS.

427. Definitions.—The **Parameter** of an ellipse, with respect to any diameter, is a third proportional to the diameter and its conjugate. Thus, if a', b' denote the lengths of any two conjugate semi-diameters, we shall have, for the value of the corresponding parameter,

$$\text{parameter} = \frac{(2b')^2}{2a'} = \frac{2b'^2}{a'}.$$

The parameter with respect to the axis major, is called the *principal parameter;* or, the *parameter of the curve.* We shall represent its length by the symbol $4p$.

428. From the definition above, we have, for the value of the parameter of the Ellipse,

$$4p = \frac{2b^2}{a}.$$

Hence, (Art. 357, Cor.,) the principal parameter is identical with the line which we named the *latus rectum;* that is, it is *the double ordinate drawn through the focus to the axis major.*

429. In Art. 415, we proved that the focal chord (or double ordinate) parallel to any diameter is a third proportional to the axis major and the diameter. Now (Art. 366) the axis major is greater than any other diameter — greater, therefore, than the diameter conjugate to that of which the focal chord is a parallel, unless the chord is the *latus rectum*. Hence,

Theorem XLIX.—*No parameter of an ellipse, except the principal, is equal in value to the corresponding focal double ordinate.*

POLE AND POLAR.

430. We can now show that the reciprocal relation of points and right lines which we established (Arts. 318 — 321) in the case of the Circle, is a property of the generic curve of which the Circle is only a particular case. We shall develop the conception of the polar line in the Ellipse by the same steps as in the former investigation.

431. Chord of Contact in the Ellipse.—Let $x'y'$ be the fixed point from which the two tangents that determine the chord are drawn, and x_1y_1, x_2y_2 their respective points of contact. Their equations (Art. 421) will be

$$\frac{x_1 x}{a'^2} + \frac{y_1 y}{b'^2} = 1, \quad \frac{x_2 x}{a'^2} + \frac{y_2 y}{b'^2} = 1.$$

But $x'y'$ being upon both tangents, we have

$$\frac{x_1 x'}{a'^2} + \frac{y_1 y'}{b'^2} = 1, \quad \frac{x_2 x'}{a'^2} + \frac{y_2 y'}{b'^2} = 1.$$

That is, the co-ordinates of both points of contact satisfy the equation

$$\frac{x'x}{a'^2} + \frac{y'y}{b'^2} = 1.$$

This is therefore the *equation to the chord of contact.*

432. Locus of the Intersection of Tangents to the Ellipse.—Let $x'y'$ be the fixed point through which the chord of contact belonging to two intersecting tangents is drawn, and x_1y_1 the intersection of the tangents. The equation to the chord (Art. 431) will be

$$\frac{x_1x}{a'^2} + \frac{y_1y}{b'^2} = 1,$$

and, as $x'y'$ is on the chord, we shall have the condition

$$\frac{x_1x'}{a'^2} + \frac{y_1y'}{b'^2} = 1,$$

irrespective of the direction of the chord. The co-ordinates of intersection for the two tangents drawn at its extremities must therefore always satisfy the equation

$$\frac{x'x}{a'^2} + \frac{y'y}{b'^2} = 1:$$

which is for that reason the equation to the required locus.

433. Tangent and Chord of Contact included in the wider conception of the Polar.—The equations to the tangent, to the chord of contact, and to the locus of the intersection of tangents drawn at the extremities of chords that pass through a fixed point, are thus seen to be identical in form. These three lines are therefore only different expressions of a common formal law; and, inasmuch as the fixed point $x'y'$, in the case of the chord of contact, is restricted to being *without* the curve; and, in that of the tangent, to being *on* the curve; while, in the case of the locus in question, it is not restricted at all: it follows that the tangent and chord of contact are cases of the locus, due to bringing the point $x'y'$ upon the curve or outside of it. Moreover, the relation between the locus which thus absorbs the tangent and chord of contact, and the fixed point $x'y'$, is that of *polar reciprocity*. For, by precisely the same argument as that used (Art. 321) in the case of the Circle, we have the twofold theorem:

I. *If from a fixed point chords be drawn to any ellipse, and tangents to the curve be formed at the extremities of each chord, the intersections of the several pairs of tangents will lie on one right line.*

II. *If from different points lying on one right line pairs of tangents be drawn to any ellipse, their several chords of contact will meet in one point.*

It thus appears that the Ellipse imparts to every point in its plane the power of determining a right line; and reciprocally.

434. Equation to the Polar with respect to an Ellipse.—From what has been shown in the preceding articles, it is evident that this equation, referred to any pair of conjugate diameters, is

$$\frac{x'x}{a'^2} + \frac{y'y}{b'^2} = 1,$$

$x'y'$ being the point to which the polar corresponds. Consequently, the equation referred to the axes of the curve will be

$$\frac{x'x}{a^2} + \frac{y'y}{b^2} = 1.$$

435. Definitions.—The **Polar** of any point, with respect to an ellipse, is the right line which forms the locus of the intersection of the two tangents drawn at the extremities of any chord passing through the point.

The **Pole** of any right line, with respect to an ellipse, is the point in which all the chords of contact corresponding to different points on the line intersect each other.

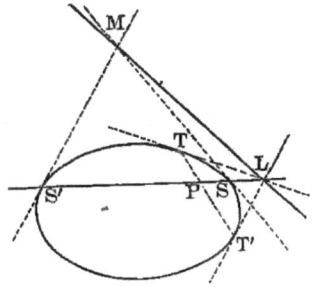

From these definitions, we obtain the following constructions :—When the pole P is given, draw through it any two chords $T'T$, $S'S$, and form the corresponding pairs of tangents, $T'L$ and TL, $S'M$ and SM. Join the intersection of the first pair to that of the second, forming the line LM: this will be the polar of P. When the polar is given, take any two points upon it, as L and M, and from each draw a pair of tangents to the curve: the point P, in which the corresponding chords of contact $T'T$, $S'S$ intersect, will be the pole of LM.

In case the pole is *without* the curve, as at L, the polar is the chord of contact of the two tangents from L; and, when the pole is *on* the curve, as at T, the polar is the tangent at T. In either case, then, the construction may be made in the way these facts require.

436. Direction of the Polar.—The equation to the polar of any point $x'y'$, namely (Art. 434),

$$\frac{x'x}{a'^2} + \frac{y'y}{b'^2} = 1,$$

when compared with that of the diameter conjugate to
the same point, namely (Arts. 372, 418),

$$\frac{x'x}{a'^2} + \frac{y'y}{b'^2} = 0,$$

shows (Art. 98, Cor.) that the polar and the diameter
are parallel. We have, then, the following extension
of the property reached in Art. 389:

Theorem L.—*The polar of any point, with respect to
an ellipse, is parallel to the diameter conjugate to that
which passes through the point.*

437. Polars of Special Points.—It is easy to see,
by comparing the equations to the polar in the Ellipse
and in the Circle (Arts. 434, 323), that the general
properties of polars proved in Art. 324 are true in the
case of the Ellipse. We leave the student to convince
himself of this, and will here present certain special
properties of polars, which depend on taking the pole
at particular points.

If we substitute for $x'y'$, in the equation of Art. 434,
the co-ordinates of the center, we shall get $1 = 0$: an
expression conforming to the type (Art. 110)

$$C = 0.$$

Hence, *The polar of the center is a right line at infinity.*

If in the same equation we make $y' = 0$, we shall get

$$x = \frac{a'^2}{x'}.$$

Hence, *The polar of any point on a diameter is a right
line parallel to the conjugate diameter, and its distance from
the center is a third proportional to the distance of the point
and the length of the semi-diameter.*

Similary, *The polar of any point on the axis major is the perpendicular whose distance from the center is a third proportional to the distance of the point and the length of the semi-axis.*

Corollary.—The second of these properties obviously leads to the following construction of the polar:—Join the given point with the center of the curve, and, from the latter as origin, lay off upon the resulting diameter a third proportional to the distance of the point and the length of the semi-diameter. Through the point thus reached, draw a parallel to the conjugate diameter, which will be the polar required.

438. Polar of the Focus.—The equation to the polar of either focus, by substituting $(\pm ae, 0)$ for $x'y'$ in the second equation of Art. 434, is found to be

$$x = \pm \frac{a}{e}.$$

Hence, *The polar of either focus in an ellipse is the perpendicular which cuts the axis major at a distance from the center equal to* a : e, *measured on the same side as the focus.*

439. The distance of any point P of the curve from either focal polar, say DR, is evidently equal to the distance of that polar from the center, diminished by the abscissa of the point. Thus,

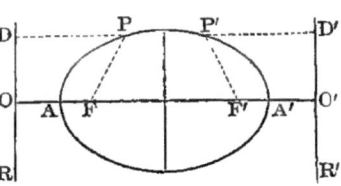

$$PD = \frac{a}{e} - x = \frac{a - ex}{e}.$$

Now (Art. 360) $a - ex = FP$. Therefore,

$$\frac{FP}{PD} = e,$$

An. Ge. 33.

and we have the remarkable property, which will hereafter be found to characterize all the Conics,

Theorem LI.—*The distance of any point on an ellipse
from the focus is in a constant ratio to its distance from
the polar of the focus, the ratio being equal to the eccentricity of the curve.*

Corollary 1.—Upon this theorem is founded the following method for constructing any arc of an ellipse. The
process is not simple enough for extensive use, but is
interesting as exhibiting the analogy between the Ellipse
and the other two Conics in regard to the important
property just established.

Take any point F, and any fixed
right line DR. Draw FR perpendicular to DR, and, at any convenient
point of the latter, as D, make DP
parallel to FR and greater than FP,
to express the property that the eccentricity of the Ellipse (which, by the
theorem above, equals $FP: PD$) is less
than unity. On PD describe a semicircle, and from P as center, with a

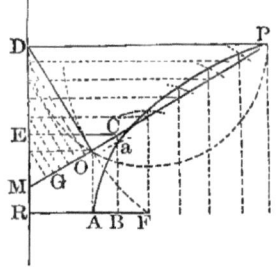

radius FP, form an arc cutting the semicircle in O. Join DO, PO,
producing the latter to meet DR in M: the triangle DOP will be
right-angled at O, being inscribed in a semicircle. Now divide FR
in the ratio $FP: PD$, suppose at A: then will A be the vertex of
the ellipse of which P is a point, F the focus, and DR the polar
of the focus. At any point on the line RF to the right of A, as
at B, erect a perpendicular, meeting PM in C. Draw CE parallel
to PD, and EG parallel to DO. Next, from F as center, with a
radius equal to CG, describe an arc cutting BC in the point a:
then will a be a point on the curve. For the triangle CEG is by
construction similar to PDO, so that $CG: CE = PO: PD$; or,
$Fa: CE = FP: PD$. Thus the focal distance of the point is to its
distance from the line which may now be considered the focal polar,
in a constant ratio less than unity. By repeating the process just
described, other points of the curve may be found in sufficient

numbers to determine its outline, and it can then be drawn through them.

Corollary 2.—The fact has been brought to light (Arts. 431—433), that the positions of the pole and polar may be as near to or as remote from each other as we please. It is therefore not only true that an ellipse imparts to every point in its plane the power of determining a right line, but *any* given right line is the polar of *any* given point, with respect to some ellipse. Now the construction just explained rests upon this principle, and the given line *DR* is therefore constantly taken as the boundary against which, as the polar of the given point *F*, the horizontal distances of the points on the curve are measured. From this constant relation to the figure of the resulting ellipse, the polar of the focus is called the *directrix* of the curve.

Corollary 3.—The theorem of this article invests the term *ellipse* with a new meaning. We now see that the name of the curve may be interpreted as signifying *the conic in which the constant ratio between the focal and polar distances* falls short *of unity*.

440. Focal Angle subtended by any Tangent.—By this is meant the angle *PFT* included between two focal radii *FT*, *FP*: one drawn to the point of contact of the tangent passing through any fixed point *P*; and the other, to such fixed point itself. The determination of this angle involves a remarkable relation, which, although not depending upon the properties of conjugate diameters, we shall nevertheless present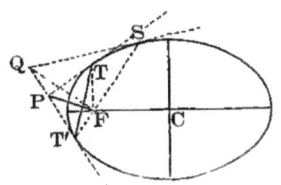
here, because the proper statement of it implies that the reader is acquainted with the definition of the pole and the polar.

Let xy be the arbitrary point P through which a tangent PT is drawn, and $x'y'$ the corresponding point of contact T. Also, let

$$\theta = CFP, \qquad \theta' = CFT,$$
$$\rho = FP, \quad \text{and} \quad \rho' = FT.$$

We shall then have

$$\cos \theta = \frac{ae - x}{\rho}, \qquad \sin \theta = \frac{y}{\rho};$$
$$\cos \theta' = \frac{ae - x'}{\rho'}, \qquad \sin \theta' = \frac{y'}{\rho'}.$$

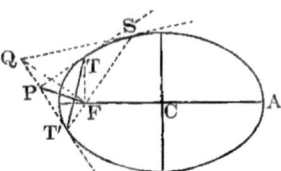

Therefore, putting $\phi = PFT$, we shall get (Trig., 845, IV)

$$\cos \phi = \frac{(ae - x)\,(ae - x') + y'y}{\rho\rho'}$$

Now, from the equation to the tangent, $y'y = b^2 - \dfrac{b^2}{a^2} x'x.$ Hence,

$$\rho\rho' \cos \phi = a^2 - ae\,x - ae\,x' + e^2 x'x = (a - ex')\,(a - ex).$$

But (Art. 360) $\rho' = a - ex'$. Hence, finally,

$$\cos \phi = \frac{a - ex}{\rho}.$$

441. This expression, being independent of the point of contact $x'y'$, must be true for either of the tangents drawn from P. Hence $\phi = PFT = PFT'$ Therefore, with respect to the whole angle TFT', we have

Theorem LII.—*The right line that joins the focus of an ellipse to the pole of any chord, bisects the focal angle which the chord subtends.*

Corollary.—Since the angle subtended by any focal chord is 180°, we obtain the further special property: *The line that joins the focus to the pole of any focal chord is perpendicular to the chord.*

III. THE CURVE REFERRED TO ITS FOCI.

442. We are now prepared to attach the proper meanings to the constants which enter the equations of Art. 152 and the subjoined Remark; and it may deserve to be mentioned in passing, that the phrases *polar equation, polar co-ordinates*, etc., have no reference to the polar *relation* lately developed as a property of the Ellipse.

443. From the preceding discussions, then, we are henceforth to understand that in the polar equations

$$\rho = \frac{a\,(1 - e^2)}{1 - e\cos\theta}, \quad \rho = \frac{a\,(1 - e^2)}{1 + e\cos\theta},$$

the constant a is the semi-axis major of the given ellipse, and the constant e its eccentricity.

Further : If we replace $1 - e^2$ by its value (Art. 151) $b^2 : a^2$, we may write these equations

$$\rho = \frac{b^2}{a} \cdot \frac{1}{1 - e\cos\theta}, \quad \rho = \frac{b^2}{a} \cdot \frac{1}{1 + e\cos\theta}.$$

Now (Art. 428) $b^2 : a$ is half the parameter of the curve.

$$\therefore \; \rho = \frac{2p}{1 \pm e\cos\theta},$$

the upper or lower sign being used according as the right-hand or left-hand focus is taken for the pole.

444. Polar Equation to the Tangent.—We shall obtain this most readily by transforming the equation

$$\frac{x'x}{a^2} + \frac{y'y}{b^2} = 1$$

from rectangular to polar co-ordinates, at the same time removing the pole to the left-hand focus, whose co-ordinates are $- ae, 0$. We have (Art. 58) $x' = \rho'\cos\theta' - ae$, $x = \rho\cos\theta - ae$, $y' = \rho'\sin\theta'$, $y = \rho\sin\theta$. Hence, the transformed equation is

$$\frac{(\rho'\cos\theta' - ae)\,(\rho\cos\theta - ae)}{a^2} + \frac{\rho'\rho\cos\theta\cos\theta'}{b^2} = 1.$$

But $\rho'\theta'$ is on the curve : therefore, (Art. 443,)

$$\rho' = \frac{a\,(1 - e^2)}{1 - e\cos\theta'} = \frac{b^2}{a\,(1 - e\cos\theta')}.$$

Substituting these values in the first and second terms of the last equation respectively, and reducing,

$$(\cos\theta'-e)\,(\rho\cos\theta-ae)+\rho\sin\theta\sin\theta'=a\,(1-e\cos\theta') \quad (1).$$

Therefore, (Trig., 845, IV,) the required equation is

$$\rho = \frac{a\,(1-e^2)}{\cos(\theta-\theta')-e\cos\theta}\,.$$

Corollary.—Since the equation to the diameter conjugate to $x'y'$ (Art. 372), differs from that of the tangent at $x'y'$ only in having 0 for its constant term, we have, by putting 0 instead of $a\,(1-e\cos\theta')$ in (1), and reducing,

$$\rho = \frac{ae\,(\cos\theta'-e)}{\cos(\theta-\theta')-e\cos\theta}\,,$$

as the *polar equation to the diameter conjugate to that which passes through* $\rho'\theta'$.

445. These polar equations afford a proof of Theorem XIX so much simpler than the one given in Art. 378, that we present it here expressly to invite comparison.

Let $\rho'\theta'$ be the extremity D of any diameter DD'. The equation to its conjugate SS' is

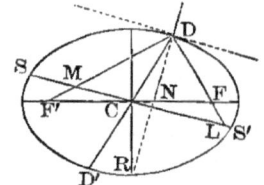

$$\rho = \frac{ae\,(\cos\theta'-e)}{\cos(\theta-\theta')-e\cos\theta}\,.$$

In this, making $\theta=\theta'$, we obtain

$$\rho = F'M = \frac{ae\,(\cos\theta'-e)}{1-e\cos\theta'}\,.$$

Now, from the polar equation to the curve, as given in Art. 152,

$$\rho' = F'D = \frac{a\,(1-e^2)}{1-e\cos\theta'}\,.$$

Hence,

$$DM = F'D - F'M = a.$$

Remark.—The geometric proof of this theorem is perhaps still simpler. Thus: — The normal DN bisects the angle $F'DF$ (Art. 403), and is perpendicular to SS' (Art. 389). Hence, the triangle MDL is isosceles, and $DM=DL$. But, by drawing a parallel to LM through F, it becomes apparent, since $CF'=CF$, that $F'M=FL$. Therefore, $DM+DL=F'D-F'M+FD+FL=F'D+FD$. That is, (Art. 355),

$$2DM=A'A \ . \ \ DM=a.$$

We have given these three proofs of the same proposition for the purpose of illustrating the importance of a proper selection of methods, even in the elementary work of the beginner. In attempting to establish theorems, several methods are often available to the student, and he should select that which will combine rigor, simplicity, and elegance, in the highest degree. While analytic processes are generally able to satisfy this condition the best, it nevertheless sometimes happens, that the proof from pure geometry is superior. In such a case, of course, the latter is to be preferred.

IV. AREA OF THE ELLIPSE.

446. The area of any ellipse whose axes are given, may be determined by the following application of the geometric *method of infinitesimals.*

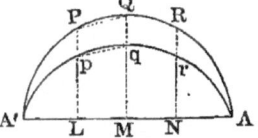

$A'A$ being the axis major of the curve, describe a circle upon it, and divide it into any number of equal parts at L, M, N, etc. Erect the ordinates LP, MQ, NR, etc., cutting the ellipse in p, q, r, etc. Join PQ, pq: then, since $Lp:LP=Mq:MQ=b:a$ (Art. 359), the trapezoids Lq, LQ are in the ratio $b:a$. Now the same must be true of *any* two corresponding trapezoids: therefore, the area of the polygon inscribed in the ellipse is to the area of the corresponding polygon inscribed in the circle as b is to a. Hence, as this proportion holds true, no matter how many sides the inscribed polygons may have, and as we can make the number of sides as great as we please by continually subdividing the axis major, in the limiting case where the polygons vanish into their respective curves, we shall have *area of ellipse : area of circumscribed circle* = $b : a$. Now (Geom., 500) the area of the circle = πa^2. Therefore, putting A = the required area,

$$A = \pi ab.$$

That is,

Theorem LIII.— *The area of an ellipse is equal to π times the rectangle under its semi-axes.*

Corollary.—Since $\pi ab = \sqrt{\pi a^2 . \pi b^2}$, we have the additional property : *The area of an ellipse is a geometric mean between the areas of its circumscribed and inscribed circles.*

EXAMPLES ON THE ELLIPSE.

1. Find the equations to the tangent and normal at the extremity of the latus rectum, and determine the eccentricity of the ellipse in which the normal mentioned passes through the extremity of the axis minor.

2. Find the equations to the diameter passing through the extremity of the latus rectum, and the chord joining the extremities of the axes; and determine the eccentricity of the ellipse in which these lines are parallel.

3. A point P is so taken on the normal of an ellipse, that its distance from the foot of the normal is in a constant ratio to the length of the normal: find the locus of P, and prove that when P is the *middle* point of the normal, its locus is an ellipse whose eccentricity e' is connected with that of the given one by the condition

$$(1 - e'^2)(1 + e^2)^2 = 1 - e^2.$$

4. Prove that two ellipses of equal eccentricity and parallel axes can have only two points in common. Also, show that if three such ellipses intersect, their three common chords will meet in one point.

5. If two parallels be drawn, one from an extremity A of the axis major, and the other from the adjacent focus, meeting the axis minor in M and N, the circle described from N as a center, with a radius equal to MA, will either *touch* the ellipse or fall entirely *outside* of it.

6. A circle is inscribed in the triangle formed by the axis major and any two focal radii: find the locus of its center.

7. Find a point on an ellipse, such that the tangent there may be equally inclined to both axes. Also, a point such that the tangent may form upon the axes intercepts proportional to them.

8. Prove that the circle described on any focal radius of an ellipse touches the circumscribed circle.

9. From the vertex of an ellipse a chord is drawn to any point on the curve, and the parallel diameter is also drawn: the locus of the intersection of this diameter and the tangent at the extremity of the chord, is the tangent at the opposite vertex.

10. From the center of an ellipse, two radii vectores are drawn at right angles to each other, and tangents to the curve are formed at their extremities: the tangents intersect on the ellipse

$$\frac{x^2}{a^4} + \frac{y^2}{b^4} = \frac{1}{a^2} + \frac{1}{b^2} \, .$$

11. Two ellipses have a common center, and axes coincident in direction, while the sum of the squares on the axes is the same in both: find the equation to a common tangent.

12. The ordinate of any point P on an ellipse is produced to meet the circumscribed circle in Q: the focal perpendicular upon the tangent at Q is equal to the focal distance of P.

13. The lines which join transversely the foci and the feet of the focal perpendiculars on any tangent, intersect on the corresponding normal, and bisect it.

14. If a right line drawn from the focus of an ellipse meets the tangent at a constant angle θ, the locus of its foot is a circle, which touches the curve or falls entirely outside of it according as $cos\,\theta$ is less or greater than e.

15. When the angle between a tangent and its *focal* radius of contact is least, the radius $= a$; and when the angle between a tangent and its *central* radius of contact is least, the radius $= \frac{1}{2}\sqrt{a^2 + b^2}$.

16. The locus of the foot of the central perpendicular upon any tangent to an ellipse, is the curve

$$\rho^2 = a^2 \cos^2 \theta + b^2 \sin^2 \theta.$$

17. The locus of the variable intersection of two circles described on two conjugate semi-diameters of an ellipse, is the curve

$$2\rho^2 = a^2 \cos^2 \theta + b^2 \sin^2 \theta.$$

18. If lines drawn through any point of an ellipse to the extremities of any diameter meet the conjugate CD in M and N, prove that $CM . CN = CD^2$.

19. In an ellipse, the rectangle under the central perpendicular upon any tangent and the part of the corresponding normal intercepted between the axes, is constant, and equal to $a^2 - b^2$.

20. The condition that two diameters of an ellipse may be conjugate, referred to a pair of conjugates as axes of co-ordinates, is

$$\tan \theta \tan \theta' = -\frac{b'^2}{a'^2}.$$

21. Normals at P and D, the extremities of conjugate diameters, meet in Q: prove that the diameter CQ is perpendicular to PD, and find the locus of its intersection with the latter.

22. Given any two semi-diameters, if from the extremity of each an ordinate be drawn to the other, the triangles so formed will be equal in area. Also, if tangents be drawn at the extremity of each, the triangles so formed will be equal in area.

23. Find the locus of the intersection of the *focal* perpendicular upon any tangent with the radius vector from the *center* to the point of contact. Also, the locus of the intersection of the *central* perpendicular with the radius vector from the *focus* to the point of contact.

24. The equi-conjugates being taken for axes, find the equation to the normal at P, and prove that the normal bisects the line joining the feet of the perpendiculars dropped from P upon the equi-conjugates.

25. Find the locus of the intersection of tangents drawn through the extremities of conjugate diameters.

26. Putting ρ, ρ' to denote the focal radii of any point on an ellipse, and ϕ for its eccentric angle, prove that

$$\rho = a (1 - e \cos \phi), \qquad \rho' = a (1 + e \cos \phi).$$

27. Express the lengths of two conjugate semi-diameters in terms of the eccentric angle, namely, by

$$a'^2 = a^2 \cos^2 \phi + b^2 \sin^2 \phi, \quad b'^2 = a^2 \sin^2 \phi + b^2 \cos^2 \phi.$$

28. The ordinate MP of an ellipse being produced to meet the circumscribed circle in Q, find the locus of the intersection of the radius CQ with the focal radius FP.

29. Normals to the ellipse and circumscribed circle pass through the points P and Q just mentioned, and intersect in R: find the locus of R.

[First show that the equation to the normal of the ellipse is

$$\frac{ax}{\cos \phi} - \frac{by}{\sin \phi} = c^2,$$

ϕ being the eccentric angle of the point of contact.]

30. Prove that the area of any parallelogram circumscribed about an ellipse may be expressed by

$$\text{area} = \frac{4ab}{\sin (\phi - \phi')},$$

where ϕ, ϕ' are the eccentric angles corresponding to the points of contact of the adjacent sides. Show that this area is least when the points of contact are the extremities of conjugates.

31. Upon the axis major of an ellipse, two supplemental chords are erected, and perpendiculars are drawn to them from the vertices: show that the locus of the intersection of these perpendiculars is another ellipse, and find its axes.

32. Let CP, CD be any two conjugate semi-diameters: the supplemental chords from P to the extremities of any diameter are parallel to those from D to the extremities of the conjugate.

33. The rectangle under the segments of any focal chord is to the whole chord in a constant ratio.

34. The sum of two focal chords drawn parallel to two conjugate diameters is constant.

35. The sum of the reciprocals of two focal chords at right angles to each other is constant.

36. To a series of confocal ellipses, tangents are drawn from a fixed point on the axis major: find the locus of the points of contact.

37. Tangents to two confocal ellipses are drawn to cut each other at right angles: the locus of their intersection is a circle concentric with the ellipses.

38. Find the sum of the focal perpendiculars upon the polar of $x'y'$.

39. The intercept formed on any variable tangent by two fixed tangents, subtends a constant angle at the focus. Also, the line which joins the focus to the point in which any chord cuts the directrix, is the external bisector of the focal angle subtended by the chord.

40. One vertex of a circumscribed parallelogram moves along one directrix of an ellipse: prove that the opposite vertex moves along the other, and that the two remaining vertices move upon the circumscribed circle.

CHAPTER FOURTH.

THE HYPERBOLA.

I. THE CURVE REFERRED TO ITS AXES.

447. In discussing the Hyperbola by means of its equation (Art. 167)

$$\frac{x^2}{a^2} - \frac{y^2}{b^2} = 1,$$

we shall avoid the repetition of much that has already been said in connection with the Ellipse, by considering that the similarity of the equations to these two curves makes most of the arguments used in the foregoing pages at once applicable to the Hyperbola. We shall therefore avail ourselves of the principle developed in the corollary to Art. 167, and, for details, shall refer the student to the proper article in the preceding Chapter.

For the sake of bringing out the antithesis between the Ellipse and Hyperbola, alluded to in the Remark under Art. 167, the theorems of this Chapter are numbered like the corresponding ones of the preceding.

THE AXES.

448. Making y and x successively equal to zero in the equation of Art. 167, we get, for the intercepts of the Hyperbola upon the lines termed its *axes*,

$$x = \pm a, \quad y = \pm b \sqrt{-1}.$$

Hence, the curve cuts the transverse axis in two *real* points equally distant from the focal center, and the conjugate axis in two *imaginary* points situated on opposite sides of that center at the distance $b\sqrt{-1}$. Assuming, then, the conjugate axis to be measured by the imaginary unit $\sqrt{-1}$, we may infer

Theorem I.—*The focal center of any hyperbola bisects the transverse axis, and also the conjugate.*

Corollary.—In the light of the analysis leading to this theorem, we should therefore interpret the constants a and b in the equation

$$\frac{x^2}{a^2} - \frac{y^2}{b^2} = 1$$

as respectively denoting half the transverse axis and half the modulus of the imaginary conjugate axis.

449. At the outset (see Art. 166), we arbitrarily used the phrase *conjugate axis* to denote the *whole* line drawn through the center C at right angles to the transverse axis $A'A$. We now see that the phrase in strictness means an imaginary *portion* of that line, of the length $= 2b\sqrt{-1}$.

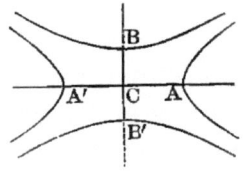

But, as was promised in Art. 166, we shall now show

that a certain *real* portion of this line has a most significant relation to the Hyperbola, on account of which it is
by universal consent taken for the conjugate axis. This
relation depends on the companion-curve called the *conjugate hyperbola*, whose equation we developed in Art. 168.

By referring to the close of Art. 168, it will be seen
that the equations to two conjugate hyperbolas, when
referred to their common center and axes, differ only in
the sign of the constant term. The equation to the curve
whose branches lie one above and the other below the
line $A'A$ in the diagram, may therefore be written

$$\frac{x^2}{a^2} - \frac{y^2}{b^2} = -1.$$

Now, if in *this* we make $x = 0$, we get

$$y = \pm b.$$

Hence, the conjugate hyperbola has a *real* axis, identical
in direction with the imaginary axis of the primary curve,
whose length is the same multiple of 1 that the length
of the imaginary is of $\sqrt{-1}$. Moreover, it is found that
this real axis of the conjugate hyperbola, when used instead of the imaginary one of the primary curve, enables
us to state the properties of the latter in complete analogy
to those of the Ellipse. It is customary, therefore, to lay
off CB, CB' each equal to b, and to treat the resulting
line $B'B$ as the *conjugate* axis of the original hyperbola,
though in fact it is only the *transverse* axis of the conjugate curve.

Adopting this convention, the statement in Theorem I
is to be taken without reference to imaginary quantities,
and the constants a and b in the equations

$$\frac{x^2}{a^2} - \frac{y^2}{b^2} = \pm 1$$

are henceforth to be interpreted as denoting the semi-axes of the curve.

450. If in the equation of Art. 167, which may be written

$$y = \pm \frac{b}{a} \sqrt{x^2 - a^2},$$

we suppose $x < a$ or $> -a$, the corresponding values of y are imaginary; so that no point of the curve is nearer to the origin, either on the right or on the left, than the extremities of the transverse axis. But (Art. 171), for the distance from the origin to either focus, we have

$$c^2 = a^2 + b^2.$$

Hence, c can not be less than a, though it may approach infinitely near to the value of a, as b diminishes toward zero. Therefore,

Theorem II.—*The foci of any hyperbola fall* without *the curve.*

451. Moreover, $c - a$ measures the distance of either focus from the adjacent vertex; while the distance of either from the remote vertex $= c + a$. Hence,

Theorem III.—*The vertices of the curve are equally distant from the foci.*

452. From Art. 448, the length of the transverse axis $= 2a$. But (Art. 167) $2a =$ the constant difference of the focal radii of any point on the curve. That is,

Theorem IV.—*The difference of the focal radii of any point on an hyperbola is equal to the length of its transverse axis.*

Corollary.—We may therefore construct the curve by points as follows:—From either focus, as F', lay off

$F'M$ equal to the transverse axis. Then from F'' as a center, with any radius $F'R$ greater than $F'A$, describe two small arcs, one above the axis, and the other below it. From the remaining focus F as a center, with a radius MR, describe two other arcs, intersecting the former in P and P': these points will be upon

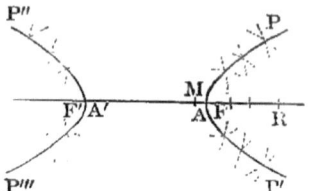

the required hyperbola; for $F'P - FP = F'R - MR = A'A = F'P' - FP'$. By using the radius $F'R$ from F, and MR from F', two points, P'' and P''', may be found upon the second branch of the curve. The operation must be repeated until the outline of the two branches is distinctly marked, when the curve may be drawn through the points determined. The conjugate curve may be formed in the same way, at the same time, if desired.

453. The abbreviation $b^2 = c^2 - a^2$ adopted (Art. 167) for the Hyperbola, gives us

$$b = \sqrt{(c+a)(c-a)}.$$

Hence, attributing to a, b, c the meanings now known to belong to them, we have

Theorem V.—*The conjugate semi-axis of any hyperbola is a geometric mean between the segments formed upon the transverse axis by either focus.*

Corollary.—Transposing in the abbreviation above, we get $c^2 = a^2 + b^2$. But, from the diagram, $a^2 + b^2 = AB^2$. Therefore, *The distance from the center to either*

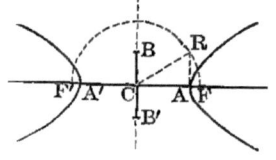

focus of an hyperbola is equal to the distance between the extremities of its axes. Hence, when the axes are given, we may construct the foci as follows: — From the center C, with a radius equal to the diagonal of the rectangle under the semi-axes, describe an arc cutting the transverse axis produced in F and F': the two points of intersection will be the foci sought.

454. By an analysis similar to that of Art. 357, the details of which the student must supply,* we obtain

Theorem VI.—*The squares on the ordinates drawn to either axis of an hyperbola are proportional to the rectangles under the corresponding segments of that axis.*

Corollary.—For the ordinate passing through either focus, we shall therefore have

$$y'^2 = \frac{b^2}{a^2}(c^2 - a^2).$$

But $c^2 - a^2 = b^2$. Hence, doubling FP or $F'P'$,

$$\text{latus rectum} = \frac{2b^2}{a} = \frac{(2b)^2}{2a}.$$

That is, *The latus rectum of any hyperbola is a third proportional to the transverse axis and the conjugate.*

455. Throwing the equations to the Hyperbola and its conjugate into the forms

$$\frac{y^2}{(x+a)(x-a)} = \frac{b^2}{a^2}, \qquad \frac{x^2}{(y+b)(y-b)} = \frac{a^2}{b^2},$$

we at once obtain

Theorem VII.—*The squares on the axes of any hyperbola are to each other as the rectangle under any two segments of either is to the square on the ordinate which forms the segments.*

---

* When seeking properties of the *conjugate* axis, we must of course use the equation to the *conjugate hyperbola.*

An. Ge. 34.

456. If we put the equation to the Hyperbola into the form

$$y^2 = \frac{b^2}{a^2}(x^2 - a^2) \qquad (1),$$

and compare it with that of the circle described upon the transverse axis, namely, with

$$y^2 = a^2 - x^2,$$

we see that the ordinates of the two curves, corresponding to a common abscissa, have only an imaginary ratio. The analogy between the Hyperbola and the Ellipse, so far as concerns the circle mentioned, is therefore defective. If, however, we suppose $b = a$ in (1), we get

$$y^2 = x^2 - a^2 \qquad (2),$$

and, if we now divide (1) by (2), we obtain

$$y_h : y_r :: b : a.$$

Now equation (2) evidently represents the curve which in Art. 177 we named a *rectangular* hyperbola, but which we may henceforth call an *equilateral* hyperbola, since its equation is obtained from that of the ordinary curve by supposing the axes equal. We have therefore proved

Theorem VIII.—*The ordinate of any hyperbola is to the corresponding ordinate of its equilateral, as the conjugate semi-axis is to the semi-transverse.*

Remark.—The peculiarity in the figure of the Equilateral Hyperbola is, that *the curve is identical in form with its conjugate.* For the equation to its conjugate (Art. 449) is

$$x^2 - y^2 = -a^2;$$

and if we transform this to the conjugate axis as the axis of x, by revolving the

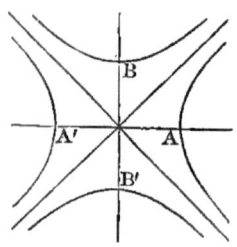

refercnce-axes through 90°, and therefore (Art. 56, Cor. 3) replacing
x by $-y$, and y by x, we obtain

$$x^2 - y^2 = a^2;$$

so that the conjugate curve, when referred to its own transverse
axis, is represented by the same equation as its primary, and is
therefore the same curve. The diagram presents a pair of conjugate
equilaterals.

Corollary.—Notwithstanding the defective analogy between the Ellipse and the Hyperbola with respect to the
circle formed upon the transverse axis, this curve still
aids us in fixing the meaning of the abbreviation

$$\frac{a^2 + b^2}{a^2} = e^2$$

adopted in Art. 170, and warrants us in calling e the
eccentricity of the Hyperbola. For, as in the case of the
Ellipse, since $a^2 + b^2 = c^2$, we learn that e is *the ratio
which the distance from the center to either focus of an
hyperbola bears to its transverse semi-axis.* Let us, then,
suppose a series of ellipses and hyperbolas to be described
upon a common transverse
axis: we saw (Art. 359,
Cor. 2) that, as the varying ellipse of such a series
deviates more and more
from the circle formed
upon the same axis, and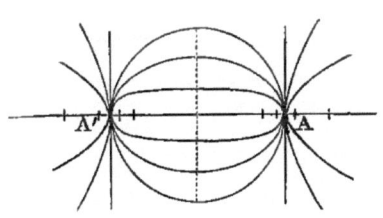
approaches nearer and nearer to coincidence with the
axis $A'A$, the eccentricity e advances nearer and nearer
to the limit 1. Assuming, then, that e actually reaches
this limit, the corresponding ellipse must vanish into the
line $A'A$, which forms the common axis. Now, from the
abbreviation above, the e of the Hyperbola lies between
the limits 1 and ∞: hence the series of hyperbolas may
be said to arise out of the common axis $A'A$ at the

instant when the series of ellipses vanishes into it, and to recede farther and farther from the axis as e advances from 1 toward ∞. But since e, with respect to the ellipses and the hyperbolas both, is at all times the ratio between the same elements of the curves; and since, taking the circle described upon the common axis as the starting-point, this ratio steadily advances from 0 through 1 toward ∞; we may regard the recession of the hyperbolas from the axis $A'A$ as a farther deviation from the curvature of the circle mentioned, and consequently call e, which measures this deviation, the *eccentricity*.

We may therefore interpret the name *hyperbola* (derived from the Greek ὑπερβάλλειν, *to exceed*) as signifying, that, in this curve, the eccentricity is *greater than unity*.

Since e increases as b increases, it follows that the greater the eccentricity, the more obtuse will be the branches of the corresponding hyperbola. In case the curve is *equilateral*, or $b = a$, we shall have

$$e = \sqrt{2}.$$

457. The distance of any point on an hyperbola from either focus, may be expressed in terms of the abscissa of the point. For, putting ρ to denote any such focal distance, we have (Art. 167)

$$\rho^2 = (x \pm c)^2 + y^2.$$

Substituting for y^2 from the equation to the curve, and reducing by means of the relations in Art. 171, we get

$$\rho = ex \pm a,$$

where the upper sign corresponds to the left-hand focus, and the lower to the right-hand one. Hence,

Theorem IX.—*The focal radius of any point on an hyperbola is a linear function of the corresponding abscissa.*

Remark.—The expression obtained in this article, like that found for the *Ellipse in Art. 360, is accordingly known as the *Linear Equation to the Hyperbola.*

458. By reasoning similar to that employed in Art. 361, we may verify the figure of the Hyperbola, as drawn in Art. 165. We leave the student to show, by interpreting the equation

$$y^2 = \frac{b^2}{a^2} (x^2 - a^2),$$

that the curve consists of two infinite branches, separated by the transverse axis $= 2a$, facing in opposite directions, and symmetric to both axes.

DIAMETERS.

459. Equation to any Diameter.—To obtain an expression for the locus of the middle points of chords in an hyperbola which have a common inclination θ' to the transverse axis, we write (Art. 167, Cor.) $-b^2$ for b^2 in the final equation of Art. 362. Hence, the required equation is

$$y = \frac{b^2}{a^2} x \cot \theta'.$$

Corollary.—Putting $\theta =$ the inclination of the diameter itself, we obtain (Art. 78, Cor. 1), as the *condition connecting the inclination of any diameter with that of the chords which it bisects,*

$$\tan \theta \tan \theta' = \frac{b^2}{a^2} .$$

460. Since the equation to a diameter conforms to the type $y = mx$, we at once infer

Theorem X.—*Every diameter of an hyperbola is a right line passing through the center.*

Corollary.—The angle θ' being arbitrary, it follows from the above condition, that θ^* is also arbitrary. Hence the converse theorem : *Every right line that passes through the center of an hyperbola is a diameter.*

461. Eliminating, then, between the equation

$$y = x \tan \theta$$

and the equation to the Hyperbola, we get, for the abscissas of intersection between the curve and any diameter,

$$x = \pm \frac{ab}{\sqrt{(b^2 - a^2 \tan^2 \theta)}}.$$

Now these abscissas evidently become imaginary when $a^2 \tan^2 \theta > b^2$. Hence,

Theorem XI.—*The proposition that every diameter cuts the curve in two real points, is not true of the Hyperbola.*

Corollary 1.—It is obvious, however, that the intersections will be real, and at a finite distance from the center, so long as $a^2 \tan^2 \theta < b^2$. Hence, the diameters corresponding to $a^2 \tan^2 \theta = b^2$, that is, the two diameters whose tangents of inclination are respectively

$$\tan \theta = \frac{b}{a}, \quad \tan \theta' = -\frac{b}{a},$$

form the limits between those diameters which have *real* intersections with the curve and those which have not. But, from the values of their tangents of inclination, these two diameters are the diagonals of the rectangle contained by the axes. We learn, then, that diameters which cut the Hyperbola in real points must either make with the transverse axis an angle *less* than is made by the *first* of these diagonals, or *greater* than is made by the *second.*

It deserves notice, that the condition $a^2 \tan^2 \theta = b^2$ renders the abscissas of intersection, as expressed above, *infinite*. The two limiting diameters therefore meet the curve at infinity: and we have come upon the analogue of the equi-conjugates in the Ellipse. We shall soon find that these lines are the most remarkable elements of the Hyperbola, giving it a series of properties in which the other Conics do not share.

Corollary 2.—Eliminating between $y = x \tan \theta$ and the equation to the *conjugate* hyperbola, we get

$$x = \pm \frac{ab}{\sqrt{(a^2 \tan^2 \theta - b^2)}}.$$

Here, then, the condition of *real* intersection is $a^2 \tan^2 \theta > b^2$. Hence, *Every diameter that cuts an hyperbola in two imaginary points, cuts its conjugate in two real ones.*

462. Length of any Diameter.—This being double the central radius vector of the curve, may be determined (Art. 170) by

$$\rho^2 = \frac{b^2}{e^2 \cos^2 \theta - 1};$$

or, if the diameter meets the conjugate curve instead of the primary, by

$$\rho^2 = \frac{b^2}{1 - e^2 \cos^2 \theta}.$$

an expression readily obtained by transforming to polar co-ordinates (Art. 57, Cor.) the equation to the conjugate hyperbola, found in Art. 449.

463. The first of the above expressions is least when $\theta = 0$; and the second, when $\theta = 90°$. Hence,

Theorem XII.—*Each axis is the minimum diameter of its own curve.*

Remark.—We see, then, that the terms *major* and *minor* are not applicable to the axes of an hyperbola.

464. By the same argument as in Art. 367, we obtain

Theorem XIII.—*Diameters which make supplemental angles with the transverse axis of an hyperbola are equal.*

Corollary.—It also follows, as in the corollary to Art. 367, that we can construct the axes when the curve is given. The diagram illustrates the process in the case of the Hyperbola, and the student may transfer the statements of Art. 367, Cor., to this figure,

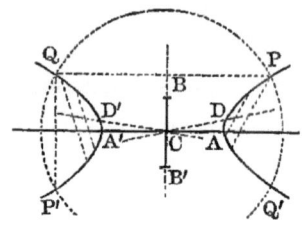

letter by letter. We deem it unnecessary to repeat them.

465. The inclinations of two diameters being represented by θ and θ', the argument of Art. 368 obviously applies to the Hyperbola, with respect to the condition (Art. 459, Cor.)

$$\tan \theta \tan \theta' = \frac{b^2}{a^2} .$$

Hence,

Theorem XIV.—*If one diameter of an hyperbola bisects chords parallel to a second, the second bisects chords parallel to the first.*

466. Two diameters of an hyperbola which are thus related, are called *conjugate* diameters, as in the case of the Ellipse. The phrase *ordinates to any diameter* is also used in connection with the Hyperbola, to signify the halves of the chords which the diameter bisects; or, the right lines drawn from the diameter, parallel to its conjugate, to meet the curve.

Corollary.—The construction of a pair of conjugates in an hyperbola, may therefore be effected exactly in the manner

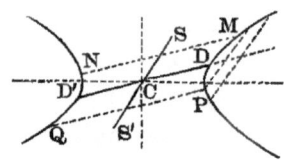

described in the corollary to Art. 369. The details may be gathered by applying to the parts of the annexed diagram, the statements of that corollary.

467. Equation of Condition for Conjugates in the Hyperbola.—The conjugate of any diameter being parallel to the chords which the diameter bisects, the inclinations of two conjugates must be connected in the same way as those of a diameter and its ordinates. Hence, if θ and θ' represent the inclinations, the required condition (Art. 459, Cor.) is

$$\tan \theta \tan \theta' = \frac{b^2}{a^2}.$$

Corollary.—Hence, if $\tan \theta < b : a$, $\tan \theta' > b : a$; and if $\tan \theta > -b : a$, $\tan \theta' < -b : a$. Therefore (Art. 461, Cors. 1, 2), *If one of two conjugates meets an hyperbola, the other meets the conjugate curve.*

Remark.—The condition of this article might have been obtained from that of Art. 370, by merely changing b^2 into $-b^2$.

468. The preceding condition shows that the tangents of inclination have like signs. Hence, the angles made with the transverse axis by two conjugates are either both *acute*, or else both *obtuse*. That is,

Theorem XV.—*Conjugate diameters of an hyperbola lie on the same side of the conjugate axis.*

469. Equation to a Diameter conjugate to a Fixed Point.—Making the requisite change of sign (Art. 167, Cor.) in the equation of Art. 372, we get the one now sought, namely,

$$\frac{x'x}{a^2} - \frac{y'y}{b^2} = 0.$$

An. Ge. 35.

Corollary.—The diameter conjugate to that which passes through $(a, 0)$ is therefore $x = 0$, that is, the conjugate axis. Hence, *The axes of an hyperbola constitute a case of conjugate diameters.*

470. Problem.—*Given the co-ordinates of the extremity of a diameter, to find those of the extremity of its conjugate.*

By the extremities of the conjugate diameter, are meant the points in which the conjugate cuts the conjugate hyperbola. The required co-ordinates are therefore found by eliminating between the equation of Art. 469 and

$$\frac{x^2}{a^2} - \frac{y^2}{b^2} = -1.$$

They are

$$x_c = \pm \frac{ay'}{b}, \quad y_c = \pm \frac{bx'}{a}.$$

Remark.—By comparing these expressions with those of Art. 373, we notice that the abscissa and ordinate of the conjugate diameter in the Ellipse have *opposite* signs, but in the Hyperbola *like* signs. This agrees with the properties developed in Arts. 371, 468.

471. The equations of Art. 470, like those of Art. 373, give rise to

Theorem XVI.—*The abscissa of the extremity of any diameter is to the ordinate of the extremity of its conjugate, as the transverse axis is to the conjugate axis.*

472. By following, with respect to the second expression of Art. 470, the steps indicated in Art. 375, excepting that we *subtract* the y'^2, we arrive at

Theorem XVII.—*The difference of the squares on the ordinates of the extremities of conjugate diameters is constant, and equal to the square on the conjugate semi-axis.*

Remark.—We leave the student to prove the analogous property : *The difference of the squares on the abscissas of the extremities of conjugate diameters is constant, and equal to the square on the transverse semi-axis.*

473. Problem.—*To find the length of a diameter in terms of the abscissa of the extremity of its conjugate.*

Let $x'y'$ be the extremity of any diameter, a' half its length, and b' half the length of its conjugate. Then $a'^2 = x'^2 + y'^2$, and we get (Art. 470)

$$a'^2 = \frac{a^2}{b^2} y_c^2 + \frac{b^2}{a^2} x_c^2 = (x_c^2 + a^2) + \frac{b^2}{a^2} x_c^2,$$

since x_c and y_c must satisfy the equation to the conjugate hyperbola. Hence, (Art. 171,)

$$a'^2 = e^2 x_c^2 + a^2.$$

By performing similar operations with respect to b', we should get

$$b'^2 = e^2 x'^2 - a^2.$$

474. Between these results and those of Art. 376, there is a striking difference ; and, as only the value of b'^2 equals (Art. 457) the rectangle of the focal radii drawn to $x'y'$, it appears as if the property proved of the Ellipse in Art. 377 were only true of the Hyperbola with respect to those diameters which meet the conjugate curve instead of the primary. But when we reflect that it is entirely arbitrary which of two conjugate hyperbolas we consider the primary, it becomes evident that the property of Art. 377 is also true of the diameters which meet the curve hitherto *called* the primary, provided we suppose the focal radii in question to be drawn from the foci of the *conjugate* curve. With this understanding, then, we may state

Theorem XVIII.—*The square on any semi-diameter of an hyperbola is equal to the rectangle under the focal radii drawn to the extremity of its conjugate.*

475. It is evident on inspection, that the fourth formula in Art. 378 will not be altered by changing b^2 into $-b^2$. Hence, in the Hyperbola as well as in the Ellipse, we have

$$\delta^2 = DM^2 = \frac{a^2(a^2c^2 - 2aex' + x'^2 + y'^2)}{(a - cx')^2} \ ,$$

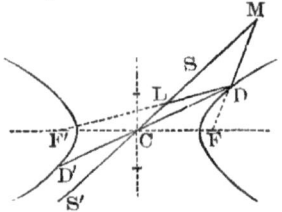

in which $x'y'$ is the extremity D of any diameter $D'D$. But $x'^2 + y'^2 = a'^2 = $ (Art. 473) $e^2x_c^2 + a^2 = $ (Art. 472, Rem.) $e^2(x'^2 - a^2) + a^2$. Hence, after substituting and reducing,

$$\delta = DM = a;$$

or, in the Hyperbola as well as in the Ellipse, we have

Theorem XIX.—*The distance from the extremity of any diameter to its conjugate, measured upon the corresponding focal radius, is constant, and equal to the transverse semi-axis.*

476. Let a', b' denote the lengths of any two conjugate semi-diameters in an hyperbola. Then (Art. 473)

$$a'^2 = e^2x_c^2 + a^2 \qquad (1).$$

Also, $b'^2 = x_c^2 + y_c^2 = x_c^2 + \{b^2(x_c^2 + a^2) : a^2\}$, since x_c and y_c satisfy the equation to the conjugate hyperbola. Hence, (Art. 171,)

$$b'^2 = e^2x_c^2 + b^2 \qquad (2).$$

Subtracting (2) from (1), member by member,

$$a'^2 - b'^2 = a^2 - b^2.$$

Hence, as the antithesis of Art. 379,

Theorem XX.—*The* difference *of the squares on any two conjugate diameters of an hyperbola is constant, and equal to the difference of the squares on the axes.*

477. Angle between two Conjugates.—Using the same symbols as in Art. 380, it is plain that, in the Hyperbola also, we shall have

$$\sin \varphi = \frac{x'y_c - y'x_c}{a'b'}.$$

Substituting for x_c and y_c from Art. 470, reducing, and remembering that $b^2x'^2 - a^2y'^2 = a^2b^2$, we get

$$\sin \varphi = \frac{ab}{a'b'}.$$

478. Clearing this expression of fractions, we have

$$a'b' \sin \varphi = ab.$$

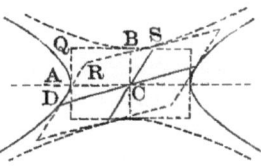

The first member of this equation obviously expresses the area of the parallelogram $CDRS$; and the second, that of the rectangle $CAQB$. Therefore,

Theorem XXI.—*The parallelogram under any two conjugate diameters is constant, and equal to the rectangle under the axes.*

Remark.—The diagram represents the parallelogram and rectangle as *inscribed* in the pair of conjugate hyperbolas. The figure will in due time be justified. Also, as in the case of the Ellipse, the theorem might have been stated thus: *The triangle formed by joining the extremities of any two conjugate diameters is constant, and equal to that included between the semi-axes.*

Corollary 1.—If we suppose $\varphi = 90°$, then $\sin \varphi = 1$; and we get

$$a'b' = ab.$$

Now (Art. 476),

$$a'^2 - b'^2 = a^2 - b^2.$$

Solving these equations for a' and b', we find, as the only *real* values,

$$a' = a, \quad b' = b.$$

Therefore, *In any hyperbola there is but one pair of conjugate diameters at right angles to each other, namely, the axes.*

Corollary 2.—We saw (Arts. 381, Cor. 1; 382, Cor.) that, in the Ellipse, $\sin \varphi$ lies between the limits 1 and $2ab : (a^2 + b^2)$. But (Art. 476), $a'^2 = b'^2 + constant$, in the Hyperbola: whence a' and b' must increase or diminish *together*. Therefore, as any diameter (Art. 461, Cor. 1) tends toward an infinite length the nearer its inclination approaches the limit $\theta = \tan^{-1} b : a$, the semi-conjugates a' and b' must advance *together* toward the value ∞, and the product $a'b'$ tends toward ∞ for its maximum. That is, $\sin \varphi$ tends toward the limit 0; or, *The angle between two conjugates in an hyperbola diminishes without limit.*

But though the conjugates thus tend to final coincidence as each tends to an infinite length, the relation $a'^2 - b'^2 = constant$ renders it impossible that the condition $a' = b'$ shall ever arise in the Hyperbola, unless the curve is equilateral. The infinite diameters that form the limit of the ever-approaching conjugates are therefore not *equal* infinites, and the conception of equi-conjugates is not in general present in the curve. However, from the equation of condition for conjugate diameters, namely,

$$\tan\theta \, \tan \theta' = \frac{b^2}{a^2},$$

it is plain that when the conjugates finally coincide, each makes with the transverse axis an angle whose tangent is either $b : a$ or else $- b : a$. Hence, the two right lines which pass through the center with the respective inclinations

$$\theta = \tan^{-1}\frac{b}{a}\,, \quad \theta = \tan^{-1}-\frac{b}{a}\,,$$

may each be regarded as the limiting case of a pair of conjugate diameters; or, each may be called a diameter *conjugate to itself.* The curve, then, replaces the conception of *equi-conjugates* by that of *self-conjugates.*

479. From what has just been shown, it follows that the inclinations of the self-conjugate diameters to the transverse axis are determined by the formula

$$\tan\theta = \pm\frac{b}{a}\,.$$

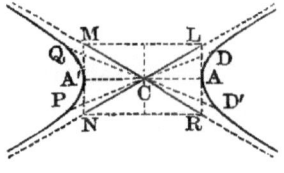

By drawing the rectangle of the axes, *LMNR*, it becomes evident that the first of the values here expressed corresponds to the angle *ACL*; and the second, to the angle *ACM*. Hence,

Theorem XXII.—*The self-conjugates of an hyperbola are the diagonals of the rectangle contained under its axes.*

Corollary.—Hence, further, *An hyperbola has two, and only two, self-conjugates.* Their mutual inclination *LCM*, or *LCR*, as we readily find, is determined by

$$\sin\varphi = \frac{2ab}{a^2 + b^2}\,.$$

480. We have thus found the two lines of the Hyperbola which, in Art. 383, we said were foreshadowed by the equi-conjugates of the Ellipse. That the two self-conjugates are in reality the analogue of the equi-conjugates, we can easily show : for though it is true, as we saw in the second corollary to Art. 478, that the two infinitely long conjugates which unite in either of the self-conjugates are *not* equal infinites, still the two self-conjugates, when compared with each other, *are* equal infinites. For, since they make equal angles *ACL*, *ACR* with the transverse axis, they are the limiting case to which two equal diameters *DP*, *D'Q* necessarily tend as their extremities *D* and *D'* move along the curve in opposite directions from the vertex *A*.

But the chief interest of the self-conjugates is due to a property in which the equi-conjugates of the Ellipse have no share, and in virtue of which they are called the *asymptotes* of the Hyperbola. From this property are derived several others, peculiar to the latter curve, which will receive a separate consideration in the proper place.

THE TANGENT.

ᵛ **481. Equation to the Tangent.**—To obtain this for the Hyperbola, we simply change b^2 into $-b^2$ in the equation of Art. 385. We thus get

$$\frac{x'x}{a^2} - \frac{y'y}{b^2} = 1.$$

482. Condition that a Right Line shall touch an Hyperbola.—Making the characteristic change of sign in the condition of Art. 386, we have

$$n = \sqrt{m^2a^2 - b^2}$$

as the condition that the line $y = mx + n$ may touch the curve

$$\frac{x^2}{a^2} - \frac{y^2}{b^2} = 1.$$

Corollary.—Hence, every line whose equation is of the form

$$y = mx + \sqrt{m^2 a^2 - b^2}$$

is a tangent to the hyperbola whose semi-axes are a and b. Like the similar expressions found in treating the Circle and the Ellipse, an equation of this form is called the *Magical Equation to the Tangent.*

483. The Eccentric Angle.—The expression of any point on an hyperbola in terms of a single variable, is effected by employing an angle analogous to that whose use in connection with the Ellipse was explained in Art. 387. If from the foot of the ordinate corresponding to any point P of an hyperbola, we draw MQ tangent to the inscribed

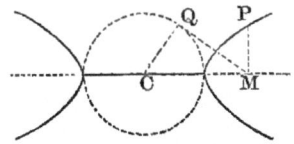

circle at Q, and join Q to the center C, QCM is called the *eccentric angle* of P.

Now (Trig., 860) $CM = CQ \sec QCM$. Also, from the equation to the Hyperbola, combined with this value of CM,

$$MP^2 = \frac{b^2}{a^2}(CM^2 - a^2) = b^2 \tan^2 QCM.$$

Hence, if we represent the arbitrary point P by $x'y'$,

$$x' = a \sec \phi, \quad y' = b \tan \phi.$$

Substituting for x' and y' in Art. 481, we may write the equation to the tangent, in this notation,

$$\frac{x}{a} \sec \phi - \frac{y}{b} \tan \phi = 1.$$

The analogy of the angle QCM, as formed in the case of the Hyperbola, to the similarly named angle in the Ellipse, may perhaps be obscure to the beginner; but it will become apparent when we reach the conception of a hyperbolic subtangent.

484. Problem.—*If a tangent to an hyperbola passes through a fixed point, to find the co-ordinates of contact.*

Let $x''y''$ be the fixed point, and $x'y'$ the required point of contact. Then, changing the sign of b^2 in the results of Art. 388, we get

$$x' = \frac{a^2 b^2 x'' \mp a^2 y'' \sqrt{a^2 y''^2 - b^2 x''^2 + a^2 b^2}}{b^2 x''^2 - a^2 y''^2},$$

$$y' = \frac{a^2 b^2 y'' \pm b^2 x'' \sqrt{a^2 y''^2 - b^2 x''^2 + a^2 b^2}}{b^2 x''^2 - a^2 y''^2}.$$

Corollary 1.—The form of these values indicates that from any given point *two* tangents can be drawn to an hyperbola: *real* when $a^2 y''^2 - b^2 x''^2 + a^2 b^2 > 0$, that is, when the point is *inside* of the curve; *coincident* when $a^2 y''^2 - b^2 x''^2 + a^2 b^2 = 0$, that is, when the point is *on* the curve; *imaginary* when $a^2 y'' - b^2 x''^2 + a^2 b^2 < 0$, that is, when the point is *outside* of the curve.

Corollary 2.—With regard to any two *real* tangents drawn from a given point, it is evident that their abscissas of contact will have *like* signs, if they both touch the same branch of the curve, and *unlike* signs, if the two touch different branches. But, if the two values of x' above have like signs, then, merely numerical relations being considered,

$$a^2 b^2 x'' > a^2 y'' \sqrt{a^2 y''^2 - b^2 x''^2 + a^2 b^2};$$

that is, after squaring, transposing, and reducing,

$$y'' < \frac{b}{a} x''.$$

Hence, as $y = (b : a) x$ is the equation to the diagonal of the rectangle formed upon the axes (Art. 461, Cor. 1),

the ordinate of the point from which two tangents can
be drawn to the *same* branch of an hyperbola must be
less than the corresponding ordinate of the diagonal;
that is, the point itself must lie
somewhere within the space in-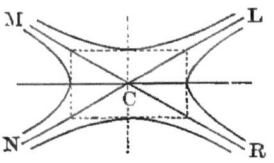
cluded between the self-conjugates
CL, CR (or *CM, CN*) and the
adjacent branch of the curve.
Hence, generally, *The two tan-*
gents which can be drawn to an hyperbola from any
point inside of the curve, will touch the same *branch or*
different *branches, according as the point is taken* within
or without *the angle of the self-conjugates which incloses*
the two branches.

485. The argument of Art. 389 will be seen, on a
moment's inspection, to hold good when hyperbolic
equations are substituted for the elliptic. Therefore,

Theorem XXIII.—*The tangent at the extremity of any*
diameter of an hyperbola is parallel to the conjugate
diameter.

Corollary.—*Tangents at the extremities of a diameter*
are parallel to each other.

Remark.—By drawing any diameter and its conjugate,
and passing a parallel to the latter through the extremity
of the former, we can readily form a tangent to a given
hyperbola. If we construct tangents at the extremities
of both diameters, we shall have an *inscribed parallelo-*
gram. Thus the diagram of Art. 478 is verified; for,
as only one parallel to a given line can be drawn
through a given point, lines drawn through the extremities of conjugate diameters so as to form their parallelogram must be tangents to the curve.

486. Let PT be a tangent to an hyperbola at any point P, and FP, $F'P$ its focal radii of contact. From the equations

$$b^2x'x - a^2y'y = a^2b^2 \quad (PT),$$

$$y'(x-c)-(x'-c)y=0 \quad (FP),$$

$$y'(x+c)-(x'+c)y=0 \quad (F'P),$$

we readily find, by the same steps as in Art. 390,

$$\tan FPT = \frac{b^2}{cy'}, \quad \tan F'PT = \frac{b^2}{cy'}.$$

Hence, $FPT = F'PT$; or, we have

Theorem XXIV.—*The tangent of an hyperbola bisects the internal angle between the focal radii drawn to the point of contact.*

Corollary 1.—We therefore obtain the following solution of the problem : *To construct a tangent to an hyperbola at a given point.* Draw the focal radii FP, $F'P$ to the given point P. On the longer, say $F'P$, lay off $PQ = FP$, and join QF. Through P draw SPT at right angles to QF: then will SPT be the tangent sought. For QPF is by construction an isosceles triangle; and SPT, the perpendicular from its vertex to its base, must therefore bisect the angle $F'PF$.

Corollary 2.—Hence, all rays emanating from F, and striking the curve, will be reflected in lines which, if traced backward, converge in F'; and reciprocally. Accordingly, to suggest the resemblance between these points and the corresponding ones of the Ellipse, they are called the *foci*, or *burning points*, of the Hyperbola.

487. Let us suppose $y = 0$ in the equation

$$b^2 x' x - a^2 y' y = a^2 b^2.$$

We shall thus find, as the value of the intercept which the tangent makes upon the transverse axis,

$$x = CT = \frac{a^2}{x'}.$$

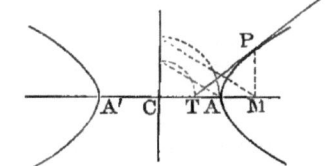

In the Hyperbola, then, as well as in the Ellipse, this intercept is a third proportional to the abscissa of contact and the transverse semi-axis, and we have the same constructions for the tangent at any point P of the curve, or from any point T of the transverse axis, as are described in Art. 391.

488. The Subtangent.—For the length of the *subtangent of the curve* in the Hyperbola, we have $MT = CM - CT$; or, by the preceding article,

$$\text{subtan} = \frac{x'^2 - a^2}{x'} = \frac{(x' + a)(x' - a)}{x'}.$$

But $x' + a = A'M$, and $x' - a = MA$. Hence,

Theorem XXV.—*The subtangent of an hyperbola is a fourth proportional to the abscissa of contact and the two segments formed upon the transverse axis by the ordinate of contact.*

Corollary 1.—Let x_c' be the abscissa of contact for any tangent to the circle described on the transverse axis of an hyperbola, and x_h' that of any tangent to the hyperbola itself. Then (Art. 311),

$$\text{subtan circ.} = \frac{a^2 - x_c'^2}{x_c'}.$$

Suppose, now, that $x_c' = a^2 : x_h'$; that is (Art. 487), that the abscissa of contact in the circle is the intercept of a tangent to the hyperbola. We at once get

$$\text{subtan circ.} = \frac{x_h'^2 - a^2}{x_h'} = \text{subtan hyp.}$$

We see, then, that if from the foot T of any tangent to an hyperbola an ordinate TQ be drawn to the in-scribed circle, the tangent to this circle at Q will pass through M, the foot of the ordinate of contact in the hyperbola; or, *If tangents be drawn to an hyperbola and its inscribed circle from the head and foot of any ordinate to either, the resulting sub-tangents will be identical.*

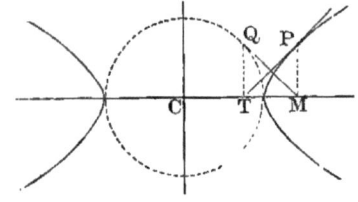

We thus learn that the corresponding points of an hyperbola and its inscribed circle are *those which have a common subtangent.* And, in fact, by turning to the dia-gram of Art. 392, it will be seen that the corresponding points of an ellipse and its circumscribed circle may be defined in the same way. Hence, the defect in the anal-ogy between the two curves with respect to those circles, which came to light in Art. 456, can now be supplied.

Corollary 2.—Accordingly, we can construct the tan-gent by means of the inscribed circle as follows:—When the point of contact P is given, draw the ordinate PM, and from its foot M make MQ tangent to the inscribed circle at Q. Let fall the circular ordinate QT, and join its foot T with the given point P. PT will be the required tangent, by the property established.

When T the foot of the tangent is given, erect the

circular ordinate TQ, and draw the corresponding tangent QM. From M, the foot of this, erect the hyperbolic ordinate MP, and join its extremity P with the given point T.

··· **Remark.**—By comparing the diagrams of Arts. 387, 483 with those of Art. 392 and the present article, the complete analogy of the *eccentric angles* in the two curves will, as we stated in Art. 483, become apparent. The eccentric angle of any point on either curve, may be defined as *the central angle determined by the corresponding point of the circle described upon the transverse axis*, it being understood that the "corresponding" points are those which have a common subtangent.

489. Perpendicular from the Center to any Tangent.—The length of the perpendicular from the origin upon the line

$$b^2x'x - a^2y'y = a^2b^2,$$

(Art. 92, Cor. 2) must be

$$p = \frac{a^2b^2}{\sqrt{(b^4x'^2 + a^4y'^2)}} = \frac{ab}{\sqrt{(e^2x'^2 - a^2)}} \,.$$

Now (Art. 473) $e^2x'^2 - a^2 = b'^2$. Therefore,

$$p = \frac{ab}{b'} \,;$$

or, as in the Ellipse, we have

Theorem XXVI.—*The central perpendicular upon any tangent of an hyperbola is a fourth proportional to the parallel semi-diameter and the semi-axes.*

490. Central Perpendicular in terms of its inclination to the Transverse Axis.—Changing the sign of b^2 in the formula of Art. 394, we get

$$p = \sqrt{a^2\cos^2\theta - b^2\sin^2\theta}.$$

491. Making the same change in the final equation of Art. 395, we obtain, as the equation to the locus of the intersection of tangents to an hyperbola which cut at right angles,

$$x^2 + y^2 = a^2 - b^2.$$

From this (Art. 136) we at once get

Theorem XXVII.—*The locus of the intersection of tangents to an hyperbola which cut each other at right angles, is the circle described from the center of the hyperbola, with a radius* $= \sqrt{a^2 - b^2}$.

492. Perpendiculars from the Foci to any Tangent.—For the length of the perpendicular from the right-hand focus $(ae, 0)$ upon $b^2x'x - a^2y'y = a^2b^2$, we have (Art. 105, Cor. 2)

$$p = \frac{b^2x'ae - a^2b^2}{\sqrt{(b^4x'^2 + a^4y'^2)}} = \frac{b(ex' - a)}{\sqrt{(e^2x'^2 - a^2)}};$$

or, since (Arts. 457, 473) $ex' - a = \rho$, and $e^2x'^2 - a^2 = b'^2$,

$$p = \frac{b\rho}{b'}.$$

And, in like manner, for the perpendicular from the left-hand focus,

$$p' = \frac{b\rho'}{b'}.$$

Corollary.—Since $b'^2 = \rho\rho'$ (Art. 474), we may also write

$$p^2 = \frac{b^2\rho}{\rho'}, \quad p'^2 = \frac{b^2\rho'}{\rho}.$$

493. Upon dividing the value of p by that of p', we obtain

Theorem XXVIII.—*The focal perpendiculars upon any tangent of an hyperbola are proportional to the adjacent focal radii of contact.*

And if we multiply these values together, $pp' = b^2$; or, we have

Theorem XXIX.—*The rectangle under the focal perpendiculars upon any tangent is constant, and equal to the square on the conjugate semi-axis.*

494. Changing the sign of b^2 in the first two equations of Art. 399, we get

$$y - mx = \sqrt{m^2a^2 - b^2},$$
$$my + x = \sqrt{a^2 + b^2},$$

as the equations to any hyperbolic tangent and its focal perpendicular. Adding the squares of these together, we eliminate m, and obtain

$$x^2 + y^2 = a^2$$

as the constant relation between the co-ordinates of intersection belonging to these lines. Hence, (Art. 136,)

Theorem XXX.—*The locus of the foot of the focal perpendicular upon any tangent of an hyperbola, is the circle inscribed within the curve.*

Corollary.—We may therefore apply in the case of the Hyperbola, the construction given in the corollary to Art. 399, as follows:

To draw a tangent to an hyperbola through any given point: — Join the given point P with either focus F, and upon PF describe a circle cutting the inscribed circle in Q and Q'. The

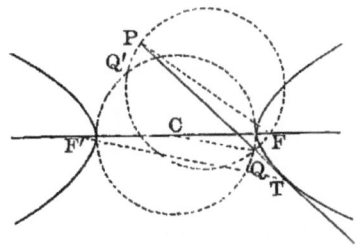

line which joins P to either of these points, for example

the line PQ, will touch the hyperbola at some point T; for the angles PQF, $PQ'F$ being inscribed in a semi-circle, Q and Q' are the feet of focal perpendiculars.

When P is on the curve, and PF consequently a focal radius, we can prove, as in Ex. 8, p. 359, that the circle described on PF will *touch* the inscribed circle. The foot of the focal perpendicular must then be found by joining the middle point of PF with the center C, and noting the point in which the resulting line cuts the inscribed circle.

495. We see, then, that if an *hyperbola* is given, every chord drawn from the focus to meet the inscribed circle must be a focal perpendicular to some tangent of the hyperbola. On the other hand, it is obvious that any point outside of a given *circle*, may be considered the focus of some *circumscribed* hyperbola. Hence,

Theorem XXXI.—*If from any point* without *a circle a chord be drawn, and a perpendicular to it at its extremity, the perpendicular will be tangent to the circumscribed hyperbola of which the point is a focus.*

Corollary.—Since this is equivalent to saying that the hyperbola is the *envelope* of the perpendicular, we may approximate the outline of an hyperbola, as is done in the annexed figure, by drawing chords to a circle from a fixed point P outside of it, and forming perpendiculars at their extremities. It should be noticed, that only the parts of these perpendiculars which lie on *opposite* sides of the chord that determines them, enter into the formation of the curve; in the Ellipse, on the contrary, the perpendiculars lie on the *same* side of the determining chords. When the chords assume the

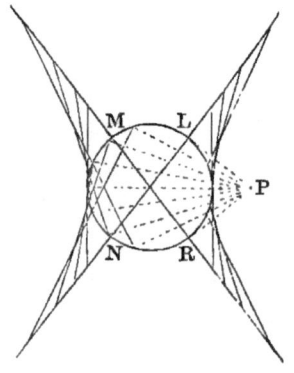

limiting positions *PL, PR,* so as to *touch* the circle at *L* and *R,* the corresponding perpendiculars *LN, MR* are the two lines which we have named the self-conjugates.

496. A little inspection of the equations in Art. 401, after the sign of b^2 has been changed in the first and second, will show that the reasoning of that article is entirely applicable to the Hyperbola. Hence,

Theorem XXXII.—*The diameters which pass through the feet of the focal perpendiculars upon any tangent of an hyperbola, are parallel to the corresponding focal radii of contact.*

Corollary.—Hence, also, as in the case of the Ellipse, we have the converse theorem, *Diameters parallel to the focal radii of contact meet the tangent at the feet of its focal perpendiculars.* Con-sequently, after finding the foot *Q* of the focal perpen-dicular, we can determine the point of contact *T*, if we wish to do so, by sim-ply drawing *F'T* parallel to *CQ*.

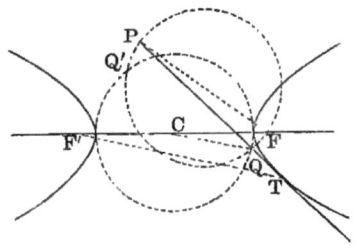

It follows, also, that *the distance between the foot of the perpendicular drawn from either focus to a tangent, and the foot of the perpendicular drawn from the remaining focus to the parallel tangent, is constant, and equal to the length of the transverse axis.*

THE NORMAL.

497. Equation to the Normal.—From the equa-tion of Art. 402, by changing the sign of b^2, we have

$$\frac{a^2x}{x'} + \frac{b^2y}{y'} = c^2.$$

498. Let PN be the normal to an hyperbola at any point P; and FP, $F'P$ the corresponding focal radii. The equations to the latter (Art. 95) are

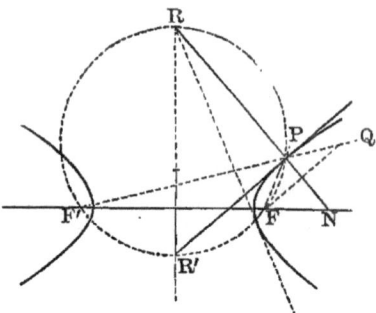

$$y'(x-c)-(x'-c)y=0 \ (\ FP),$$
$$y'(x+c)-(x'+c)y=0 \ (F'P).$$

Combining the equation to the normal with each of these in succession, we get (Art. 96)

$$\tan FPN = \frac{cy'}{b^2}, \quad \tan F'PN = -\frac{cy'}{b^2}.$$

Hence, $FPN = 180° - F'PN = QPN$; and we have

Theorem XXXIII.—*The normal of an hyperbola bisects the* external *angle between the focal radii of contact.*

Corollary 1.—Comparing Theorems XXIV, XXXIII of the Hyperbola with the same of the Ellipse, we at once infer: *If an ellipse and an hyperbola are confocal, the normal of the one is the tangent of the other at their intersection.*

Corollary 2.—To construct a normal at any point P of the curve, we draw the focal radii FP, $F'P$, produce one of them, as $F'P$, until $PQ = FP$, and join QF: then will PN, drawn through P at right angles to QF, be the required normal. For it will bisect the angle FPQ, according to the well-known properties of the isosceles triangle.

Corollary 3.—To draw a normal through any point R on the conjugate axis, we pass a circle $RF'R'F$ through the given point and the foci, and join the point where this circle cuts the hyperbola with the given point by the line RPN: this line will bisect the angle FPQ, because R is the middle point of the arc $F'RF$.

It is important to notice, however, that the auxiliary circle cuts each branch of the curve in *two* points, as P and P', and that only one of these (P, in the diagram) answers the conditions of the present construction. For the line joining R to the other, as RP', will bisect the *internal* angle between the focal radii, instead of the *external*. We thus see that we can use this method for drawing a *tangent* from any point in the conjugate axis: a statement which applies to the Ellipse also, provided the point R is outside of the curve.

499. Intercept of the Normal.—Making $y = 0$ in the equation of Art. 497, we obtain

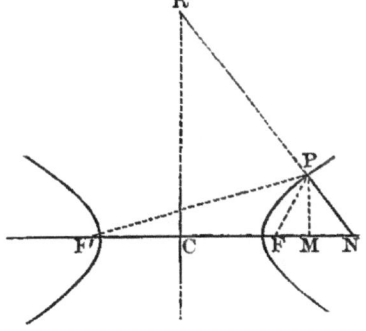

$$x = CN = \frac{c^2}{a^2} x' = e^2 x'.$$

We can therefore, as in the case of the Ellipse (Art. 404), construct a normal at any point P of the curve, or one from any point N of the transverse axis.

500. By an argument in all respects similar to that of Art. 405, we have $F'N : FN = F'P : FP$; that is,

Theorem XXXIV.—*The normal of an hyperbola cuts the distance between the foci in segments proportional to the adjacent focal radii of contact.*

501. Length of the Subnormal.—For the portion of the transverse axis included between the foot of the

normal and that of the ordinate of contact, we have
$MN = CN - CM = e^2 x' - x' = (e^2 - 1) x'$. Hence,

$$\text{subnor} = \frac{b^2}{a^2} x'.$$

502. Comparing the results of Arts. 499 and 501,
$CN : MN = c^2 : b^2$. Or, since $c^2 = a^2 + b^2$, we have

Theorem XXXV.—*The normal of an hyperbola cuts the
abscissa of contact in the constant ratio* $(a^2 + b^2) : b^2$.

503. Length of the Normal.—Changing the sign
of b^2 in the first formula
of Art. 408, and then
applying the formula of
Art. 473, we get

$$PN = \frac{bb'}{a},$$

$$PR = \frac{ab'}{b}.$$

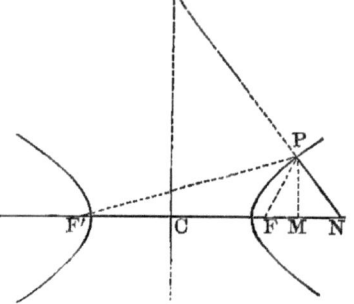

504. Hence, $PN.PR = b'^2$; and we have

Theorem XXXVI.—*The rectangle under the segments
formed by the two axes upon the normal is equal to the
square on the semi-diameter conjugate to the point of
contact.*

Corollary.—Hence, too, (Art. 474) $PN.PR = \rho\rho'$; or,
*The rectangle under the segments of the normal is equal
to the rectangle under the focal radii of contact.*

505. Also (Art. 489), putting Q for the foot of the
central perpendicular on the tangent at P, $CQ.PR = a^2$,
and $CQ.PN = b^2$. That is,

Theorem XXXVII.—*The rectangle under the normal and the central perpendicular upon the corresponding tangent is constant, and equal to the square on the semi-axis other than the one to which the normal is measured.*

SUPPLEMENTAL AND FOCAL CHORDS.

506. Condition that Chords of an Hyperbola be Supplemental.—Let φ, φ' denote the inclinations of any two supplemental chords DP, $D'P$. Then, from Art. 412, by the characteristic change of sign, the required condition will be

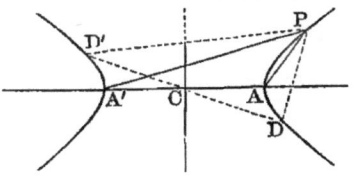

$$\tan \varphi \tan \varphi' = \frac{b^2}{a^2}.$$

507. Hence, the argument of Art. 413 applies directly to the Hyperbola, and we have

Theorem XXXVIII.—*Diameters of an hyperbola which are parallel to supplemental chords are conjugate.*

Corollary 1.—*To construct a pair of conjugate diameters at a given inclination.* The method of solving this problem in the Hyperbola being identical with that given for the Ellipse in the first corollary to Art. 413, we do not consider it necessary to repeat the details here.

Corollary 2.—*To construct a tangent parallel to a given right line.* Let LM be the given line. Draw any diameter QR, and through its extremity Q pass the chord QS

parallel to *LM*. Form the supplemental chord *SR* and its parallel diameter *DP*: the latter, by the present theorem, will be conjugate to that drawn parallel to *LM*; and (Art. 485) the line *PT*, drawn through its extremity *P*, and

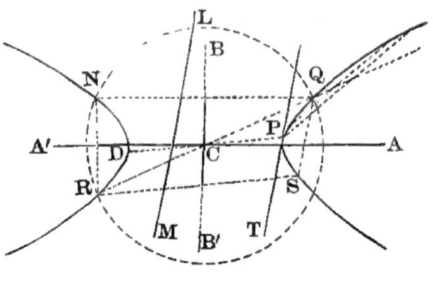

parallel to *LM*, will be the tangent required.

Corollary 3.—*To construct the axes in the empty curve.* Draw any two parallel chords, bisect them, and form the corresponding diameter, say *QR*. On the latter, describe a semicircle cutting the hyperbola in *N*. Join *RN*, *NQ*, and through the middle point of *QR* draw *A'A*, *B'B* parallel to them : the latter will be the axes, by the same reasoning as that used in Art. 413, Cor. 3.

508. Focal Chords.—The properties of these chords presented in Exs. 33—35, p. 361, are as true for the Hyperbola as for the Ellipse. The reader can easily convince himself of this by looking over his solutions of those examples, and making such changes in the formulæ as the equation to the Hyperbola requires. We shall here consider only that single property, proved for the Ellipse in Art. 415, which serves to characterize the parameter of the curve.

509. For the length of any focal chord in an hyperbola, we have, by changing the sign of b^2 in the formula at the foot of p. 334,

$$\text{cho} = \frac{2}{a} \cdot \frac{b^2}{e^2 \cos^2 \theta - 1},$$

in which $\theta =$ the inclination of the chord. Hence (Art. 462), putting $a' =$ the semi-diameter parallel to the chord,

$$\text{cho} = \frac{2a'^2}{a} = \frac{(2a')^2}{2a}.$$

That is,

Theorem XXXIX.—*Any focal chord of an hyperbola is a third proportional to the transverse axis and the diameter parallel to the chord.*

Remark.—The *latus rectum* is the focal chord parallel to the conjugate axis, and its value (Art. 454, Cor.) exemplifies this theorem.

II. THE CURVE REFERRED TO ANY TWO CONJUGATES.

DIAMETRAL PROPERTIES.

510. Equation to the Hyperbola, referred to any two Conjugate Diameters.—The equation to the *primary* curve, transformed to two conjugates whose respective inclinations are θ and θ', is found by simply changing the sign of b^2 in the equation at the middle of p. 336. It is

$$(a^2\sin^2\theta - b^2\cos^2\theta)\, x^2 + (a^2\sin^2\theta' - b^2\cos^2\theta')\, y^2 = -a^2b^2.$$

Hence, by changing the sign of the constant term, the equation to the *conjugate* hyperbola, referred to the same pair of diameters, is

$$(a^2\sin^2\theta - b^2\cos^2\theta)\, x^2 + (a^2\sin^2\theta' - b^2\cos^2\theta')\, y^2 = a^2b^2.$$

Now let a', b' denote the lengths of the semi-diameters

of reference: we shall get, by making $y = 0$ in the first of these equations, and $x = 0$ in the second,

$$a^2\sin^2\theta - b^2\cos^2\theta = -\frac{a^2b^2}{a'^2}, \quad a^2\sin^2\theta' - b^2\cos^2\theta' = \frac{a^2b^2}{b'^2}.$$

Substituting in the first equation above, we obtain

$$\frac{x^2}{a'^2} - \frac{y^2}{b'^2} = 1.$$

Corollary 1.—The transformed equation to the *conjugate* curve is therefore

$$\frac{x^2}{a'^2} - \frac{y^2}{b'^2} = -1.$$

Moreover (since $a'^2 - b'^2 = a^2 - b^2$), in the Equilateral Hyperbola we have $b' = a'$: hence, the equations to that curve and its conjugate, referred to any two conjugate diameters, are

$$x^2 - y^2 = \pm\, a'^2.$$

Corollary 2.—The new equation to the Hyperbola differs from the analogous equation to the Ellipse (Art. 417), only in the sign of b'^2. Hence, *Any function of* b' *that expresses a property of the Ellipse, will be converted into one expressing a corresponding property of the Hyperbola by merely replacing its* b' *by* b' $\sqrt{-1}$.

511. The remarks of Art. 418 evidently apply to the equations

$$\frac{x^2}{a^2} - \frac{y^2}{b^2} = \pm\, 1, \quad \frac{x^2}{a'^2} - \frac{y^2}{b'^2} = \pm\, 1.$$

Hence, we have the following extensions of Theorems VI, VII:

Theorem XL.—*The squares on the ordinates to any diameter of an hyperbola are proportional to the rectangles under the corresponding segments of the diameter.*

Theorem XLI.—*The square on any diameter of an hyperbola is to the square on its conjugate, as the rectangle under any two segments of the diameter is to the square on the corresponding ordinate.*

512. Writing the equation of Art. 510 in the form

$$y^2 = \frac{b'^2}{a'^2} (x^2 - a'^2),$$

and comparing it with that of the Equilateral Hyperbola, namely, with

$$y^2 = x^2 - a'^2,$$

we get $y_h : y_r = b' : a'$. That is, as the extension of Theorem VIII,

Theorem XLII.—*The ordinate to any diameter of an hyperbola is to the corresponding ordinate of its equilateral, as the conjugate semi-diameter is to the semi-diameter.*

Remark.—We may take the *corresponding* ordinate of the equilateral as signifying either the *oblique* ordinate of the equilateral described upon the same transverse axis as the given hyperbola, or the *rectangular* ordinate of the equilateral described upon the diameter selected for the axis of x. For the equation $x^2 - y^2 = a'^2$ will denote either of these equilaterals, according as it is supposed to refer to oblique or rectangular axes. Only we must understand that, in either interpretation, the corresponding ordinates are those which have a common abscissa.

It is evident, also, that the ratio between the corresponding ordinates of the hyperbola and the circle $x^2 + y^2 = a'^2$, described on any diameter of the curve, is imaginary. Hence, with respect to this circle, there is a defect in the analogy between the Ellipse and the Hyperbola: a defect that will be supplied, however, as soon as we develop the conception of the *subtangent to any diameter.*

513. We leave the student to show, by interpreting the equation

$$y^2 = \frac{b'^2}{a'^2}\,(x^2 - a'^2),$$

that, with reference to *any* diameter, the Hyperbola consists of two infinite branches, extending in opposite directions, and both symmetric to the diameter.

CONJUGATE PROPERTIES OF THE TANGENT.

514. Equation to the Tangent, referred to any two Conjugate Diameters.—By changing the sign of b'^2 (Art. 510, Cor. 2) in the equation of Art. 421, the equation now sought is seen to be

$$\frac{x'x}{a'^2} - \frac{y'y}{b'^2} = 1.$$

515. Intercept of the Tangent on any Diameter.—Making $y = 0$ in the equation just found, we get, for the intercept in question,

$$x = CT' = \frac{a'^2}{x'}.$$

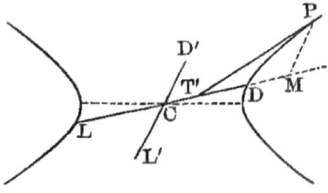

Hence, as the extension of Art. 487,

Theorem XLIII.— *The intercept cut off by a tangent upon any diameter of an hyperbola is a third proportional to the abscissa of contact and the semi-diameter.*

Corollary.— *To construct a tangent from any given*

point. The method of the corollary to Art. 422 applies directly to the Hyperbola, and the student may interpret the statements there made, as referring to the present diagram letter by letter.

516. The properties of tangential intercepts, proved in Art. 423 with respect to the Ellipse, are also true of the Hyperbola. We shall merely restate them here, leaving the reader to make such simple modifications of the analyses in I, II, III of the article mentioned, as may be necessary to establish them. 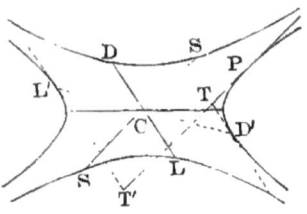 To aid him in this, the parts of the annexed diagram are lettered identically with the corresponding parts of that in Art. 423.

I. **Theorem XLIV.**—*The rectangle under the intercepts cut off upon two fixed parallel tangents by any variable tangent of an hyperbola is constant, and equal to the semi-diameter parallel to the two tangents.*

II. **Theorem XLV.**—*The rectangle under the intercepts cut off upon any variable tangent of an hyperbola by two fixed parallel tangents is variable, being equal to the square on the semi-diameter parallel to the tangent.*

III. **Theorem XLVI.**—*The rectangle under the intercepts cut off upon any variable tangent of an hyperbola by two conjugate diameters is equal to the square on the semi-diameter parallel to the tangent.*

Corollary 1.—By the same reasoning as in the first corollary to III of Art. 423, we have: *Diameters drawn through the intersections of any tangent with two parallel tangents are conjugate.*

Corollary 2.—The problem, *Given two conjugate diameters of an hyperbola in position and magnitude, to construct the axes,* is solved by the same process as the corresponding one on p. 342; excepting that the point *P* must be taken on the side of *D* next to *C,* instead of on the side remote from it.

517. Subtangent to any Diameter.—For the length of this, we have $MT' = CM - CT'$. Hence, putting $x' = CM$, and substituting the value of CT' from Art. 515,

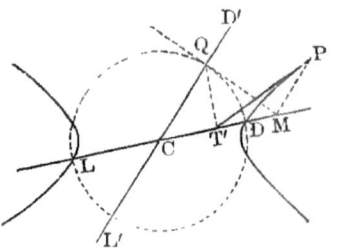

$$\text{subtan}' = \frac{x'^2 - a'^2}{x'}.$$

That is, since $x' + a' = LM$, and $x' - a' = MD$, *The subtangent to any diameter of an hyperbola is a fourth proportional to the abscissa of contact and the corresponding segments of the diameter.*

Corollary.—If we compare this value of the general subtangent with that of the *subtangent of the curve* (Art. 488), we see at once that the argument used in Art. 488, Cor. 1, with respect to the Hyperbola and its inscribed circle, applies to the curve and the circle described upon *any* of its diameters. Hence, *If through the head and foot of an ordinate to any diameter of an hyperbola tangents be drawn to the curve and to the circle described upon the diameter, they will have a common subtangent.*

In other words, if Q is the point in which a rectangular ordinate drawn through the foot of a tangent to the hyperbola pierces the circle mentioned, the tangent to this circle at Q passes through M, the foot of the ordinate of contact for the tangent to the hyperbola. The defect noticed in the Remark under Art. 512, is therefore supplied; and we may employ the circle in question, to solve the following problem:

To draw a tangent to an hyperbola from any given point. Let T' be the given point. Draw the diameter $DT'L$, and form the corresponding circle C-DQL. At the given point, set up $T'Q$ a rectangular ordinate to this circle, and through its extremity Q draw the tangent QM. Then, through the foot M of this tangent, pass MP parallel to the diameter conjugate to DL: the point P in which this parallel cuts the hyperbola, will be the point of contact of the required tangent, which may be obtained by joining $T'P$.

Remark.—To form a tangent at any point P of the curve, we draw the ordinate PM, and, through its foot, the circular tangent MQ. Then, if QT' be drawn at right angles to the diameter DL, T' will be the foot of the required tangent.

518. By the same reasoning as in Art. 425, we get

Theorem XLVII.—*The rectangle under the subtangent and the abscissa of contact is to the square on the ordinate of contact, as the square on the corresponding diameter is to the square on its conjugate.*

519. Changing the sign of b'^2 in the equations of Art. 426, and then taking the steps indicated there, we obtain

Theorem XLVIII.—*Tangents at the extremities of any chord of an hyperbola meet on the diameter which bisects that chord.*

PARAMETERS.

520. Definitions.—The **Parameter** of an hyperbola, with respect to any diameter, like the parameter of an ellipse, is a third proportional to the diameter and its conjugate. Thus,

$$\text{parameter} = \frac{(2b')^2}{2a'} = \frac{2b'^2}{a'}.$$

The parameter with respect to the transverse axis, is called the *principal parameter;* or, the *parameter of the curve.* We shall denote its length by $4p$.

521. For the value of the parameter of the Hyperbola, we accordingly have

$$4p = \frac{2b^2}{a}.$$

Thus (Art. 454, Cor.) the principal parameter is identical with the *latus rectum*, and may therefore be described as *the double ordinate to the transverse axis, drawn through the focus.*

522. In Art. 509, we proved that the focal double ordinate parallel to any diameter is a third proportional to the transverse axis and the diameter. Now (Art. 463) the transverse axis is less than any other diameter — less, therefore, than the diameter conjugate to that of which the focal chord is a parallel, unless the chord is the *latus rectum.* Hence,

Theorem XLIX.—*No parameter of an hyperbola, except the principal, is equal in value to the corresponding focal double ordinate.*

POLE AND POLAR.

523. We now proceed to develop the polar relation as a property of the Hyperbola; and shall follow the steps already twice taken, in connection with the Circle and the Ellipse.

524. Chord of Contact in the Hyperbola.—Let $x'y'$ be the fixed point from which the two tangents that

determine the chord are drawn. Then, by merely changing the sign of b'^2 in the equation of Art. 431, the equation
to the hyperbolic chord of contact will be

$$\frac{x'x}{a'^2} - \frac{y'y}{b'^2} = 1.$$

**525. Locus of the Intersection of Tangents to the
Hyperbola.**—Let $x'y'$ denote the fixed point through
which the chord of contact belonging to any two of the
intersecting tangents is drawn, and change the sign of
b'^2 in the equation of Art. 432: the equation to the locus
now considered will then be

$$\frac{x'x}{a'^2} - \frac{y'y}{b'^2} = 1.$$

**526. Tangent and Chord of Contact taken up
into the wider conception of the Polar.**—From the
identity in the form of the last two equations with the
form of the equation to the tangent, we see that, in the
Hyperbola also, the law which connects the tangent with
its point of contact, and the chord of contact with the
point from which its determining tangents are drawn, is
the same that connects the locus of the intersection of
tangents drawn at the extremities of chords passing
through a fixed point, with that point.

In short, the three right lines represented by these
equations are only different expressions of the same formal
law: a law, moreover, of which the locus mentioned is the
generic expression. For, in the case of the tangent, the
point $x'y'$ is restricted to being *on* the curve; and, in that
of the chord of contact, to being *within;* while, in that of
the locus, it is unrestricted: so that the tangent and the
chord of contact are cases of the locus, due to bringing

the point $x'y'$ upon the curve or within it. Moreover, the formal law which connects the locus with the fixed point is the law of *polar reciprocity*. For, by its equation, the locus is a right line; and, if we suppose the point $x'y'$ to be any point on a *given* right line, the co-efficients of the equation in Art. 524 will be connected by the relation $Ax' + By' + C = 0$; whence (Art. 117) we have the twofold theorem:

I. *If from a fixed point chords be drawn to any hyperbola, and tangents to the curve be formed at the extremities of each chord, the intersections of the several pairs of tangents will lie on one right line.*

II. *If from different points lying on one right line pairs of tangents be drawn to any hyperbola, their several chords of contact will meet in one point.*

The Hyperbola, then, imparts to every point in its plane the power of determining a right line; and reciprocally.

527. Equation to the Polar with respect to an Hyperbola.—From the conclusions now reached, this equation, referred to any two conjugate diameters, must be

$$\frac{x'x}{a'^2} - \frac{y'y}{b'^2} = 1;$$

or, referred to the axes of the curve,

$$\frac{x'x}{a^2} - \frac{y'y}{b^2} = 1,$$

$x'y'$ being the point to which the polar corresponds.

528. Definitions.—The **Polar** of any point, with respect to an hyperbola, is the right line which forms the locus of the intersection of the two tangents drawn at the extremities of any chord passing through the point.

The **Pole** of any right line, with respect to an hyperbola, is the point in which all the chords of contact corresponding to different points on the line intersect.

Hence the following constructions : — When the pole P is given, draw through it any two chords $T'T$, $S'S$, and form the corresponding pairs of tangents, $T'L$ and TL, $S'M$ and SM: the line LM, which joins

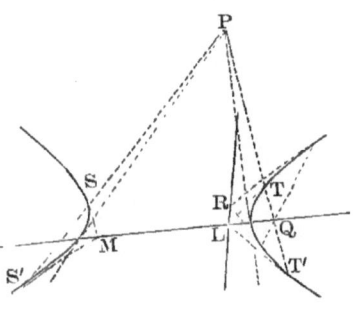

the intersection of the first pair to that of the second, will be the polar of P. When the polar is given, take upon it any two points, as L and M, and draw from each a pair of tangents, LT and LT', MS and MS': the point P, in which the corresponding chords of contact $T'T$, $S'S$ intersect, will be the pole of LM.

This construction is applicable in all cases; and, when the pole is *without* the curve, as at Q, it *must* be used. But if the pole is *within* the curve, as at P, the polar LM may be obtained by drawing the chord of contact of the two tangents from P; and if it is *on* the curve, as at T, the polar is the corresponding tangent LT.

529. Direction of the Polar.—By changing the sign of b'^2 in the equations of Art. 436, and then using the principle of inference employed there, we obtain

Theorem L.—*The polar of any point, with respect to an hyperbola, is parallel to the diameter conjugate to that which passes through the point.*

530. Polars of Special Points.—A comparison of the equation to the polar in an hyperbola with its equation as related to the Circle (Art. 323), will show that the general properties proved of polars in Art. 324 are true for the Hyperbola. We therefore pass at once to those special properties which characterize the polars of certain particular points.

Applying the processes of Art. 437 to the equation of Art. 527, we get

I. *The polar of the center is a right line at infinity.*

II. *The polar of any point on a diameter is a right line parallel to the conjugate diameter, and its distance from the center is a third proportional to the distance of the point and the length of the semi-diameter.*

III. *The polar of any point on the transverse axis is the perpendicular whose distance from the center is a third proportional to the distance of the point and the length of the semi-axis.*

Corollary.—From II it follows, that the construction for the polar, given under Art. 437 with respect to the Ellipse, is entirely applicable to the Hyperbola.

531. Polar of the Focus.—The equation to this is found by putting $(\pm ae, 0)$ for $x'y'$ in the second equation of Art. 527, and is therefore

$$x = \pm \frac{a}{e}\,.$$

Hence, *The polar of either focus in an hyperbola is the perpendicular which cuts the transverse axis at a distance from the center equal to* a : e, *measured on the same side as the focus.*

Remark.—Since the *e* of the Hyperbola is *greater* than unity, the distance of the focal polar from the center is in that curve *less* than *a*. In the Ellipse, on the contrary, this distance is *greater* than *a*, because the *e* of that curve is *less* than unity. Hence, in the Ellipse, the polar of the focus is *without* the curve; but, in the Hyperbola, it is situated *within*.

532. The distance of any point P of an hyperbola from either focal polar, for instance from DR, is obviously equal to the abscissa of the point, diminished by the distance of the polar from the center. That is,

$$PD = x - \frac{a}{e} = \frac{ex - a}{e}.$$

But (Art. 457) $ex - a = FP$. Therefore,

$$\frac{FP}{PD} = e.$$

In other words, the property of Art. 439 re-appears, and we have

Theorem LI.—*The distance of any point on an hyperbola from the focus is in a constant ratio to its distance from the polar of the focus, the ratio being equal to the eccentricity of the curve.*

Corollary 1.—We may therefore describe an hyperbola by a continuous motion, as follows:

Take any point F, and any fixed right line DR. Against the latter, fasten a ruler DD', and place a second ruler NQL (right-angled at L) so that its edge LN may move freely along DD'. At F fasten one end of a thread equal in length to the edge NQ of this last ruler, to whose extremity Q the other end must be attached. Then, with the point P of a pencil, stretch this thread against the edge NQ, and move the pencil so that the thread shall be kept stretched while the ruler NQL slides along DD': the path of P will be an hyperbola. For, by the conditions

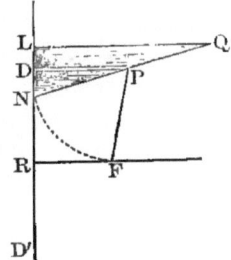

named, FP must equal PN in every position of the pencil: whence $FP \cdot PD = NQ : QL$. That is, since DR must be the polar of F with respect to *some* hyperbola, the focal distance of P is in a constant ratio to its distance from the focal polar.

This construction derives interest from a comparison with that of the Ellipse in Art. 439, Cor. 1. It will be seen that the essential principle is the same in both, namely, the use of the parts of a right triangle to determine a constant ratio between the distances of points from a fixed point and a fixed right line. It is noticeable, that, in the Ellipse, this constant ratio is that of the base to the hypotenuse; while, in the Hyperbola, it is that of the hypotenuse to the base; thus illustrating the inverse relation existing between the two curves.

Corollary 2.—In the construction just explained, the polar of the focus is used as the directing line of the motion which generates the curve. For this reason, it is called the *directrix* of the corresponding hyperbola.

Corollary 3.—In the light of the present theorem, we may interpret the name *hyperbola* as denoting *the conic in which the constant ratio between the focal and polar distances* exceeds *unity.*

533. **Focal Angle subtended by any Tangent.**—By examining the investigation conducted in Art. 440, the student will see that it is applicable to the Hyperbola, with the single exception of a change in the sign of the final result. Hence, if $\rho =$ the focal distance of any given point from which a tangent is drawn to an hyperbola, and $x =$ the abscissa of the point, the angle φ which the portion of the tangent intercepted between the given point and the point of contact subtends at the focus, will be determined by the formula

$$\cos \phi = \frac{\epsilon x - a}{\rho}.$$

534. This expression, being independent of the point of contact $x'y'$, would seem to indicate that both of the tangents that can be drawn from a given point to the curve subtend the same focal angle. It is found, however, as in the case presented in the diagram, that when the given point P is taken under such conditions (Art. 484, Cor. 2) as fix the two points of contact T and T' on opposite branches of the curve, the angles PFT, PFT' are not *equal*, but *supplemental.* But, whether they be the one or the other, the line FP must bisect the whole angle $T'FT$ subtended by the chord of contact, either internally or externally. Hence,

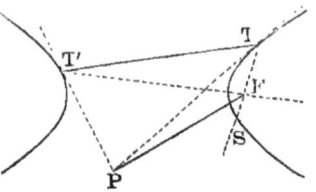

Theorem LII.—*The right line that joins the focus to the pole of any chord, bisects the focal angle which the chord subtends.*

Corollary.—The angle subtended by a *focal* chord being 180°, we have, as a special case of the preceding: *The line that joins the focus to the pole of any focal chord is perpendicular to the chord.*

III. The Curve referred to its Foci.

535. In the polar equations of Art. 172 and the subjoined Remark, namely, in

$$\rho = \frac{a\,(1-e^2)}{1-e\cos\theta}, \quad \rho = \frac{a\,(e^2-1)}{1-e\cos\theta},$$

we now know that the constant a is the transverse semi-axis of the corresponding hyperbola, and the constant e its eccentricity.

Replacing, then, $e^2 - 1$ by its value (Art. 171) $b^2 : a^2$, we may write these equations

$$\rho = -\frac{b^2}{a}\cdot\frac{1}{1-e\cos\theta}, \quad \rho = \frac{b^2}{a}\cdot\frac{1}{1-e\cos\theta}.$$

But (Art. 521) $b^2 : a$ is half the parameter of the curve.

$$\therefore\ \rho = \frac{\pm\,2p}{1-e\cos\theta},$$

the upper sign corresponding to the right-hand focus, and the lower to the left-hand.

536. Polar Equation to the Tangent.—By an analysis exactly similar to that in Art. 444, we find this to be

$$\rho = \frac{a\,(1 - e^2)}{\cos(\theta - \theta') - e\cos\theta}.$$

Corollary.—The equation to the diameter conjugate to $x'y'$ (Art. 469) differs from that of the tangent at $x'y'$ only in having 0 for its constant term. Hence, as in the corollary to Art. 444,

$$\rho = \frac{ae\,(\cos\theta' - e)}{\cos(\theta - \theta') - e\cos\theta}$$

is the *polar equation to the diameter conjugate to that which passes through $\rho'\theta'$.*

IV. The Curve referred to its Asymptotes.

537. Hitherto, the properties established for the Hyperbola have had a fixed relation, either of identity or of antithesis, to those of the Ellipse. We now come, however, to a series of properties peculiar to the Hyperbola, arising from the presence of the two lines which we have named the self-conjugate diameters. We might proceed at once to transform the equation of Art. 167 to these diameters as axes of reference; but, before doing so, let us subject the self-conjugates themselves to a more minute examination.

538. Definition.—An **Asymptote** of any curve is a line which continually approaches the curve, but meets it only at infinity.

Asymptotes are either *curvilinear* or *rectilinear*. The term *asymptote* is derived from the Greek *a privative*, and συμπίπτειν, *to coincide*, and may be taken as signifying that the line to which it is applied *never meets* the curve which it forever approaches.

539. We have already once or twice spoken of the self-conjugates of the Hyperbola as its *asymptotes*. We now proceed to show that they are such. We have proved (Arts. 461, Cor. 1; 480) that they meet the curve only at infinity: it remains to show that they draw nearer and nearer to the curve the farther they recede from the center.

Let CM be any common abscissa of an hyperbola and its self-conjugate diameter CL. The equations to CL and the curve being respectively

$$y = \frac{b}{a} x, \qquad y = \frac{b}{a} \sqrt{x^2 - a^2},$$

we get, for the difference between any two corresponding ordinates,

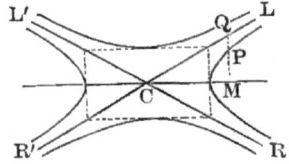

$$PQ = \frac{b}{a}\left(x - \sqrt{x^2 - a^2}\right) = \frac{ab}{x + \sqrt{x^2 - a^2}}.$$

Hence, as x increases, PQ diminishes; so that, if we suppose x to be increased without limit, or the point P of the curve to recede to an infinite distance from the origin, PQ will converge to the limit 0. Now the distance of any point P of the curve from the self-conjugate CL is equal to $PQ \sin PQC$: therefore, as the angle PQC is the same for every position of P, this distance diminishes continually as P recedes from the center; or, we have

An. Ge. 38.

Theorem LIII.—*The self-conjugate diameters of an hyperbola are asymptotes of the curve.*

Remark.—We have inferred this theorem with respect to $L'R$ as well as LR', although the preceding investigation is conducted in terms of LR' only. But it is manifest that a similar analysis applies to $L'R$; and it can be shown, in like manner, that LR' and $L'R$ are asymptotes of the *conjugate* hyperbola. We leave the proof of this, however, as an exercise for the student.

540. Angle between the Asymptotes.—From Art. 479, we have $\tan LCM = b : a$, and $\tan L'CM = -b : a$; hence $LCM = 180° - L'CM = RCM$. If, then, we put $\varphi =$ the required angle LCR, and $\theta = LCM$, we shall get $\varphi = 2\theta$; or, since $\tan \theta = b : a$, and therefore $\cos \theta = a : c$,

$$\varphi = 2 \sec^{-1} e.$$

Hence, if the eccentricity of an hyperbola is given, the inclination of its asymptotes is also given; for it is *double the angle whose secant is the eccentricity.* Conversely, when the inclination of the asymptotes is known, the eccentricity is found by taking the secant of half the inclination.

Thus, in the case of an *equilateral* hyperbola, whose eccentricity (Art. 456, Cor.) $= \sqrt{2}$, we have

$$\varphi = 2 \sec^{-1} \sqrt{2} = 90°:$$

which agrees with the property by which (Art. 177, Cor.) we originally distinguished this curve.

541. Equations to the Asymptotes.—These are respectively (Art. 479) $y = (b : a) x$, $y = - (b : a) x$. Or we may write them

$$\frac{x}{a} - \frac{y}{b} = 0, \quad \frac{x}{a} + \frac{y}{b} = 0.$$

Hence, (Art. 124,) the equation to *both* asymptotes is

$$\frac{x^2}{a^2} - \frac{y^2}{b^2} = 0.$$

542. Let CD, CD' be any two conjugate semi-diameters. Then, from the fact that the equation to the Hyperbola, when referred to these, is identical in form with its equation as referred to the axes, we may at once infer that

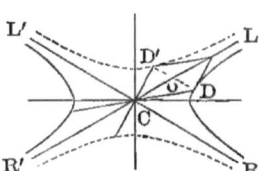

$$\frac{x}{a'} - \frac{y}{b'} = 0, \quad \frac{x}{a'} + \frac{y}{b'} = 0$$

are the equations to the asymptotes, referred to any pair of conjugates.

Now the first of these lines (Art. 95, Cor. 2) passes through the point $a'b'$, that is, through the vertex of the parallelogram formed on CD and CD'; while the second (Art. 98, Cor.) is parallel to the line

$$\frac{x}{a'} + \frac{y}{b'} = 1,$$

that is, to the diagonal $D'D$ of the same parallelogram. Hence, the asymptotes have the same direction as the diagonals of this parallelogram; or, extending the property to the figure of which this parallelogram is the fourth part, we get

Theorem LIV.—*The asymptotes are the diagonals of every parallelogram formed on a pair of conjugate diameters.*

Corollary.—If, then, we have any two conjugate diameters given, we can find the asymptotes; and, conversely, given the asymptotes and any diameter CD, we can find its conjugate by drawing DO parallel to CR, and producing it till $OD' = OD$, when D' will be the extremity of the conjugate sought.

543. From the equation to the tangent (Art. 481) we get

$$y = \frac{b^2 x' x}{a^2 y'} - \frac{b^2}{y'};$$

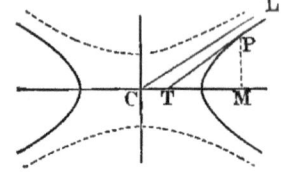

or, after substituting for y' from the equation to the curve, and factoring,

$$y = \frac{b}{a} x \cdot \frac{1}{\left\{ 1 - \dfrac{a^2}{x'^2} \right\}^{\frac{1}{2}}} - \frac{b^2}{y'}.$$

Supposing, then, that x' and y' are increased without limit, or that the point of contact P recedes to an infinite distance from the origin, the limiting form to which this equation tends is

$$y = \frac{b}{a} x.$$

But this (Art. 541) is the equation to CL; and a like result can be readily obtained with respect to the other asymptote. Hence,

Theorem LV.—*The asymptotes are the limits to which the tangents of an hyperbola converge as the point of contact recedes toward infinity.*

Remark.—We might therefore define the asymptotes as *the right lines which meet the hyperbola in two consecutive points at infinity.*

544. Accordingly, by Theorem XXIX (Art. 493), the product of the focal perpendiculars upon an asymptote must be equal to b^2. But, since the asymptote passes through the center, these focal perpendiculars must be equal to each other, and therefore each equal to b. That is,

Theorem LVI.—*The perpendicular from either focus to an asymptote is equal to the conjugate semi-axis.*

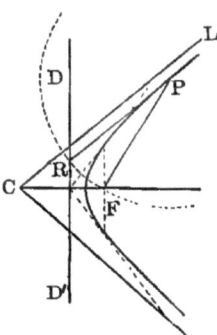

545. Let FP be the focal distance of any point on an hyperbola, and PD its distance from the directrix DR. By Art. 532, $FP = e.PD$. But (Art. 540), $e = \sec LCF$. Hence, $FP = PD \sec LCF = PD \operatorname{cosec} PRD$, if PR be drawn parallel to the asymptote CL. That is (Trig., 859), $FP = PR$; or, we have

Theorem LVII.—*The focal distance of any point on an hyperbola is equal to its distance from the adjacent directrix, measured on a parallel to either asymptote.*

Corollary.—We here find a new reason for the method of generating an hyperbola, given in the first corollary to Art. 532. For, by the requirements of the method, $FP = PR = e.PD$. Now, by the diagram, $PR = PD \sec RPD$. Hence, the method makes $\sec RPD = e$; that is (Art. 540), it makes the angle RPD equal to the inclination of the asymptote, and PR therefore parallel to that line.

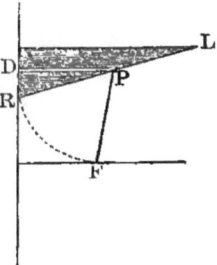

546. Equation to the Hyperbola, referred to its Asymptotes.—The equation to the Hyperbola, transformed to a pair of oblique axes whose inclinations to the transverse axis are respectively θ and θ', is found

by changing the sign of b^2 in the first equation of p. 336, and is

$$(a^2 \sin^2 \theta - b^2 \cos^2 \theta)\, x^2 + (a^2 \sin^2 \theta' - b^2 \cos^2 \theta')\, y^2$$
$$+ 2\,(a^2 \sin \theta \sin \theta' - b^2 \cos \theta \cos \theta')\, xy = - a^2 b^2.$$

If, then, the new axes are the asymptotes, and therefore (Art. 479) $\tan^2 \theta = b^2 : a^2 = \tan^2 \theta'$, we shall have $a^2 \sin^2 \theta - b^2 \cos^2 \theta = 0 = a^2 \sin^2 \theta' - b^2 \cos^2 \theta'$; and the equation will become

$$2\,(a^2 \sin \theta \sin \theta' - b^2 \cos \theta \cos \theta')\, xy = - a^2 b^2.$$

In this, again, since * $\sin \theta = - b : \sqrt{a^2 + b^2} = -\sin \theta'$, and $\cos \theta = a : \sqrt{a^2 + b^2} = \cos \theta'$, it is evident that we have $a^2 \sin \theta \sin \theta' - b^2 \cos \theta \cos \theta' = - 2\,a^2 b^2 : (a^2 + b^2)$. Hence, the required equation is

$$xy = \frac{a^2 + b^2}{4};$$

and putting k^2 to represent the constant in the second member, we may write it in the form in which it is usually quoted, namely,

$$xy = k^2.$$

Corollary.—Hence, the equation to the *conjugate* hyperbola will be

$$xy = - k^2;$$

and, in the case of an *equilateral* hyperbola, we shall have

$$xy = \pm \frac{a^2}{2}.$$

---

* It must be remembered that $\theta =$ the inclination of tho new axis of x; and, in our investigation, the axis of x is that asymptote which corresponds to $\theta = \tan^{-1} - b : a.$

547. If $\varphi =$ the angle LCR, the parallelogram $CMPN$, contained by the asymptotic co-ordinates of any point P, will be expressed by $xy \sin \varphi$. There-fore, from the equation of the preceding article, this parallelo-gram is equal to $\frac{1}{4}(a^2 + b^2) \sin \varphi$.

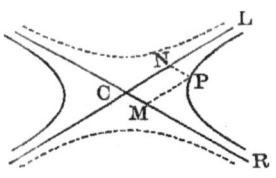

But (Art. 479, Cor.) $\sin \varphi = 2ab : (a^2 + b^2)$. Hence, the parallelogram is in fact equal to $\frac{1}{2}ab$; and we have, as the geometric interpretation of the equation $xy = k^2$,

Theorem LVIII.—*The parallelogram under the asymptotic co-ordinates of an hyperbola is constant, and equal to half the rectangle under the semi-axes.*

548. Equation to any Chord, referred to the Asymptotes.—Let $x'y'$, $x''y''$ be the extremities of the chord. Then (Art. 95), the equation will be of the form

$$\frac{y - y'}{x - x'} = \frac{y'' - y'}{x'' - x'} .$$

Now, since the extremities of the chord are in the curve, $x' = k^2 : y'$, and $x'' = k^2 : y''$. Substituting these values, and reducing, we get for the required equation

$$y'x + x''y = x''y' + k^2 \qquad (1);$$

or, after dividing through by $k^2 = x'y' = x''y''$, the more symmetric form

$$\frac{x}{x'} + \frac{y - y'}{y''} = 1 \qquad (2).$$

549. Let Q $(x'y')$ and S $(x''y'')$ be any two *fixed* points on an hyperbola, and P $(\alpha\beta)$ a *variable* point. Then (Art. 548) the equations to PQ and PS will be

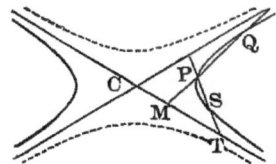

$$y'x + \alpha y = \alpha y' + k^2,$$
$$y''x + \alpha y = \alpha y'' + k^2.$$

Making $y = 0$ in each of these, we get $CM = a + x'$, $CT = a + x''$. Hence, $MT = x'' - x'$, which being independent of $\alpha\beta$, we have

Theorem LIX.—*The right lines which join two fixed points of an hyperbola to any variable point on the curve, include a constant portion of the asymptote.*

550. Equation to the Tangent, referred to the Asymptotes.—Assuming that the point $x''y''$ in the final equation of Art. 548 becomes coincident with $x'y'$, we get, for the equation now sought,

$$\frac{x}{x'} + \frac{y}{y'} = 2.$$

551. Equations to Diameters, referred to the Asymptotes.—The diameter which passes through a fixed point $x'y'$ (Art. 95, Cor. 2), is represented by $y'x - x'y = 0$; and this equation, when $x'y'$ is on the curve, may be written (Art. 546)

$$\frac{x}{k^2} - \frac{y}{y'^2} = 0 \qquad\qquad (1).$$

The equation to the diameter conjugate to $x'y'$ must have 0 for its absolute term (Art. 63); and, as the diameter is parallel to the tangent at $x'y'$, the variable part of its equation (Art. 98, Cor.) must be identical with that of the tangent. Hence, the equation is

$$\frac{x}{x'} + \frac{y}{y'} = 0 \qquad\qquad (2).$$

The transverse axis bisects the angle between the asymptotes, and therefore, at its extremity, $x' = y'$. Hence, the equations to the *axes*, referred to the asymptotes, are

$$x - y = 0, \quad x + y = 0 \qquad\qquad (3).$$

552. Eliminating between (2) of the preceding article, and the equation to the conjugate hyperbola, we get, for the co-ordinates of the extremity of the diameter conjugate to $x'y'$,

$$x_c = \mp x', \quad y_c = \pm y'.$$

553. From the equation to the tangent (Art. 550), we get, by making y and x successively equal to zero, $CT = 2x'$, and $CS = 2y'$. Hence, P is the middle point of ST; and we have

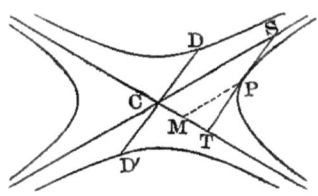

Theorem LX.—*The portion of the tangent included between the asymptotes, is bisected at the point of contact.*

Corollary.—Since (Art. 542) S is a vertex of the parallelogram formed on the conjugate semi-diameters CP and CD, we have $PS = CD$. Hence, $ST = 2PS = DD'$. That is, *The segment cut from the tangent by the asymptotes is equal to the diameter conjugate to the point of contact.*

Remark.—Theorem LX might have been obtained geometrically, as a corollary to Theorem LIV.

554. From the preceding article, we at once obtain

$$CT \cdot CS = 4x'y' = a^2 + b^2,$$

since $4x'y' = 4k^2$. That is,

Theorem LXI.—*The rectangle under the intercepts cut off upon the asymptotes by any tangent is constant, and equal to the sum of the squares on the semi-axes.*

An. Ge. 39

555. For the area of the triangle SCT, we have

$$T = 2\,x'y' \sin \varphi = ab,$$

since $2\,x'y' = 2k^2$, and $\sin \varphi = 2ab : (a^2 + b^2)$. Hence,

Theorem LXII.—*The triangle included between any tangent and the asymptotes is constant, and equal to the rectangle under the semi-axes.*

556. The equations to the tangents at the extremities of two conjugate diameters (Arts. 550, 552) are

$$\frac{x}{x'} + \frac{y}{y'} = 2, \quad \frac{x}{x'} - \frac{y}{y'} = -2.$$

Adding these together, we get $x = 0$, the equation to the line CL. Hence,

Theorem LXIII—*Tangents at the extremities of conjugate diameters meet on the asymptotes.*

557. The equation to $Q'Q$, an ordinate to any diameter VD, will only differ from that of the conjugate diameter $V'D'$ by some constant, which we may call $2c$. The equation will therefore [Art. 551, (2)] be

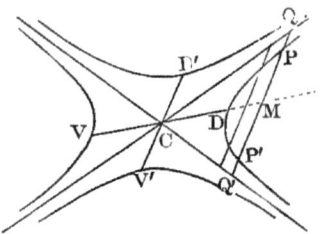

$$\frac{x}{x'} + \frac{y}{y'} = 2c.$$

Now this, when combined with the equation to VD,

$$\frac{x}{x'} - \frac{y}{y'} = 0,$$

gives $x = cx'$, $y = cy'$ as the co-ordinates of M, the point in which $Q'Q$ cuts VD. But the intercepts of $Q'Q$ upon the asymptotes are obviously $CQ' = 2cx'$, $CQ = 2cy'$. Hence, M is the middle point of $Q'Q$; or, we have

Theorem LXIV.—*The segments formed by the asymptotes upon an ordinate to any diameter are equal.*

Corollary 1.—By the definition of a diameter, M is the middle point of $P'P$; so that $PQ = P'Q'$, and we get the property: *The portions of any chord that are intercepted between the curve and the asymptotes, are equal.*

Corollary 2.—We can now readily solve the problem, *Given the asymptotes and one point, to form the curve.* Let CL, CR be the given asymptotes, and P the given point. Through P draw any right line $Q'Q$, cutting the asymptotes in Q' and Q. On its longer segment, lay off $Q\,1$ equal to the shorter segment PQ': then will (1) be a point on the curve, by Cor. 1 above. In the same manner, other points, (2), (3), etc., may be obtained; and, when enough are found, the curve can be drawn through them. The given point P may be *any* point of the curve; but in practice it is usually the *vertex*, and is so represented in the diagram.

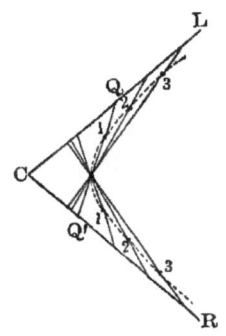

558. The equation to any chord $Q'Q$, in terms of the extremity $x'y'$ of its bisecting diameter VD, being (Art. 557)

$$\frac{x}{x'} + \frac{y}{y'} = 2c,$$

the abscissas of P' and P, the points in which it cuts the curve, will be found by eliminating between this equation and $xy = k^2$. But, as D is on the curve, $x'y' = k^2$: whence, by combining with $xy = k^2$,

$$\frac{y}{y'} = \frac{x'}{x}.$$

Substituting this value in the equation to the chord, we get

$$\frac{x}{x'} + \frac{x'}{x} = 2c \quad \cdot \quad x^2 - 2cx'x = - x'^2$$

as the quadratic determining the required abscissas. Hence,

$$(P') \quad x = x'(c + \sqrt{c^2-1}), \quad (P) \quad x = x'(c - \sqrt{c^2-1}).$$

Now, ϕ being the angle between the asymptotes, and θ the inclination of $Q'Q$, the distance from Q to any other point on $Q'Q$ is determined (Arts. 101, Cor. 2; 102) by the formula

$$l = \frac{x - x_1}{h} = \frac{(x - x_1)\sin\phi}{\sin(\phi-\theta)}$$

Now the x_1 of $Q = 0$; and (Art. 553, Cor.) $\sin\phi : \sin(\phi-\theta) = b' : x'$, where x' is the abscissa of D, and b' the semi-diameter conjugate to D. Hence,

$$P'Q = b'(c + \sqrt{c^2-1}), \qquad QP = b'(c - \sqrt{c^2-1}).$$

Therefore, $P'Q \cdot QP = b'^2$; and we have

Theorem LXV.—*The rectangle under the segments formed upon parallel chords by either asymptote is constant, and equal to the square on the semi-diameter parallel to the chords.*

V. Area of the Hyperbola.

559. The area of the segment *ALMP*, included between the curve, the asymptote, the ordinate of the vertex, and the ordinate of any given point P, is by general consent called the *area of the hyperbola*. Its value may be determined as follows : *

Let $x' =$ the abscissa CM of the point P to which the area is to be computed. Since the co-ordinates of the vertex are equal to each other, we shall have (from the equation $xy = k^2$) $CL = k$.

---

* See Hymers' *Conic Sections*, p. 121, 3d edition.

It is customary to take the quantity k as the unit in this computation. Adopting this convention, let the distance LM be so subdivided at n points $R, S, \ldots, M$, that the abscissas $CL, CR, CS, \ldots, CM$ may increase by geometric progression. Then, if $CR = x$, we shall have

$$CL = 1, \quad CR = x, \quad CS = x^2, \ldots, CM = x^n.$$

Thus $x' = x^n$; or, $x = x'^{\frac{1}{n}}$: so that, as n increases, x diminishes, and converges toward 1 as n converges to infinity. Now, at $R, S, \ldots, M$, erect n ordinates, and form n corresponding parallelograms $RA, Sa, \ldots, Mo$, situated as in the figure. Then

area $RA = RL \cdot LA \sin \phi = (CR - CL) \, LA \sin \phi = (x-1) \sin \phi,$

" $\quad Sa = SR \cdot Ra \, \sin \phi = (x^2 - x) \dfrac{1}{x} \sin \phi = (x-1) \sin \phi,$

" $\quad Te = TS \cdot Se \, \sin \phi = (x^3 - x^2) \dfrac{1}{x^2} \sin \phi = (x-1) \sin \phi,$

and so on for the whole series of n parallelograms. Hence, replacing x by its value $x'^{\frac{1}{n}}$, and putting $\varSigma =$ the sum of the n parallelograms, we get

$$\varSigma = n \left(x'^{\frac{1}{n}} - 1 \right) \sin \varphi.$$

But an inspection of the diagram shows that the greater the number of the parallelograms, the more nearly does the sum of their areas approach the area $ALMP$. And since x, the ratio of the successive abscissas, tends to 1 as n tends to ∞, we can make the number of the equal parallelograms as great as we please; in other words, the true value of the area $ALMP$ is *the limit toward which $\varSigma$ converges as* n *converges to* ∞. Hence,

area $ALMP = \underset{n = \infty}{n \left(x'^{\frac{1}{n}} - 1 \right) \sin \phi} = \underset{n = \infty}{n \left[\{ 1 + (x' - 1) \}^{\frac{1}{n}} - 1 \right] \sin \phi}$

area $ALMP = \left\{ (x'-1) + \frac{1}{2}\left(\frac{1}{n}-1\right)(x'-1)^2 \right.$

$\left. + \frac{1}{2\cdot 3}\left(\frac{1}{n}-1\right)\left(\frac{1}{n}-2\right)(x'-1)^3 + \ldots \right\} \sin \varphi_{n\,=\,x}$

$= \left\{ \frac{x'-1}{1} - \frac{(x'-1)^2}{2} + \frac{(x'-1)^3}{3} - \frac{(x'-1)^4}{4} + \ldots \right\} \sin \varphi.$

Now [Alg., 373, (5)] the series in the braces denotes the Naperian logarithm of x'. Therefore, calling this logarithm lx', and the hyperbolic area A, we obtain

$$A = \sin \varphi \cdot lx'.$$

But (Alg., 376) $\sin \varphi \cdot lx' =$ the logarithm of x' in a system whose modulus $= \sin \varphi$. Hence,

Theorem LXVI.—*The area of any hyperbolic segment is equal to the logarithm of the abscissa of its extreme point, taken in a system whose modulus is equal to the sine of the angle between the asymptotes.*

Corollary.—In an equilateral hyperbola, since $\varphi = 90°$, $\sin \varphi = 1$; and we get $A = lx'$. That is, *The area of an equilateral hyperbola is equal to the Naperian logarithm of the abscissa of the extreme point.*

For this reason, Naperian logarithms are called *hyperbolic*. But, as we have just seen, the title belongs with equal propriety to logarithms with any modulus.

EXAMPLES ON THE HYPERBOLA.

1. Prove that the middle points of a series of parallels intercepted between an hyperbola and its conjugate, lie on the curve

$$4\left\{ \frac{x^2}{a'^2} - \frac{y^2}{b'^2} \right\} = \frac{b'^2}{y^2}.$$

2 Find the several loci of the centers of the circles inscribed and escribed to the triangle $F'PF$, F' and F being the foci of any hyperbola, and P any point on the curve.

3. An ellipse and a pair of conjugate hyperbolas are described upon the same axes, and, at the points where any line through the center meets the ellipse and one of the hyperbolas, tangents are drawn : find the locus of their intersection.

4. In any triangle inscribed in an equilateral hyperbola, the three perpendiculars from the vertices to the sides, converge in a point upon the curve.

5. To determine the hyperbola which has two given lines for asymptotes, and passes through a given point.

6. Between the sides of a given angle, a right line moves so as to inclose a triangle of constant area: the locus of the center of gravity in the triangle is the hyperbola represented by

$$9\,xy\,\sin\phi = 2k^2,$$

where $\phi =$ the given angle, and $k^2 =$ the constant area.

7. QQ' is a double ordinate to the axis major $A'A$ of an ellipse; QA, $A'Q'$ are produced to meet in P: find the locus of P.

8. To a series of confocal ellipses, tangents are drawn having a constant inclination to the axes: the locus of the points of contact is an hyperbola concentric with the ellipses.

9. The radius of the circle which touches an hyperbola and its asymptotes, is equal to that part of the latus rectum produced which is intercepted between the asymptote and the curve.

10. About the focus of an hyperbola, a circle is described with a radius equal to the conjugate semi-axis, and tangents are drawn to it from any point on the curve : their chord of contact is tangent to the inscribed circle.

11. Tangents to an hyperbola are drawn from any point on either branch of the conjugate curve: their chord of contact touches the opposite branch.

12. In any equilateral hyperbola, let $\phi =$ the inclination of a diameter passing through any point P, and $\phi' =$ that of the polar of P, the transverse axis being the axis of x: then will

$$\tan\phi\,\tan\phi' = 1.$$

13. The circle which passes through the center of an equilateral hyperbola and any two points A and B, passes also through the intersection of two lines drawn the one through A parallel to the polar of B, and the other through B parallel to the polar of A.

14. The locus of a point such that the rectangle under the focal perpendiculars upon its polar with respect to a given ellipse shall be constant, is an ellipse or an hyperbola according as the foci are on the *same* side or on *opposite* sides of the polar.

15. A line is drawn at right angles to the transverse axis of an hyperbola, meeting the curve and its conjugate in P and Q: show that the *normals* at P and Q intersect upon the transverse axis. Also, that the *tangents* at P and Q intersect on the curve

$$\frac{y^4}{4} \left\{ \frac{y^2}{b^2} - \frac{x^2}{a^2} \right\} = \frac{b^4 x^2}{a^2}$$

16. Given two unequal circles: the locus of the center of the circle which touches them both externally, is an hyperbola whose foci are the centers of the given circles.

17. Every chord of an hyperbola bisects the portion of either asymptote included between the tangents at its extremities.

18. If a pair of conjugate diameters of an ellipse be the asymptotes of an hyperbola, to prove that the points of the hyperbola at which its tangents will also touch the ellipse, lie on an ellipse concentric and of the same eccentricity with the given one.

19. A tangent is drawn at a point P of an hyperbola, cutting the asymptote CY in E; from E is drawn any right line EKH cutting one branch of the curve in K, H; and Kk, PM, Hh are drawn parallel to CY cutting the asymptote CX in k, M, h: to prove that

$$Hh + Kk = 2MP.$$

20. Three hyperbolas have parallel asymptotes: show that the three right lines which join two and two the intersections of the hyperbolas, meet in one point.

CHAPTER FIFTH.

THE PARABOLA.

I. The Curve referred to its Axis and Vertex.

560. In discussing the Parabola, we shall find it most convenient to transform its equation, as found in

Art. 181, to a new set of reference-axes. Before doing so, however, we may deduce an important property, which will enable us to give the constant p, involved in that equation, a more significant interpretation.

THE AXIS.

561. Making $y = 0$ in the equation mentioned, namely, in

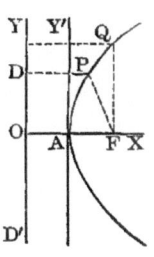

$$y^2 = 4p \, (x - p),$$

we obtain $x = OA = p$. Now (Art. 181) $2p = OF$; hence, $OA = \frac{1}{2} OF$; or, we have

Theorem I.—*In any parabola, the vertex of the curve bisects the distance between the focus and the directrix.*

Corollary.—Whenever, therefore, the constant p presents itself in a parabolic formula, we may interpret it as denoting *the distance from the focus to the vertex of the curve.*

Remark.—We might also infer the theorem of this article directly from the definition of the curve in Art. 179.

562. Since AF is thus equal to p, the distance of the focus from the vertex will converge to 0 whenever $2p = OF$ converges to that limit, but will remain finite as long as $2p$ remains so. In other words (since we may take the focus F as close to the directrix $D'D$ as we please), the focus may approach infinitely near to the vertex, but can not pass beyond it. Hence,

Theorem II.—*The focus of a parabola falls* within *the curve.*

563. Let us now transform the equation $y^2 = 4p \, (x-p)$, by moving the axis of y parallel to itself along OF to the vertex A.   This we accomplish (Art. 55) by putting $x + p$ for x, and thus get

$$y^2 = 4px.$$

564. This equation asserts that the ordinate of the Parabola is a geometric mean of the abscissa and four times the focal distance of the vertex.

It therefore leads directly to the fol-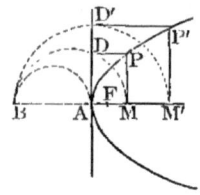
lowing construction of the curve by
points, when the focus and the vertex
are given:—Through the focus F draw
the axis $M'A$, and produce it until
$AB = 4AF$.   Through the vertex A
draw AD' perpendicular to the axis.   At any convenient points on the axis, erect perpendiculars of indefinite lengths, as MP, $M'P'$; and upon the distances BM, BM', etc., as diameters, describe circles BDM, $BD'M'$, etc.   From D, D', . . . , where these circles cut the perpendicular AD', draw parallels to the axis: the points P, P', . . . , in which these meet the perpendiculars MP, $M'P'$, . . . , will be points of the parabola required.   For we shall have $PM = AD = \sqrt{AB.AM} = \sqrt{4AF.AM}$, and a similar relation for $P'M'$ and all other ordinates formed in the same way.

565. The property asserted by the equation $y^2 = 4px$, and involved in the foregoing construction, may be otherwise stated as

Theorem III.—*The square on any ordinate of a parabola is equal to four times the rectangle under the corresponding abscissa and the focal distance of the vertex.*

Corollary.—Since p is constant for any given parabola, y^2 will increase or diminish directly as x does.   That is,

The squares on the ordinates of any parabola vary as the corresponding abscissas.

566. The double ordinate $L'L$ passing through the focus, in the Parabola also, is called the *latus rectum.* Making $x = p$ in the equation to the curve, we obtain the value of FL, namely, $y = 2p$. Hence,

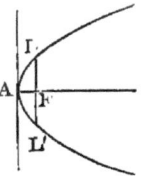

$$\text{latus rectum} = 4p.$$

Corollary.—This result would lead us to state Theorem III as follows: *The square on any ordinate is equal to the rectangle under the corresponding abscissa and the latus rectum.*

567. It is now important to show that a certain relation in form exists between the Parabola and the Ellipse, such that we may consider a parabola as the limiting shape to which an ellipse approaches when we conceive its axis major to increase continually, while its focus and the adjacent vertex remain fixed. By establishing this, we shall be enabled to bring the symbol e, arbitrarily written $= 1$ in Art. 184, under the conception which gives it meaning in the other two conics.

The equation we are now using for the Parabola being referred to the vertex, we must refer the Ellipse to its vertex, if we desire to exhibit the relation mentioned. Transforming, then, the equation of Art. 147 by putting $x - a$ for x, we

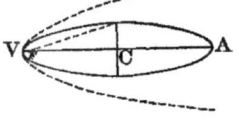

get, as the equation to the Ellipse referred to its vertex V,

$$y^2 = \frac{b^2}{a^2} (2ax - x^2).$$

Putting p as an arbitrary symbol for the distance VF between the vertex and the adjacent focus, we have (Art. 151) $p = a - \sqrt{a^2 - b^2}$: whence

$$\frac{b^2}{a^2} = \frac{2p}{a} - \frac{p^2}{a^2},$$

and the above equation becomes

$$y^2 = 4px - \left\{ \frac{2p^2}{a} x + \left(\frac{2p}{a} - \frac{p^2}{a^2} \right) x^2 \right\}.$$

Suppose, now, that the distance VF remains fixed, while the whole axis major VA increases to infinity: in the limit, where $a = \infty$, we get

$$y^2 = 4px,$$

the equation to the Parabola: which proves our proposition.

Corollary 1.—Since we may thus regard a parabola as an ellipse with an infinitely long axis, that is, with a center infinitely distant from its vertex and its focus, it deserves to be considered whether the Parabola has any element analogous to the so-called *circumscribed circle* of the Ellipse. To settle this point, we only need to consider, that, if a tangent be drawn to a circle, and the radius of the circle be then continually increased without changing the point of contact, the circle will tend more and more nearly to coincidence with its tangent the greater the radius becomes; so that, if we were to suppose the radius *infinitely* great, the circle would become straight by actually coinciding with the tangent. If, then, we draw at the vertex of an ellipse a common tangent to the curve and its circumscribed circle, and subject the axis major to continual increase under the conditions which will cause the curve to assume the form of a parabola in the limit where $a = \infty$, the circumscribed circle will continually approach the common tangent as its center recedes from the vertex, and, in the limit where the ellipse vanishes into a parabola, will coincide with the tangent. We learn, then, that the Parabola *has* an analogue of the circumscribed circle; that this analogue is in fact the tangent of the curve at its vertex; and that it is also the line used as the axis of y in the equation $y^2 = 4px$, since this line and the tangent are both *perpendicular* to the axis of the curve at its vertex, and must therefore coincide. All these results will soon be confirmed by analysis.

Corollary 2.—But, as a point of greater importance, the relation established above enables us to assert that the symbol $e = 1$ denotes the *eccentricity* of the corresponding parabola. For, in the Ellipse, we have

$$e^2 = \frac{a^2 - b^2}{a^2} = 1 - \frac{b^2}{a^2} \;;$$

and if in this we make $a = \infty$, or suppose the curve to become a parabola, we get $e = 1$. That is, we may consider a parabola to be an ellipse in which the eccentricity has reached the limit 1. Moreover, in the view taken of this subject in the corollaries to Arts. 359, 456, the condition $e = 1$ corresponds to that ellipse which has so far deviated from the curvature of its circumscribed circle as to vanish into the right line that forms its axis major. Now, when we recollect (Art. 195, Cor.) that a right line is a particular case of the Parabola, and that, in reaching the limit 1, e has assumed a value fixed and the same for *all* parabolas, it becomes evident that we may consider the condition $e = 1$ as marking that stage of deviation from circularity which characterizes the Parabola, and therefore, in connection with this curve also, appropriately call e the *eccentricity.*

Hence, the name *parabola* (derived from the Greek παραβάλλειν, *to place side by side, to make equal*) may be taken as signifying, that, in the curve which it denotes, the eccentricity is equal to unity.

In closing this article, we would again direct the student's attention to the fact, that all parabolas have the *same* eccentricity.

568. Let $\rho = FP$, the focal distance of any point on a parabola. Then, by the definition of the curve (BD being the directrix), $FP = PD$. Also, from the diagram, $PD = BM = BA + AM$. Now, $AM = x$, and (Art. 561) $BA = p$. Hence,

$$\rho = p + x.$$

That is,

Theorem IV.—*The focal radius of any point on a parabola is a linear function of the corresponding abscissa.*

Remark.—This expression for ρ is similar to those found in the case of the Ellipse and of the Hyperbola (Arts. 360, 457), and is called the *Linear Equation to the Parabola.*

569. The methods of drawing the curve, given in Arts. 179, 564, have already familiarized the reader with the figure of the Parabola, and suggested a tolerably clear conception of its details. Let us now see how the equation

$$y^2 = 4px$$

verifies and completes the impressions made by the diagrams:

I. No point of the curve lies on the *left* of the perpendicular to the axis, drawn through the vertex. For this line is the axis of *y*; and, if we make *x* negative in the equation, the resulting values of *y* are imaginary.

II. But the curve extends to infinity on the *right* of the perpendicular mentioned, both above and below the axis. For *y* is real for every possible positive value of *x*.

III. The curve is symmetric to the axis. For there are *two* values of *y*, numerically equal but opposite in sign, corresponding to every value of *x*.

570. The Parabola, then, differs from the Ellipse, and resembles the Hyperbola, in having infinite continuity of extent. There is, however, a marked distinction between its infinite branch and the infinite branches of the Hyperbola, which calls for a more minute examination.

In the first place, the limiting forms to which the two curves tend are essentially different: that of the Hyperbola (Art. 176) being two *intersecting* right lines, which pass through its center; while that of the Parabola (Arts. 192; 195, Cor.) is two *parallel* right lines, or, in the extreme case, a *single* right line. This, of itself, indicates a difference in the nature of the curvature in these two conics.

Secondly, the branches of the Hyperbola, in receding from the origin, tend to meet the two lines called the asymptotes in two coincident points at infinity (Arts. 539, 543). Now, if we seek the intersections of any right line with a *parabola*, by eliminating between the equations $y = mx + b$ and $y^2 = 4px$, we find that their abscissas are

$$x = \frac{(2p - mb) \pm \sqrt{p - mb}}{m^2}.$$

In order, then, that these intersections may be coincident, we must have $mb = p$: in which case, for the abscissas, we get

$$x = \frac{p}{m^2} \qquad (1),$$

and, for the equation to the given line,

$$y = mx + \frac{p}{m} \; ; \quad \text{or,} \quad m^2 x - my + p = 0 \qquad (2).$$

If, now, the coincidence takes place at infinity, we shall have, from (1), $m = 0$; and (2) will assume the form (Art. 110)

$$C = 0.$$

That is, any right line that tends to meet a parabola in two coincident points at infinity, is situated altogether at infinity; or, what is the same thing, no parabola has any tendency to approach a finite right line in the manner characteristic of the Hyperbola.

DIAMETERS.

571. Equation to any Diameter.—In a system of parallel chords in a parabola, let θ be the common inclination to the axis, xy the middle point of any member of the system, and $x'y'$ the point in which the chord cuts the curve. We have (Art. 101, Cor. 3)

$$x' = x - l \cos \theta, \quad y' = y - l \sin \theta.$$

Hence, as $x'y'$ is on the parabola mentioned,

$$(y - l \sin \theta)^2 = 4p \, (x - l \cos \theta).$$

That is, for determining l, we get the quadratic

$$l^2 \sin^2 \theta - 2 \, (y \sin \theta - 2p \cos \theta) \, l = 4px - y^2.$$

But as xy is the *middle* point of a chord, the co-efficient of l vanishes (Alg., 234, Prop. 3d), and the *locus* of the middle point is therefore represented by the equation

$$y \sin \theta - 2p \cos \theta = 0.$$

Hence, the required equation to any diameter is

$$y = 2p \cot \theta.$$

572. Since p is fixed for any given parabola, and θ for any given system of parallel chords, this equation is of the form

$$y = \text{constant}.$$

Now such an equation (Art. 25) denotes a parallel to the axis of x. Hence,

Theorem V.—*Every diameter of a parabola is a right line parallel to its axis.*

Corollary 1.—From this, we at once infer: *All the diameters of a parabola are parallel to each other.*

Corollary 2.—The constant value of y in the above equation, being dependent on the arbitrary angle θ, is itself arbitrary. The converse of our theorem is therefore true, and we have: *Every right line drawn parallel to the axis of a parabola is a diameter.*

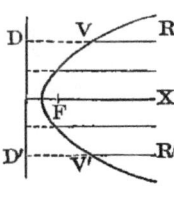

Remark.—The directrix being perpendicular to the axis, we might define a diameter of a parabola as *any right line drawn perpendicular to the directrix.* The diagram illustrates diameters from either point of view.

573. If we write (as we may) the equation $y = constant$ in the form

$$y = 0x + b,$$

An. Ge. 40.

and then eliminate y between it and $y^2 = 4px$, we get, for determining the intersections of a parabola with its diameter, the relation $0x^2 + 0x + b^2 = 4px$; or,

$$0x^2 - 4px + b^2 = 0.$$

Now (Alg., 238) the roots of this quadratic are

$$x = \frac{b^2}{4p}, \qquad x = \infty:$$

whence, observing the form of these roots, we have

Theorem VI.—*Every diameter of a parabola meets the curve in two points, one finite, the other at infinity.*

Remark.—The meaning of this theorem, in ordinary geometric language, is, of course, that a diameter only meets the curve in the one finite point whose abscissa $= b^2 : 4p$. The argument and the phraseology adopted here are only used for the purpose of completing analogies, and will seem less forced when we approach the subject from a more generic point of view.

574. From the last theorem, it follows that no chord of a parabola can be parallel to the axis, and, therefore, that no diameter can bisect a system of chords parallel to a second diameter. We thus learn, that, in the Parabola, the conception of conjugate diameters vanishes in the parallelism of all diameters.

THE TANGENT.

575. Equation to any Chord.—If $x'y'$, $x''y''$ be the extremities of any chord in a parabola, we shall have $y'^2 = 4px'$ and $y''^2 = 4px''$. Hence, $y''^2 - y'^2 = 4p(x'' - x')$, and we get

$$\frac{y'' - y'}{x'' - x'} = \frac{4p}{y' + y''}.$$

Substituting this value for the second member of the equation in Art. 95, we obtain the equation now required, namely,

$$\frac{y - y'}{x - x'} = \frac{4p}{y' + y''}.$$

576. Equation to the Tangent.—Making $y'' = y'$ in the preceding equation, reducing, and recollecting that $y'^2 = 4px'$, we get

$$y'y = 2p\,(x + x').$$

577. Condition that a Right Line shall touch a Parabola.—We have seen, in the second part of Art. 570, that $y = mx + b$ will meet the parabola $y^2 = 4px$ in two coincident points, whenever $mb = p$. The condition now required is therefore

$$b = \frac{p}{m}.$$

Corollary.—Every right line, then, whose equation is of the form

$$y = mx + \frac{p}{m}$$

is a tangent to the parabola $y^2 = 4px$. We have here another instance of the so-called *Magical Equation to the Tangent.*

578. Problem.—*If a tangent to a parabola passes through a fixed point, to find the co-ordinates of contact.*

Putting $x'y'$ for the required point of contact, and $x''y''$ for the fixed point through which the tangent passes, we have (Arts. 563, 576)

$$y'^2 = 4px', \quad y'y'' = 2p\,(x' + x'').$$

Solving these conditions for x' and y', we get

$$x' = \frac{(y''^2 - 2px'') \pm y'' \sqrt{y''^2 - 4px''}}{2p},$$

$$y' = \quad y'' \pm \sqrt{y''^2 - 4px''}.$$

Corollary.—These values indicate that from any given point *two* tangents can be drawn to a parabola: *real* when $y''^2 - 4px'' > 0$, that is, when the point is *without* the curve; *coincident* when $y''^2 - 4px'' = 0$, that is, when the point is *on* the curve; *imaginary* when $y''^2 - 4px'' < 0$, that is, when the point is *within* the curve.

579. **Definition.**—The halves of the chords which any diameter of a parabola bisects, are called the *ordinates* of the diameter. The term is also applied at times to the entire chords, or to the right lines formed by prolonging them indefinitely.

580. Let $x'y'$ be the extremity of any parabolic diameter. Then, from Art. 571, $y' = 2p \cot \theta$; and we get, for determining the angle θ which the ordinates of the diameter make with the axis,

$$\tan \theta = \frac{2p}{y'}.$$

Now the equation to the tangent at $x'y'$ (Art. 576) may be written

$$y = \frac{2p}{y'} (x + x').$$

Hence, by the principle of Art. 78, Cor. 1,

Theorem VII.— *The tangent at the extremity of any diameter of a parabola is parallel to the corresponding ordinates.*

Corollary.—Hence, further, the tangent at the vertex of the curve is perpendicular to the axis; and we confirm by analysis the result of Art. 567, Cor. 1, namely, *The vertical tangent and the axis of y are identical.*

581. The equations to any tangent of a parabola and its focal radius of contact (Arts. 576, 95) may be written

$$y'y = 2p\,(x + x') \qquad (PT),$$

$$y'x + (p - x')\,y = py' \qquad (FP).$$

For the angle *FPT* between these lines, we therefore have (Art. 96, Cor. 1)

$$\tan FPT = \frac{2p\,(p - x') + y'^2}{2py' - y'\,(p - x')} = \frac{2p}{y'}.$$

But, by Art. 580, this is also the value of $\tan QPT$. Hence $FPT = QPT$; and we get

Theorem VIII.—*The tangent of a parabola bisects the internal angle between the diameter and focal radius drawn to the point of contact.*

Corollary 1.—To draw, then, a tangent to a parabola at any point *P*, we form the focal radius *PF*, and the diameter *RP*. We next prolong *RP* until *PQ* equals *PF*, join *QF*, and draw *PT* perpendicular to *QF*: it will be the tangent required, by virtue of the theorem just proved, and the isosceles triangle *FPQ*.

Corollary 2.—Since $SPR = QPT = FPT$, all rays that strike the concave of the curve in lines parallel to the axis will be reflected to *F*, which is therefore called the *focus*, as in the Ellipse and the Hyperbola.

582. Making $y = 0$ in the equation $y'y = 2p\,(x + x')$, we obtain, as the value of the intercept formed by the tangent upon the axis of a parabola,

$$x = AT = -x'.$$

The *length* of the intercept is therefore equal to that of the abscissa of contact, the sign *minus* denoting that it is measured to the left of the vertex. If, then, we add $AF = p$ to this length, we get

$$FT = p + x'.$$

But (Art. 568) $p + x' = FP$. Hence, $FT = FP$; and we have

Theorem IX.—*In any parabola, the foot of the tangent and the point of contact are equally distant from the focus.*

Corollary.—This property obviously leads to the following constructions:

I. *To draw a tangent at any point* P *of a parabola.* Join the given point P with the focus F, and from the latter as a center, with a radius equal to PF, describe an arc cutting the axis in T: the required tangent may then be formed by joining PT.

II. *To draw a tangent to a parabola from any point* T *on the axis.* From the focus F as a center, with the radius FT, describe an arc, and note the point P in which this cuts the curve: P will be the point of contact, and the corresponding tangent may be formed by joining TP.

583. The Subtangent.—For the length of the *subtangent of the curve* in the Parabola, or of that portion of the axis which is included between the foot of the tangent and that of the ordinate of contact, we have $TM = TA + AM$; or,

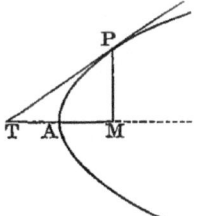

$$\text{subtan} = x' + x' = 2x'.$$

584. Thus $TA = AM = \frac{1}{2} TM$, and we get the important property,

Theorem X.—*The subtangent of a parabola is bisected in the vertex.*

Corollary 1.—We can now construct the tangent at any point of the curve or from any point on the axis, as follows:—When the point of contact P is given, draw the ordinate PM, and on the prolonged axis lay off $AT = AM$: then, by the theorem just proved, T will be the foot of the tangent at P, which is found by joining PT.

When the foot T of the tangent is given, lay off upon the axis $AM = AT$, and erect the ordinate MP. The point P in which this meets the curve, will be the point of contact sought; and the tangent is obtained by joining TP.

Corollary 2.—As the value of the subtangent is dependent upon the peculiar form of the equation to the Parabola, the present theorem is peculiar to this curve, and hence leads to the following mode of constructing it, which is often used by mechanics and draughtsmen.

Lay down two equal right lines AB, AC making any convenient angle with each other. Bisect them in E and F, join EF, BC, and draw AX perpendicular to the latter: it will bisect EF in V, and BC in X, by the well-known properties of the isosceles triangle.

Now divide AE and its equal EB into the same number of equal parts, and the equals AF, FC in the same manner: the whole lines AB, AC will thus be subdivided into equal parts at the points 1, 2, 3, E, 4, 5, 6 and 6, 5, 4, F, 3, 2, 1. Having numbered these points in reverse order upon the two lines, as in the diagram, join those which have the same numeral: the resulting lines will envelope a parabola, which we can approximate as closely as we please, by continually diminishing the 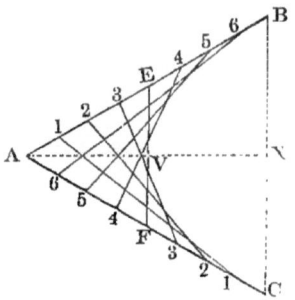 distances $A\,1$, $1 \ldots 2$, etc. For, by the construction, V is the middle point of AX; and the curve touches all the lines AB, AC, $1 \ldots 1$, $6 \ldots 6$, etc.: hence, with respect to the lines AB, AC, which may be regarded as limiting cases of all the others, it is a curve that bisects its subtangent in the vertex; that is, a parabola.

From another point of view, the curve here formed is the envelope of a line EF, which moves within the fixed lines AB, AC in such a manner that the sum of the remaining sides of the triangle EAF is constant, being equal to AB. It is therefore a parabola, by the result of the Example solved in Art. 251.

585. Perpendicular from the Focus to any Tangent.—For the length of the perpendicular from the focus $(p, 0)$ upon the line $y'y = 2p\,(x + x')$, we have (Art. 105, Cor. 2)

$$P = \frac{2p\,(p + x')}{\sqrt{(4p^2 + y'^2)}} = \frac{2p\,(p + x')}{\sqrt{\{4p\,(p + x')\}}} = \sqrt{p\,(p + x')}.$$

But (Art. 568) $p + x' = \rho$, the focal distance of the point of contact. Hence,

$$P^2 = p\rho.$$

586. In this expression, p being constant, it is evident that P^2 will change its value as ρ changes its value; or, P will change with the square root of ρ: a property usually expressed by

Theorem XI.—*The focal perpendicular upon the tangent of a parabola varies in the subduplicate ratio of the focal radius of contact.*

587. Focal Perpendicular in terms of its inclination to the Axis.—The perpendicular from $(p, 0)$ upon the line whose equation (Art. 577, Cor.) is

$$m^2x - my + p = 0,$$

will be, according to Art. 105, Cor. 2,

$$P = \frac{m^2p + p}{\sqrt{(m^4 + m^2)}} = \frac{p}{m}\sqrt{1 + m^2}.$$

Put $\theta =$ the inclination of P, measured from the axis toward the *right*: then $m = \cot\theta$, and we get

$$P = p \sec \theta.$$

588. The equation to the tangent being written (Art. 577, Cor.)

$$my - m^2x = p,$$

that of its focal perpendicular, which passes through $(p, 0)$, will be

$$my + x = p.$$

Combining these so as to eliminate m, we get

$$x = 0$$

as the equation to the locus of the point in which the focal perpendicular meets the tangent. Hence, (Art. 580, Cor.,)

Theorem XII.—*The locus of the foot of the focal perpendicular upon any tangent of a parabola, is the tangent at the vertex of the curve.*

Corollary 1.—By means of this property, we can solve in its most general form the problem,

An. Ge. 41.

To draw a tangent to a parabola through any given point. — Let P be the given point, and join it with the focus F. On PF as a diameter, describe a circle cutting the vertical tangent in Q and Q': these points will be the feet of focal perpendiculars, as the angles PQF, $PQ'F$ are inscribed in semicircles. Hence, a line joining P to either of them, for instance the line PQ, will touch the parabola in some point T.

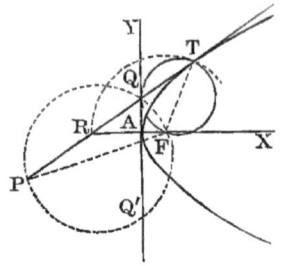

If the point of contact is required, produce the axis to meet PQ in R, and then apply the method of Art. 582, Cor. When the given point is on the curve, as at T, the auxiliary circle will *touch* the vertical tangent; but the point Q can still be found, by dropping a perpendicular from the middle point of FT upon AY.

Corollary 2.—The present theorem, and the resulting construction, completely justify the view, taken in Art. 567, Cor. 1, that the vertical tangent of the Parabola is the analogue of the circle circumscribed about the Ellipse. We thus arrive at the conclusion, often serviceable in analysis, that the Right Line may be defined as *the circle whose radius is infinite.*

589. Since the vertical tangent is the *locus* of the foot of the focal perpendicular on any other tangent, that is, the line in which the foot of *every* such perpendicular is found, it follows that every right line drawn from the focus to the vertical tangent is a focal perpendicular to some other tangent of the curve. Besides, it is obvious that *any* given point and right line may be regarded as the focus and vertical tangent of some parabola. Hence,

Theorem XIII.—*If from any point a line be drawn to a fixed right line, and a perpendicular to it be formed at the intersection, the perpendicular will be tangent to the parabola of which the point and fixed line are the focus and vertical tangent.*

Corollary.—Thus we see that a parabola is the *envelope* of such a perpendicular, and we may therefore approximate the outline of one, by drawing oblique lines to a given right line from a fixed point P, and erecting perpendiculars at their extremities. If formed close enough together, these perpendiculars will define the curve with considerable distinctness, as the diagram shows.

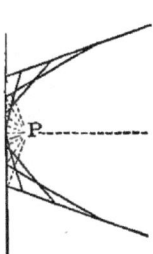

590. Let us now ascend from the theorem of Art. 588 to the general one, in which a line from the focus meets the tangent at *any* fixed angle.

Calling this angle θ, and writing the equation to the tangent (Art. 577, Cor.)

$$y = mx + \frac{p}{m} \qquad (1),$$

we obtain, for the equation to the intersecting line from the focus (Art. 103),

$$y = \frac{m + \tan \theta}{1 - m \tan \theta} (x - p) \qquad (2).$$

From (2), by clearing of fractions, expanding, and collecting terms,

$$y - mx = (1 + m^2) x \tan \theta - mp.$$

Subtracting this result from (1) member by member, and then transposing,

$$(1 + m^2) x \tan \theta = \frac{p}{m} + mp.$$

Dividing through by $(1 + m^2)$,

$$x \tan \theta = \frac{p}{m} \quad \therefore \quad m = \frac{p}{x \tan \theta} \cdot$$

Substituting this value of m in (1), we get the equation to the locus of the intersection of the tangent and the focal line, namely,

$$y = x \tan \theta + p \cot \theta.$$

But (Art. 577, Cor.) this denotes a tangent of the same parabola, inclined to the axis at the angle θ. Hence,

Theorem XIV.—*The locus of the intersection of a tangent with the focal line which meets it at a fixed angle, is the tangent which meets the axis at the same angle.*

591. **Angle between any two Tangents.**—Let $x'y'$, $x''y''$ denote the two points of contact Q, Q'. The equations to the corresponding tangents (Art. 576) will then be

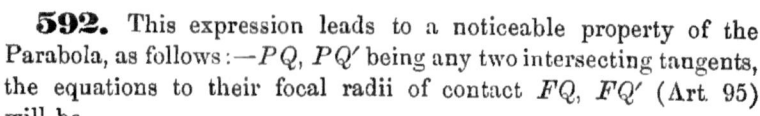

$$y'y = 2p(x + x'), \quad y''y = 2p(x + x'').$$

Hence, applying the formula of Art. 96, Cor. 1, we get

$$\tan QPQ' = \frac{2p\,(y'' - y')}{y'y'' + 4p^2}.$$

592. This expression leads to a noticeable property of the Parabola, as follows:—PQ, PQ' being any two intersecting tangents, the equations to their focal radii of contact FQ, FQ' (Art. 95) will be

$$\frac{y}{x - p} = \frac{y'}{x' - p}, \quad \frac{y}{x - p} = \frac{y''}{x'' - p}.$$

Substituting for x' and x'' their values from the equation to the Parabola, and clearing of fractions, we may write these equations,

$$(y'^2 - 4p^2)\,y = 4py'\,(x - p) \qquad (FQ),$$
$$(y''^2 - 4p^2)\,y = 4py''\,(x - p) \qquad (FQ').$$

From them, by Art. 96, Cor. 1, we get

$$\tan QFQ' = \frac{4py'\,(y''^2 - 4p^2) - 4py''\,(y'^2 - 4p^2)}{16p^2y'y'' + (y'^2 - 4p^2)\,(y''^2 - 4p^2)},$$

$$= \frac{4p\,(y'' - y')\,(y'y'' + 4p^2)}{(y'y'' + 4p^2)^2 - 4p^2\,(y'' - y')^2}.$$

Now, if we apply the formula for the tangent of a double angle (Trig., 847, III) to the angle QPQ', whose tangent we found in the preceding article, we shall get, for the tangent of $2QPQ'$, the expression just obtained. Hence $QFQ' = 2QPQ'$; and we have

Theorem XV.—*The angle between any two tangents of a parabola is equal to half the focal angle subtended by their chord of contact.*

593. The equations to any two tangents of a parabola that cut each other at right angles (Arts. 577, Cor.; 96, Cor. 3) will be

$$y = mx + \frac{p}{m}, \quad y = -\frac{x}{m} - mp.$$

Subtracting the second of these from the first, we obtain

$$x = -p$$

as the equation to the locus of the intersection. But this equation denotes a right line perpendicular to the axis at the distance p on the left of the vertex; in other words, the line called the *directrix* in our primary definitions. Hence,

Theorem XVI.—*The locus of the intersection of tangents which cut each other at right angles, is the directrix of the curve.*

Corollary.—Since this locus, in the case of the Ellipse, is a circle concentric with the curve (Art. 395), this theorem again shows us that the Circle converges to the form of the Right Line as its radius tends to infinity, and that we may therefore correctly regard the Right Line as a circle with an infinite radius.

THE NORMAL.

594. Equation to the Normal.—The equation to the perpendicular drawn through the point of contact $x'y'$ to the tangent

$$y'y = 2p(x + x'),$$

is found by Art. 103, Cor. 2, and is therefore

$$2p(y - y') = y'(x' - x).$$

595. Let P be any point $x'y'$ on a parabola, PN the corresponding normal, DP a diameter through P, and FP the focal radius of its vertex. The equation to FP (Art. 95) is

$$y'x - (x' - p)\, y = py'.$$

Comparing this with the equation to the normal, we get (Art. 96, Cor. 1)

$$\tan FPN = \frac{2py' + y'\,(x' - p)}{y'^2 - 2p\,(x' - p)} = \frac{y'}{2p}.$$

But, since every diameter is parallel to the axis, $DPN = PNX$, and we have, from the equation to the normal,

$$\tan DPN = -\frac{y'}{2p}.$$

Hence, $FPN = 180° - DPN = QPN$; and we obtain

Theorem XVII.—*The normal of a parabola bisects the external angle between the corresponding diameter and focal radius.*

Corollary.—To construct a normal at any point P of the curve, we therefore draw the focal radius PF, and the diameter DPQ, laying off upon the latter $PQ = PF$, and joining QF: then will PN, drawn perpendicular to QF, bisect the angle FPQ, and for that reason be the normal required.

596. Intercept of the Normal.—Making $y = 0$ in the equation of Art. 594, we obtain

$$x = AN = 2p + x'.$$

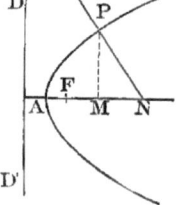

Corollary.—This result enables us to construct a normal at any point of the curve, or from any point on the axis.

For, if the point P is given, we draw the corresponding ordinate PM, and lay off MN to the right of its foot, equal to $2AF$: we thus find N, the foot of the required normal. When N is given, we lay off $NM = 2AF$ to its left, erect the ordinate MP, and join NP.

597. Since $FN = AN - AF = (2p + x') - p = p + x'$, we have (Arts. 568, 582)

Theorem XVIII.—*The foot of the normal is at the same distance from the focus as the foot and the point of contact of the corresponding tangent.*

Corollary.—This is the same as saying that the three points mentioned are on the same circle, described from the focus as center. Hence, to construct either the tangent or the normal, or both, pass a circle from the center F through either of the three points P, T, N, as one or another is given, and join the points in which it cuts the axis with the point in which it cuts the parabola.

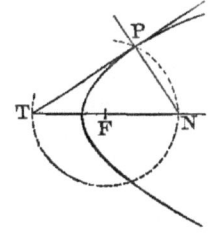

598. Length of the Subnormal.—For this (see diagram, Art. 596), we have $MN = AN - AM$. That is (Art. 596),

$$\text{subnor} = 2p.$$

In other words (Art. 561, Cor.), we have obtained

Theorem XIX.—*The subnormal of a parabola is constant, and equal to twice the distance from the focus to the vertex.*

599. Length of the Normal.—For the distance between $x'y'$ and $(2p + x', 0)$, we have (Art. 51, I, Cor. 1)

$$PN^2 = 4p^2 + y'^2 = 4p (p + x').$$

Now (Art. 568) $p + x' = \rho$. Hence,

$$PN^2 = 4p\rho.$$

But (Art. 585) $p\rho =$ the square of the focal perpendicular on the tangent at P. Therefore,

Theorem XX.—*The normal of a parabola is double the focal perpendicular on the corresponding tangent.*

II. The Curve in terms of any Diameter.

600. The equation which we have thus far employed is only a special form of a more general one, and many of the properties proved by means of it are but particular cases of generic theorems which relate, not to the axis, but to any diameter whatever. The truth of this will appear as soon as we transform $y^2 = 4px$, which is referred to the axis AX and the vertical tangent AY, to any diameter $A'X'$ and *its* vertical tangent $A'Y'$. This transformation we can effect by means of the formulæ in Art. 56, Cor. 1, observing that the 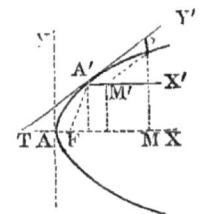 new axis of x is parallel to the primitive, and the a of the formulæ therefore equal to zero. If in addition we call the angle $Y'TX$ not β but θ, and put $x'y'$ to denote the new origin A', these formulæ of transformation will become (Art. 58)

$$x = x' + x + y \cos \theta,$$
$$y = y' + y \sin \theta.$$

601. Equation to the Parabola, referred to any Diameter and its Vertical Tangent.—Replacing the y and x of $y^2 = 4px$ by their values as given in the preceding formulæ, and collecting the terms, we get

$$y^2 \sin^2 \theta + 2 (y' \sin \theta - 2p \cos \theta) y + y'^2 - 4px' = 4px.$$

But, as the new origin $x'y'$ is on the curve, $y'^2 - 4px' = 0$. Also, since the new axis of y is a tangent, $\tan \theta = 2p : y'$; so that $y' \sin \theta - 2p \cos \theta = 0$. Hence, the transformed equation is in reality

$$y^2 \sin^2 \theta = 4px \qquad (1).$$

Putting $p : \sin^2 \theta = p'$, we may write the equation

$$y^2 = 4p'x \qquad (2).$$

This at once shows that $y^2 = 4px$ is the form which (1) assumes when the diameter chosen for the axis of x is that whose vertical tangent is perpendicular to it, so that $\sin^2 \theta = 1$.

602. Before employing our new equation itself, we may extend the property of Theorem I by means of relations derived from $y^2 = 4px$: this will better prepare the way for the use of $y^2 = 4p'x$.

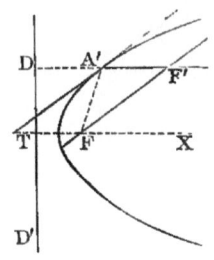

Let DA' be any diameter, and $A'T$ its vertical tangent. Parallel to the latter, draw FF' through the focus. Then, in the parallelogram $F'T$, we shall have $A'F' = FT =$ (Art. 582) FA'. But, from the definition of the curve, supposing $D'D$ to be the directrix, $FA' = A'D$. Hence, $A'D = A'F'$; or,

Theorem XXI.—*The vertex of any diameter bisects the distance from the directrix to the point in which the diameter is cut by its focal ordinate.*

603. The factor $p' = p : \sin^2 \theta$ which enters the second member of equation (2), Art. 601, may be expressed in terms of p and x', by the following process:

According to Art. 576, $\tan \theta = 2p : y'$. Hence,

$$\sin \theta = \frac{2p}{\sqrt{(4p^2 + y'^2)}} = \frac{\sqrt{p}}{\sqrt{(p + x')}}.$$

We have, then,

$$\frac{p}{\sin^2 \theta} = p + x'.$$

Now (Art. 568) $p + x' = FA'$, the focal distance of the vertex of any diameter. Hence,

Theorem XXII.—*The focal distance of the vertex of any diameter is equal to the focal distance of the principal vertex, divided by the square of the sine of the angle between the diameter and its vertical tangent.*

604. The expression just obtained aids us to interpret the new equation $y^2 = 4p'x$. For, from what precedes, the equation may be written

$$y^2 = 4 (p + x') x,$$

and we learn that the symbol p' signifies the focal distance of the vertex taken for the new origin. We therefore read from the equation at once, the following extension of Theorem III:

Theorem XXIII.—*The square on an ordinate to any diameter is equal to four times the rectangle under the corresponding abscissa and the focal distance of the vertex.*

Corollary.—Hence, the squares on the ordinates to *any* diameter vary as the corresponding abscissas.

605. In Art. 602, we found $A'F' = FA' = p + x'$.
Hence, making $x = p + x'$ in the equation of Art. 604, we get

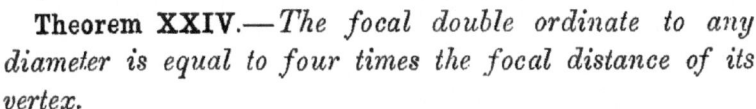

$$y = F'P = 2(p + x') = 2FA'.$$

Now, since every diameter bisects the chords parallel to its vertical tangent, $PQ = 2F'P = 4FA'$; and we have

Theorem XXIV.—*The focal double ordinate to any diameter is equal to four times the focal distance of its vertex.*

Remark.—The value of the *latus rectum* (Art. 566), furnishes a particular case of this theorem; and, as $4p$ signifies the length of the focal double ordinate to the axis, so $4p'$, by what precedes, represents that of the focal double ordinate to *any* diameter. From Arts. 429, 522, we see that this uniform analogy among the focal double ordinates to *all* diameters, is peculiar to the Parabola.

606. The reader may now interpret the equation

$$y = \pm\, 2\sqrt{p'x},$$

and show by means of it, that, with reference to *any* diameter, the Parabola consists of a single infinite branch, tending to two parallel right lines as its limiting form, and symmetric to the diameter.

DIAMETRAL PROPERTIES OF THE TANGENT.

607. Equation to the Tangent, referred to any Diameter.—From the identity of form in the equations $y^2 = 4px$, $y^2 = 4p'x$, we at once infer that this must be

$$y'y = 2p' (x + x').$$

608. Making $y = 0$ in this equation, we get, for the intercept of the tangent on *any* diameter,

$$x = -x'.$$

This shows that the tangent cuts any diameter on the left of its vertex, at a distance equal to the abscissa of contact. Thus the vertex of any diameter is situated midway between the foot of the tangent and that of its ordinate of contact, and we have, as the extension of Theorem X,

Theorem XXV.—*The subtangent to any diameter of a parabola is bisected in the corresponding vertex.*

Corollary 1.—This property enables us *to construct a tangent to a parabola from any external point whatever.* For, if T' be such point, we have only to draw the diameter $T'M'$, form the tangent $A'T$ at its vertex (by dropping a perpendicular from A' upon the axis, and setting off $AT =$ the abscissa thus determined), take $A'M' = A'T'$, draw $M'P$ parallel to the tangent $A'T$, and join PT'.

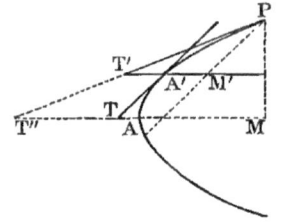

Corollary 2.—By the same property, we can *construct an ordinate to any diameter.* This is either done in the way $M'P$ was formed above, or as follows:—Take any point T'' on the axis, make $AM = AT''$, erect the perpendicular MP, and join PT''. Then, T' being the point where this tangent cuts the given diameter, take $A'M' = A'T'$, and join $M'P$.

609. Let PQ be any chord of a parabola. Then (Art. 607) the equations to the tangents at its opposite extremities P and Q, by referring them to its bisecting diameter, will be

$$2p'(x + x') - y'y = 0, \quad 2p'(x + x') + y'y = 0.$$

Subtracting the first of these from the second, we get

$$y = 0$$

as the equation to the locus of the intersection. Hence,

Theorem XXVI.—*Tangents at the extremities of any chord of a parabola meet on the diameter which bisects that chord.*

POLE AND POLAR.

610. We shall now prove that the polar relation is a property of the Parabola, following the same steps as in the Ellipse and the Hyperbola.

611. Chord of Contact in the Parabola.—Let $x'y'$ denote the point from which the two tangents that determine the chord are drawn, and x_1y_1, x_2y_2 the extremities of the chord. Since $x'y'$ is upon both tangents, we have

$$y_1y' = 2p'(x' + x_1), \quad y_2y' = 2p'(x' + x_2).$$

That is, the two extremities of the chord are on the line whose equation is

$$y'y = 2p'(x + x').$$

And this is therefore the equation to the chord itself.

612. Locus of the Intersection of Tangents to the Parabola.—Let $x'y'$ be the fixed point through which the chord of contact of the intersecting tangents is drawn. Then, if x_1y_1 be the intersection of the two tangents, since $x'y'$ is always on their chord of contact, we shall have (Art. 611)

$$y_1y' = 2p'(x' + x_1).$$

And this being true, however $x_1 y_1$ may change its position as the chord of contact revolves about $x' y'$, the co-ordinates of intersection must always satisfy the equation

$$y' y = 2p' (x + x').$$

This, therefore, is the equation to the locus sought.

613. Tangent and Chord of Contact taken up into the wider conception of the Polar.—Here, too, as well as in the Ellipse and the Hyperbola, the two equations just found are identical in form with that of the tangent. By the same reasoning, then, as in Arts. 433, 526, we learn that the tangent and chord of contact in the Parabola are particular cases of the locus just discussed. Now, too, by its equation, this locus is a right line; and, if we suppose $x' y'$ to be any point on a *given* right line, the co-efficients of the equation in Art. 611 will fulfill the condition $A x' + B y' + C = 0$, and thus (Art. 117) the chord of contact will pass through a fixed point. In the curve now before us, therefore, we have the twofold theorem:

I. *If from a fixed point chords be drawn to any parabola, and tangents to the curve be formed at the extremities of each chord, the intersections of the several pairs of tangents will lie on one right line.*

II. *If from different points lying on one right line pairs of tangents be drawn to any parabola, their several chords of contact will meet in one point.*

Thus the law that renders the locus of Art. 612 the *generic* form of which the tangent and chord of contact are special phases, is the law of *polar reciprocity:* whence the Parabola, in common with the other two Conics, imparts to every point in its plane the power of determining a right line; and reciprocally.

614. Equation to the Polar with respect to a Parabola.—This, as we gather immediately from the preceding results, is

$$y'y = 2p'(x + x'),$$

if referred to *any* diameter; or, if referred to the *axis*,

$$y'y = 2p(x + x').$$

615. Definitions.—The **Polar** of any point, with respect to a parabola, is the right line which forms the locus of the intersection of the two tangents drawn at the extremities of any chord passing through the point.

The **Pole** of any right line, with respect to the same curve, is the point in which all the chords of contact corresponding to different points on the line intersect.

We have, then, the following constructions:—When the pole P is given, draw through it any two chords $T'T$, $S'S$, and form the corresponding pairs of tangents, $T'L$ and TL, $S'M$ and SM: the line LM which joins the intersection of the first pair to that of the second, will be the polar of P. When the polar is given, take any two of its points, as L and M, draw a pair of tangents from each, and form the corresponding chords of contact, $T'T$, $S'S$: the point P in which these intersect, will be the pole of LM.

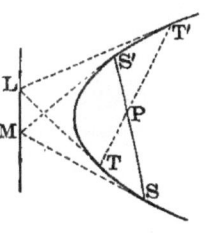

When the pole is *without* the curve, as at M, the polar is the corresponding chord of contact $S'S$; and when it is *on* the curve, as at T_0, the polar is the tangent at T. In these cases, the drawing may be made in accordance with the facts.

616. Direction of the Polar.—By referring the polar of any point to the diameter drawn through the point, the y' of its equation will become $= 0$, and the equation itself (Art. 614) will assume the form

$$x = -x'.$$

This denotes a parallel to the axis of y. Hence,

Theorem XXVII.—*The polar of any point, with respect to a parabola, is parallel to the ordinates of the diameter which passes through the point.*

Corollary 1.—The equation $x = -x'$, more exactly interpreted, gives us: *The polar of any point on a diameter is parallel to the ordinates of the diameter, and its distance from the vertex of the diameter is equal, in an opposite direction, to the distance of the point.* And, in particular, *The polar of any point on the axis is the perpendicular which cuts the axis at the same distance from the vertex as the point itself, but on the opposite side.*

Corollary 2.—To construct the polar, therefore, draw a diameter through the pole, take on the opposite side of its vertex a point equidistant with the pole, and draw through this a parallel to the corresponding ordinates.

617. Polar of the Focus.—The equation to this is obtained by putting $(p, 0)$ for $x'y'$ in the second equation of Art. 614, and is

$$x = -p.$$

The focal polar of the Parabola is therefore identical with the line which in Art. 180 we named the *directrix*, and we shall presently see that our ability to generate the curve by the means employed in Art. 179, is due to the polar relation of that line to the focus.

618. Let $D'D$ represent the polar of the focus F. Then, obviously, $PD = RA + AM$; or, the distance of any point on the curve from the polar is equal to the distance of the polar from the vertex, increased by the abscissa of the point. That is,

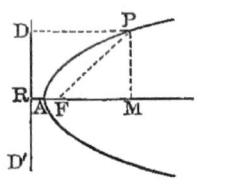

$$PD = p + x.$$

Now (Art. 568) $p + x = FP$: whence, $FP = PD$; or, since $e = 1$ in the Parabola,

$$\frac{FP}{PD} = e.$$

Here, then, the property of Arts. 439, 532 again appears, and we have

Theorem XXVIII.—*The distance of any point on a parabola from the focus is in a constant ratio to its distance from the polar of the focus, the ratio being equal to the eccentricity of the curve.*

Remark 1.—We thus complete the circuit of our analysis, and, as stated above, return upon the property from which we set forth in Art. 179. The significance of our present result consists in the fact, that *we have translated the apparently arbitrary definition of* Art. 179 *into the generic law of polarity.* And we may say that we have vindicated our method of generating the curve; because we can now see that it is the mechanical expression of the power to determine a conic, which the Point and the Right Line together possess: a power reciprocally involved, of course, in that of the Conic to bring these two Forms into the polar relation.

An. Ge. 42.

464   ANALYTIC GEOMETRY.

It may deserve mention, that the construction of Art.
179, like those of the corollaries to Arts. 439, 532, em-
ploys the parts of a right triangle in order to embody
the constant ratio of the focal and polar distances, the
constant in the case of the Parabola being the ratio of
the base to itself.

Remark 2.—The name *parabola* thus acquires a new
meaning. We may henceforth regard it as signifying
*the conic in which the constant ratio between the focal and
polar distances* equals *unity.*

619. Focal Angle subtended by any Tangent.—By an
analysis similar to that of Art. 440, x being
the abscissa of the point P from which the
tangent is drawn, ρ its radius vector FP,
and ϕ the angle PFT, we can show that

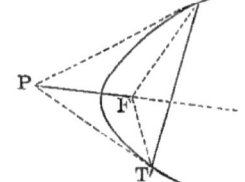

$$\cos \phi = \frac{p + x}{\rho} .$$

620. This expression, like those of Arts. 440, 533, is independent of the point of contact. Hence, the angle $PFT =$ the
angle PFT'; and, with respect to the whole angle TFT', we get

Theorem XXIX.—*The right line that joins the focus to the pole of
any chord, bisects the focal angle which the chord subtends.*

Corollary.—In particular, *The line that joins the focus to the pole
of any focal chord is perpendicular to the chord.*

Remark.—By comparing this corollary with the theorem of Art.
593, and bearing in mind that the directrix, as the polar of the
focus, is the line in which the pair of tangents drawn at the extremities of *any* focal chord will intersect, we may state the following noticeable group of related properties:
*If tangents be drawn at the extremities of any focal chord of a
parabola,*
1. *The tangents will intersect on the directrix.*
2. *The tangents will meet each other at right angles.*
3. *The line that joins their intersection to the focus will be perpendicular to the focal chord.*

621. We shall in this article solve two examples, which will show the beginner how to take advantage of such results as we have lately obtained.

I. Given two points P, Q, and their polars $T'T$, $S'S$: to determine the relation between the intercept mn, cut off on the axis by the polars, and the intercept MN, cut off by perpendiculars from the points.

Let $x'y'$, $x''y''$ denote P and Q. Then

$$MN = x'' - x'.$$

But the equations to the polars $T'T$, $S'S$ are

$$y'y = 2p(x + x'), \qquad y''y = 2p(x + x'');$$

and if we make $y = 0$ in these, and take the difference of the results,

$$mn = x'' - x'.$$

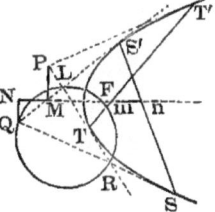

Hence, *The intercept on the axis between any two polars is equal to that between the perpendiculars from their poles.*

II. To prove that the circle which circumscribes the triangle formed by any three tangents, passes through the focus.

Let L, Q, R be the intersections of the tangents, and F the focus. By Art. 592, the angle LQR is half the focal angle subtended by $S'S$. Also, by Art. 620, the angle $LFR = LFT + TFR = 180°$ *minus* half this same focal angle. Hence, $LQR + LFR = 180°$, and the quadrilateral $LQRF$ is an *inscribed* quadrilateral. That is, F is on the circle which circumscribes the triangle LQR.

PARAMETERS.

622. Definition.—The **Parameter** of a parabola, with respect to any diameter, is a third proportional to any abscissa formed on the diameter, and the corresponding ordinate. Thus,

$$\text{parameter} = \frac{y'^2}{x'}.$$

623. From the equation to the Parabola, $y'^2 : x' = 4p'$. Also, from Art. 604, $p' =$ the distance of the vertex of a diameter from the focus. Hence,

Theorem XXX.—*The parameter of any diameter in a parabola is equal to four times the focal distance of its vertex.*

Corollary.—The parameter of the axis, or, as it is usually called, the *principal parameter*, is therefore equal to four times the distance from the focus to the vertex of the curve; that is, its value is $4p$.

Remark.—In the equations $y^2 = 4p'x$, $y^2 = 4px$, we are henceforward to understand that p' is one-fourth the parameter of the reference-diameter, and p one-fourth the parameter of the axis. Or, with greater generality, in any equation of the form

$$y^2 = Px,$$

P, the constant co-efficient of x, is to be interpreted as the parameter of the corresponding parabola, taken with respect to the diameter to which the equation is referred.

624. In Art. 605, we proved that the *focal double ordinate* to any diameter is equal to four times the focal distance of its vertex. Hence,

Theorem XXXI.—*The parameter of any diameter in a parabola is equal to the focal double ordinate of that diameter.*

Corollary.—Accordingly, the parameter of the axis is equal to the *latus rectum:* as we might also infer from the fact that both are equal to $4p$.

Remark.—From Arts. 429, 522, as already noticed in another connection, we see that the Parabola is the only conic in which the parameter of every diameter is equal to the corresponding focal double ordinate.

625. It is sometimes useful to express the parameter of any diameter in terms that refer to the axis of the curve as the axis of x. To effect this, we either use the relation

$$p' = p + x' \qquad (1),$$

where x' is the abscissa of the vertex of the diameter whose parameter is sought, measured on the axis; or else

$$p' = \frac{p}{\sin^2 \theta} \qquad (2),$$

where θ is the angle made with the axis by the tangent at the vertex of the diameter.

626. The latter of the foregoing relations may be interpreted as

Theorem XXXII.—*The parameter of any diameter varies inversely as the square of the sine of the angle which the corresponding vertical tangent makes with the axis.*

III. The Curve referred to its Focus.

627. The polar equations to the Parabola (Art. 183 cf. Rem.) being

$$\rho = \frac{2p}{1 \pm \cos \theta},$$

our present knowledge leads us to assign to p its proper meaning, and to describe the numerator in the value of ρ as *half the parameter of the curve.*

Moreover, since $e = 1$ in the Parabola, we may write

$$\rho = \frac{2p}{1 \pm e \cos \theta},$$

thus completely exhibiting the analogy of these expressions to the corresponding elliptic equations of Art. 443: an analogy which we partially established in Art. 184, and which verifies the proposition of Art. 567, that a parabola may be regarded as an ellipse in which the eccentricity has passed to the limiting value $= 1$.

628. Polar Equation to the Tangent.—In seeking this, we shall avail ourselves of the property just mentioned.

If in the equation of Art. 444 we replace $a (1 - e^2)$ by its value $2p$, as found in Art. 443, we may write the polar equation to an *elliptic* tangent

$$\rho = \frac{2p}{\cos (\theta - \theta') - e \cos \theta}.$$

Making $e = 1$ in this, we get the equation to a *parabolic* tangent, namely,

$$\rho = \frac{2p}{\cos (\theta - \theta') - \cos \theta}.$$

IV. AREA OF THE PARABOLA.

629. The area of any parabolic segment, included between the curve and any double ordinate to the axis, may be computed as follows:

Supposing A-PMQ to be the segment whose area is sought, divide its abscissa AM into any number of equal parts at B, C, D, $\dots$, erect ordinates BL, CN, DR, $\dots$ at the points of division, and through their extremities L, N, R, $\dots$ draw parallels to the axis, producing both the ordinates and the parallels until they meet as in the figure. The curve divides the circumscribed rectangle *UM* into two segments: and, by the process just described, there will be formed in both of these a number of smaller rectangles, corresponding two and two; as *UR* to *RM*, *ON* to *ND*, *EL* to *LC*, etc.

Let $x'y'$, $x''y''$ be any two successive points thus formed upon the curve; for instance, *L*, *N*. Then

$$\text{area } LC = y' \, (x'' - x') = \frac{y' \, (y''^2 - y'^2)}{4p},$$

$$\text{area } EL = x' \, (y'' - y') = \frac{y'^2 \, (y'' - y')}{4p}.$$

To express, then, the ratio of *any* interior rectangle to its corresponding exterior one, we shall have an equation of the form

$$\frac{LC}{EL} = \frac{y'' + y'}{y'} = 1 + \frac{y''}{y'}.$$

Hence, the limiting value to which this ratio tends as y'' converges to y', is evidently $= 2$: and therefore the ratio borne by the *sum* of the interior rectangles to the *sum* of the exterior, also tends to the limit 2 as y'' converges to y'. Now the condition that y'' may converge to y' is, that the subdivision of *AM* shall be continued *ad infinitum:* and if this takes place, the sum of the *interior* rectangles will converge to the area of the interior segment *APM*; and the sum of the *exterior*, to the area of the exterior segment *APU*. Therefore, $APM = 2APU$; or, putting $x = AM$, $y = MP$, and $A =$ the area of the interior segment,

$$A = \frac{2}{3} \, xy.$$

Similarly, the segment $AQM = \dfrac{2}{3} xy$: whence,

Theorem XXXIII.—*The area of a parabolic segment cut off by any double ordinate to the axis, is equal to two-thirds of the circumscribing rectangle.*

Corollary.—Since the same reasoning is obviously applicable to the equation $y^2 = 4p'x$, we may at once state the generic theorem : *The area of a parabolic segment cut off by a double ordinate to any diameter, is equal to two-thirds of the circumscribing parallelogram.*

EXAMPLES ON THE PARABOLA.

1. The extremities of any chord of a parabola being $x'y'$, $x''y''$, and the abscissa of its intersection with the axis being x, to prove that

$$x'x'' = x^2, \quad y'y'' = -4px.$$

2. Two tangents of a parabola meet the curve in $x'y'$ and $x''y''$: their point of intersection being xy, show that

$$x = \sqrt{x'x''}, \quad y = \frac{y' + y''}{2}.$$

3. The area of the triangle formed by three tangents of a parabola is half that of the triangle formed by joining their points of contact.

4. To prove that the area of the triangle included between the tangents to the parabolas

$$y^2 = mx, \quad y^2 = nx$$

at points whose common abscissa $= a$, and the portion of the corresponding ordinate intercepted between the two curves, is equal to

$$\frac{(m^{\frac{1}{2}} - n^{\frac{1}{2}}) a^{\frac{3}{2}}}{2}.$$

5. To prove that the three altitudes of any triangle circumscribed about a parabola, meet in one point on the directrix.

6. The cotangents of the inclinations of three parabolic tangents are in arithmetical progression, the common difference being $= \delta$: prove that, in the triangle inclosed by these tangents, we shall have

$$\text{area} = p^2 \delta^3.$$

7. Given the outline of a parabola: to construct the axis and the focus.

8. Find the equation to the normal of a parabola, in terms of its inclination to the axis; and prove that the locus of the foot of the focal perpendicular upon the normal is a second parabola, whose vertex is the focus of the given one, and whose parameter is one-fourth as great as that of the given one.

9. Show that the locus of the intersection of parabolic normals which cut at right angles, is a parabola whose parameter is one-fourth that of the given one, and whose vertex is at a distance $= 3p$ from the given vertex.

10. The centers of a series of circles which pass through the focus of a parabola, are situated on the curve: prove that each circle touches the directrix.

11. To find the area of the rectangle included by the tangent and normal at any point P of a parabola, and their respective focal perpendiculars; and to determine the position of P when the rectangle is a square.

12. If two parabolic tangents are intersected by a third, parallel to their chord of contact, the distances from their point of intersection to their respective points of contact are bisected by the third tangent.

13. Show that the locus of the vertex of a parabola which has a given *focus*, and touches a given right line, is a circle, of which the perpendicular from the given focus to the given line is a diameter.

14. Show that, if a parabola have a given *vertex*, and touch a given right line, its focus will move along another parabola, whose axis passes through the given vertex at right angles to the given line, and whose parameter = the distance from the given vertex to the given line.

15. TP and TQ are tangent to a given parabola at P and Q, and TF joins their intersection to the focus: prove that

$$FP.\ FQ = FT^2.$$

An. Ge. 43.

16. A right angle moves in such a manner that its sides are respectively tangent to two confocal parabolas whose axes are coincident: to find the locus of its vertex.

17. Prove that the pole of the normal which passes through one extremity of the latus rectum of a parabola, is situated on the diameter which passes through the other extremity, and find its ·exact position on that line.

18. The triangle included by two parabolic tangents and their chord of contact being of a given area $= a^2$, prove that the locus of the pole is a parabola whose equation is

$$y^2 = 4px + (2pa^2)^{\frac{2}{3}}$$

19. Prove that, if two parabolas whose axes are mutually perpendicular intersect in four points, the four points lie on a circle.

20. If two parabolas, having a common vertex, and axes at right angles to each other, intersect in the point $x'y'$· then, l denoting the latus rectum of the one, and l' that of the other,

$$l : x' :: y' : l'.$$

[This property has a special interest, on account of its connection with the ancient problem of the Duplication of the Cube. It affords, as the reader will observe, a method of determining graphically two geometric means between two given lines; and was proposed for this purpose by MENECHMUS, a geometer of the school of Plato, in connection with his attempt to solve the problem just mentioned. The graphic problem of "two means" received a variety of solutions at the hands of the Greek geometers, two of the most celebrated being discovered by DIOCLES and NICOMEDES. They are effected, respectively, with the help of the curves called the *Cissoid* and the *Conchoid*.]

CHAPTER SIXTH.

THE CONIC IN GENERAL.

630. Having in the previous Chapters become familiar with the properties of the several conics considered as separate curves, let us now ascend to the wholly generic point of view, from which we may comprehend them not as isolated Forms, but as members of a united System, and, in fact, as successive phases of a generic locus which may be called the CONIC, whose idea we sketched in the eighth Section of Part I.

We may begin by showing in what sense this name is descriptive of the system; or, how the curves may be grouped together as *sections of a cone.*

THE THREE CURVES AS SECTIONS OF THE CONE.

631. Definitions.—A **Cone** is a surface generated by moving a right line which is pivoted upon a fixed point, along the outline of any given curve whose plane does not contain the fixed point.

The fixed point is called the *vertex* of the cone; the given curve, its *directrix;* and the moving right line, its *generatrix.*

Since the generatrix extends indefinitely on both sides of the vertex, the cone will consist of two exactly similar portions, extending from the vertex in opposite directions to infinity. Of these, one is called the *upper nappe* of the cone; and the other, the *lower nappe.*

Any single position of the generatrix is called an *element* of the cone.

When the directrix is a circle, the cone is called *circular;* and the right line drawn through the vertex and the center of the directrix, is termed the *axis* of the cone.

632. Definitions.—A **Right circular Cone** is a cone whose directrix is a circle, and whose axis is perpendicular to the plane of its directrix.

The section formed with any cone by a plane, is termed a *base* of the cone; consequently, the directrix of a right circular cone may be called its base. For this reason, such a cone is often named a *right cone on a circular base.* The diagram of the next article presents an example of one.

633. We shall prove, in the proper place in Book Second, the following propositions, which we ask the student to take upon trust for the present, in order that we may use them in grouping the three curves according to their *geometric* order:

I. Every section formed by passing a plane through a right circular cone is a curve of the Second order.

II. If the angle which the secant plane makes with the base is *less* than that made by the generatrix, the section is an *ellipse.*

III. If the angle which the secant plane makes with the base is *equal* to that made by the generatrix, the section is a *parabola.*

IV. If the angle which the secant plane makes with the base is *greater* than that made by the generatrix, the section is an *hyperbola.*

These three cases are represented in the diagram: that of the Ellipse, at *AE*; that of the Parabola, at *LPR*; and that of the Hyperbola, at *HA - A'H'*. It is manifest, however, from the fact that the secant plane makes with the base an angle successively *less* than, *equal* to, and *greater* than the angle made by the generatrix (whose angle must have a fixed value for any given cone),

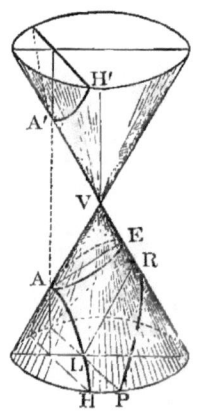

that the three sections may be formed by a single

plane, by simply revolving it on the line in which it cuts the base. Beginning with it in such a position that its inclination to the base is less than that of the generatrix (whose inclination is often called the *inclination of the side* of the cone), and revolving it upward toward the position of parallelism to the side, we shall cut out a series of ellipses of greater and greater eccentricity. When the secant plane becomes parallel to the side, the section will be a parabola. When it is pushed still farther upward, so that its angle with the base becomes greater than that of the side, it will reach across the space between the two nappes of the cone, and. pierce the upper as well as the lower one, and the section will be an hyperbola, whose two branches will lie in the two nappes respectively.

From this it appears, that, granting the proposition that the sections are the curves mentioned, the natural geometric order in which they occur is: Ellipse, Parabola, Hyperbola. This is the same as their analytic order, as we found it in Art. 200. We shall be able to give a more explicit account of their appearance as successive phases of the Conic, so soon as we have presented a fuller view of the modes in which we can represent that generic locus by analytic symbols.

VARIOUS FORMS OF THE EQUATION TO THE CONIC.

634. Equation in Rectangular Co-ordinates at the Vertex.—We have not as yet referred the three Conics to the same axes and origin: let us now do so, by transforming the central equations of the Ellipse and the Hyperbola to such a vertex of each curve as will correspond to the vertex of the Parabola.

This will require us to transform the equation

$$y^2 = \frac{b^2}{a^2}(a^2 - x^2) \qquad (1)$$

to the *left-hand* vertex of the Ellipse; and the equation

$$y^2 = \frac{b^2}{a^2}(x^2 - a^2) \qquad (2)$$

to the *right-hand* vertex of the Hyperbola. Accordingly, putting $x - a$ for x in (1), and $x + a$ for x in (2), and expanding, we get

$$y^2 = \frac{2b^2}{a}x - \frac{b^2}{a^2}x^2, \quad y^2 = \frac{2b^2}{a}x + \frac{b^2}{a^2}x^2.$$

Hence, (Arts. 428, 521,) the equations to the Ellipse, the Hyperbola, and the Parabola may be written

$$y^2 = 4px - \frac{b^2}{a^2}x^2, \quad y^2 = 4px + \frac{b^2}{a^2}x^2, \quad y^2 = 4px.$$

Remembering, now, that $b^2 : a^2 = \pm (1 - e^2)$, and that its value in the case of the Parabola must therefore be $= 0$, we learn that the equation to the Conic in General is

$$y^2 = Px + Rx^2,$$

in which P is the parameter of the curve, and R the ratio between the squares of the semi-axes; and we have the specific conditions

$$R < 0 \;\therefore\; \text{Ellipse,}$$
$$R = 0 \;\therefore\; \text{Parabola,}$$
$$R > 0 \;\therefore\; \text{Hyperbola.}$$

Corollary.—By the three equations from which $y^2 = Px + Rx^2$ was generalized, we see that in the Ellipse, the square on the ordinate is *less* than the rectangle under the abscissa and parameter; that in the Parabola, the square is *equal* to the rectangle; and that in the Hyperbola, the square is *greater* than the rectangle.

Remark.—According to PAPPUS (*Math. Coll.*, VII: c. A. D. 350), the names of the three curves were originally given to designate this property. But EUTOCIUS (A. D. 560) says that the names were derived from the fact, that, according to the ancient Greek geometers, the three sections were cut respectively from an acute-angled, a right-angled, and an obtuse-angled cone, by means of a plane always passing at right angles to the side. Thus, if the angle under the vertex of the cone were *acute*, the sum of that and the right angle made by the secant plane with the side would be *less* than two right angles, and the name *ellipse* was given, either to indicate this deficiency, or to show that the curve would then *fall short* of the upper nappe of the cone. But if the angle under the vertex were *right*, the mentioned sum of angles would be *equal* to two right angles, and the plane of the curve consequently be *parallel* to the side of the cone: to denote which facts, the name *parabola* was given. Finally, if the angle under the vertex were *obtuse*, the mentioned sum of angles would be *greater* than two right angles, and the name *hyperbola* was given, either to suggest this excess, or to indicate that the curve would then *reach over* to the upper nappe of the cone.

It is noticeable here, that the early geometers supposed the three sections to be *peculiar* respectively to an acute-angled, a right-angled, and an obtuse-angled cone. The improvement of forming them all from the same cone by merely changing the inclination of the secant plane, was introduced by APOLLONIUS OF PERGA, B. C. 250.

Which of these etymologies should have the preference, is a question among critics of mathematical history. It is remarkable, however, that the names represent equally well *all* the distinguishing properties of the curves, whether geometric or analytic: as the reader may perhaps have already observed for himself.

The name *parameter*, which simply means *corresponding measure*, or *co-efficient*, is given to the quantities $2b^2 : a$ and $y'^2 : x'$, on account of the position they occupy in the equation

$$y^2 = Px + Rx^2,$$

one or the other of them being the *first co-efficient* in the second member, according as the conic is central or non-central. Moreover, as the dimensions of the curve depend mainly upon the value of P, the ratios which it symbolizes are naturally termed *par excellence* the *measures* of the Conic. On the other hand, the name *parameter* may be defined in each conic by the value which it denotes for each, as in the preceding Chapters.

635. Equation in terms of the Focus and its Polar.—We have seen (Arts. 439, 532, 618) that in every conic the distance of any point on the curve from the focus is in a constant ratio to its distance from the polar of the focus, the ratio being equal to the eccentricity. Hence, calling the focal distance ρ, and the distance from the polar (or directrix) δ, we may write, as the equation to the Conic,

$$\rho = e.\delta :$$

which will denote an ellipse, a parabola, or an hyperbola, according as e is *less* than, *equal* to, or *greater* than unity.

636. It follows from the property mentioned above, that the Conic may be defined as *the locus of a point whose distance from a fixed point is in a constant ratio to its distance from a fixed right line.* In fact, this definition has been made the basis of several treatises upon the Conics.

Calling the fixed point $x'y'$, the fixed right line $Ax + By + C = 0$, and the variable point of the curve xy, the equation to the Conic is

$$\left[(x-x')^2 + (y-y')^2\right]^{\frac{1}{2}} = \frac{e(Ax+By+C)}{\sqrt{(A^2+B^2)}} \qquad (1).$$

If we suppose the arbitrary axes of this equation to be changed so that the given line shall become the axis of y, and a perpendicular to it through the given point the axis of x, we shall have $B = C = 0$, $y' = 0$, and a new value of x' which may be called $2p : e$. We then get

$$e^2 x^2 - y^2 = \left\{ \frac{2p - ex}{e} \right\}^2 \qquad (2).$$

This represents an ellipse when $e < 1$, a parabola when $e = 1$, an hyperbola when $e > 1$; and, as it can be written

$$e^2 y^2 = 4p (ex - p) - (1 - e^2) e^2 x^2 \qquad (3),$$

is evidently the generic relation of which the equation in Art. 181 is a particular case.

Remark.—The formulæ of this article are only modified expressions of the relation $\rho = e.\delta$; and it is obvious, on comparing this with (1) above, that *the focal distance of any point on a conic can always be expressed as a rational function of the co-ordinates of the point, in the first degree.*

We leave the student to prove the converse theorem, that *a curve must be a conic, if the distance of every point on it from a fixed point can be expressed as such a rational linear function.*

637. Linear Equation to the Conic.—It is evident from what has just been said, that this title would correctly describe either the expression of Art. 635 or the modified form of it given in (1) of Art. 636. But the phrase is in fact reserved to designate a still further modification of the same expression, which we will now obtain.

Suppose the origin of abscissas to be at the focus, and the axis of x to be the perpendicular drawn through the focus to its polar: the distance from this polar (or directrix) to the point on the curve, will then be equal to the di-tance between the directrix and the focus, increased by the abscissa of the point; or, we shall have

$$\delta = \frac{2p}{e} + x,$$

and the equation $\rho = e.\delta$ will become

$$\rho = 2p + cx.$$

Remark.—The so-called Linear Equations to the Ellipse, the Parabola, and the Hyperbola, namely (Arts. 360, 457, 568),

$$\rho = a + ex, \qquad \rho = cx - a, \qquad \rho = p + x,$$

will all assume the form just found, if we shift their respective origins to the focus, by putting $x - ae$ for x in the first, $x + ae$ for x in the second, and $x + p$ for x in the third. We leave the actual transformation to the student, only reminding him, that, in the first two curves, $\pm a(1 - e^2) = 2p$; and that, in the third, $e = 1$.

638. A form of the preceding equation with which the reader may sometimes meet, is

$$r = mx + n,$$

and any equation of this form, in which m and n are any two constants whatever, will denote a conic, whose eccentricity will $= m$, while its semi-latus rectum will $= n$.

639. Equation referred to Two Tangents.—A useful expression for the Conic may be developed as follows:

Let the equation to the curve, referred to any axes whatever, be

$$Ax^2 + 2Hxy + By^2 + 2Gx + 2Fy + C = 0 \qquad (1).$$

To determine the intercepts of the curve on the axes, we get, by making y and x successively $= 0$,

$$Ax^2 + 2Gx + C = 0, \qquad By^2 + 2Fy + C = 0.$$

But if the axes are tangents, the two intercepts on each will be equal, these quadratics will have equal roots, and we shall have

$$G^2 = AC, \qquad F^2 = BC.$$

Putting into (1) the values of A and B which these conditions give,

$$G^2x^2 + 2CHxy + F^2y^2 + 2GCx + 2FCy + C^2 = 0.$$

Whence, by adding $2FGxy - 2FGxy$, and re-arranging the terms,

$$(Gx + Fy + C)^2 = 2(FG - CH)xy.$$

This is the equation we are seeking; and, as the co-efficient of xy in it is arbitrary with respect to G, F, C, we may write it

$$(Gx + Fy + C)^2 = Mxy \qquad (2),$$

where G, F, C, M are any four constants whatever.

Corollary 1.—Making y and x successively $= 0$ in this equation, we get the distances of the two points of contact from the origin, namely,

$$x = -\frac{C}{G}, \quad y = -\frac{C}{F}.$$

Calling the first of these distances a, and the second κ, we have

$$G = -\frac{C}{a}, \quad F = -\frac{C}{\kappa};$$

and may write the equation in the more convenient form

$$\left[\frac{x}{a} + \frac{y}{\kappa} - 1\right]^2 = \mu xy \qquad (3).$$

Corollary 2.—The special modification of this which represents a parabola, deserves a separate notice. In order that (3) may denote a parabola, we must have (Art. 191)

$$\left[\frac{2}{a\kappa} - \mu\right]^2 = \frac{4}{a^2\kappa^2}:$$

a condition satisfied by either $\mu = 0$ or $\mu = 4 : a\kappa$. If $\mu = 0$, the equation becomes

$$\frac{x}{a} + \frac{y}{\kappa} = 1,$$

and denotes the chord of contact of the tangent axes. If $\mu = 4 : a\kappa$, we get, by taking the square root of both members of (3),

$$\frac{x}{a} + \frac{y}{\kappa} - 1 = 2\left[\frac{xy}{a\kappa}\right]^{\frac{1}{2}};$$

or, after transposing and again taking the square root,

$$\left[\frac{x}{a}\right]^{\frac{1}{2}} + \left[\frac{y}{\kappa}\right]^{\frac{1}{2}} = 1 :$$

an equation which is sometimes written

$$\sqrt{\kappa x} + \sqrt{ay} = \sqrt{a\kappa}.$$

640. Polar Equation to the Conic.—By comparing Arts. 443, 535, 627, it becomes evident that the Conic may be represented by the general equation

$$\rho = \frac{2p}{1 - e\cos\theta} \qquad (1).$$

In this, as the reader will see by referring to the original investigations (Arts. 152, 172, 183), the pole is at the focus, and the vectorial angle θ is reckoned from the *remote* vertex.

A more useful expression, however, and the only one *universally* applicable in Astronomy, is

$$\rho = \frac{l}{1 + e\cos\theta} \qquad (2).$$

Here, θ is reckoned from the vertex *nearest* the focus selected for the pole, and l denotes the semi-latus rectum. The conic represented is an ellipse, a parabola, or an hyperbola, according as $e < 1$, $e = 1$, or $e > 1$.

641. The Conic as the Locus of the Second Order in General.—All the *Cartesian* equations that precede, are only reduced forms of the general and unconditioned equation

$$Ax^2 + 2Hxy + By^2 + 2Gx + 2Fy + C = 0,$$

which may be converted into any one of them by a proper transformation of co-ordinates, and to whose type they all conform.

THE CONICS IN SYSTEM, AS SUCCESSIVE PHASES OF ONE FORMAL LAW.

642. That the Conics are successive phases of *some* uniform law, we have already seen in Section VII of Part I. Of the *nature* of that law, however, we were there unable to give any better account than this : that it expressed itself in the unconditioned equation of the second degree, and became visible in a threefold series of curves, determined by the successive appearance, in that equation, of the three conditions

$$H^2 - AB < 0, \quad H^2 - AB = 0, \quad H^2 - AB > 0.$$

But we have now reached a position which will enable us to state the law in *geometric* language, to exhibit the elements of *form* which it embodies, and to trace the steps by which those elements cause the three curves to appear in an unbroken series. And it deserves especial mention, that this geometric statement of the law is furnished by the *polar relation*, as expressed in the definition of Art. 636.

643. The generic *law of form* which is designated by the name of *The Conic*, may therefore be stated as follows : *The distance of a variable point from a fixed point shall be in a constant ratio to its distance from a given right line.*

Expressing this law in the equation (Art. 637)

$$\rho = 2p + ex \qquad (1),$$

let us observe the development of the system of the three curves, member after member, as the given line advances nearer and nearer to the fixed point.

We have (Art. 637), for the distance of the given line from the fixed point,

$$d = \frac{2p}{e} \qquad (2).$$

Also, supposing a perpendicular to the given line to be drawn through the fixed point, the perpendicular to be called the *axis*, and the point in which it cuts the curve to be called the *vertex*, we get, for the distance of this vertex from the fixed point, by making $x = -\rho$ in equation (1),

$$\rho' = \frac{2p}{1+e} \qquad (3).$$

Let the generation of the system begin with the given line at an infinite distance from the fixed point. In that case, from (2), we shall have $e = 0$. Under this supposition, equation (1) becomes

$$\rho = 2p,$$

which (Art. 138, Cor. 2) denotes a *circle*, described from the fixed point as a center, with a radius = the semi-latus rectum of the Conic : a result confirmed by the fact, that, under the same supposition, equation (3) becomes $\rho' = 2p$; or, the distance of the vertex from the focus becomes equal to the radius.

Now let the given line move parallel to itself along the axis, assuming successive *finite* distances from the fixed point, but with the condition that every distance shall be greater than $2p$. Then, from (2), $e < 1$; and, from (3), $\rho' < 2p$ and $> p$: so that a continuous series of *ellipses* will appear, of ever-increasing eccentricity, but with a constant latus rectum, their vertices all lying within a segment of the axis $= p$, contained between the points reached by measuring from the fixed point distances $= p$ and $2p$.

Next, let the given line have attained the distance $= 2p$ from the fixed point. We shall then have, from (2), $e = 1$; and, from (3), $\rho' = p$. That is, we shall have a *parabola*, described upon the constant latus rectum.

Finally, let the given line advance from its last position, and approach the fixed point indefinitely. Then, d being less than $2p$, we shall have, from (2), $e > 1$; and, from (3), $\rho' < p$: so that there will arise a continuous series of *hyperbolas*, with a constant latus rectum, but with an ever-increasing eccentricity; with their vertices all lying within a segment of the axis $= p$, measured from the fixed point, and with their branches tending to coincide with the given line as that line tends toward the fixed point. When the given line attains the particular distance $= p\sqrt{2}$ from the fixed point, we shall have $e = \sqrt{2}$; or, the hyperbola (Art. 456, Cor.) will be *rectangular*.

Thus, the order of the curves, as foreshadowed by their analytic criteria, is verified by a systematic generation.

644. The results of the preceding article may be tabulated as follows:

THE CONIC
- Semi-latus rectum < Distance of Focal Polar ∴ ELLIPSE.
  - $e = 0$ ∴ Circle.
  - $e < 1$ ∴ Eccentric.
- Semi-latus rectum = Distance of Focal Polar ∴ PARABOLA.
  - $e = 1$.
- Semi-latus rectum > Distance of Focal Polar ∴ HYPERBOLA.
  - $e > 1$ ∴ Oblique.
  - $e = \sqrt{2}$ ∴ Rect'r.

PROPERTIES OF THE CONIC IN GENERAL.

645. The views thus far taken of the Conic in the present Chapter, although generic, have nevertheless been obtained from a standpoint not strictly analytic. For our results have been derived from a comparison of the properties in which the three curves, after separate treatment by means of equations based upon certain *assumed* properties, have been found to agree. But we shall now, for a few pages, ascend to the strictly analytic point of view, and, beginning with the unconditioned equation

$$Ax^2 + 2Hxy + By^2 + 2Gx + 2Fy + C = 0,$$

shall show how the properties common to the System of the Conics, or peculiar to its several members, may be developed from this abstract symbol, without assuming a single one of them.

THE POLAR RELATION.

646. Intersection of the Conic with the Right Line.—If we eliminate between

$$Ax^2 + 2Hxy + By^2 + 2Gx + 2Fy + C = 0 \qquad (1)$$

and $y = mx + b$, we shall obviously get a quadratic in x to determine the abscissa of the point in which the Conic cuts any right line. Hence, *Every right line meets the Conic in* two *points, real, coincident, or imaginary.*

In particular, for the points in which the curve meets the axes of reference, we get, by making y and x in (1) successively $= 0$, the determining quadratics

$$Ax^2 + 2Gx + C = 0, \quad By^2 + 2Fy + C = 0 \qquad (2).$$

647. The Chord of the Conic.—If a right line meets the Conic in two *real* points, $x'y'$ and $x''y''$, we may write its equation

$$A(x-x')(x-x'') + 2H(x-x')(y-y'') + B(y-y')(y-y'')$$
$$= Ax^2 + 2Hxy + By^2 + 2Gx + 2Fy + C.$$

For this is the equation to *some* right line, since, upon expansion, its terms of the second degree destroy each other ; and to a line that passes through the points $x'y'$, $x''y''$ of the curve, because if x' and y' or x'' and y'' be substituted for x and y in it, we either get

$$Ax'^2 + 2Hx'y' + By'^2 + 2Gx' + 2Fy' + C = 0$$

or else

$$Ax''^2 + 2Hx''y'' + By''^2 + 2Gx'' + 2Fy'' + C = 0,$$

which are simply the conditions that $x'y'$, $x''y''$ may be on the curve.

648. The Tangent of the Conic.—Making $x'' = x'$, and $y'' = y'$, in the preceding equation to the chord, we get the equation to the tangent,

$$A(x-x')^2 + 2H(x-x')(y-y') + B(y-y')^2$$
$$= Ax^2 + 2Hxy + By^2 + 2Gx + 2Fy + C,$$

which, after expansion, assumes the form

$$2Ax'x + 2H(x'y + y'x) + 2By'y + 2Gx + 2Fy + C$$
$$= Ax'^2 + 2Hx'y' + By'^2.$$

Adding $2Gx' + 2Fy' + C$ to both members of this, and remembering that the point of contact $x'y'$ must satisfy the equation to the Conic, we get the usual form of the equation to the tangent, namely,

$$Ax'x + H(x'y + y'x) + By'y$$
$$+ G(x + x') + F(y + y') + C = 0 \qquad (1).$$

By expanding, and re-collecting the terms, this may be otherwise written

$$(Ax' + Hy' + G)x + (Hx' + By' + F)y$$
$$+ Gx' + Fy' + C = 0 \qquad (2).$$

An. Ge. 44.

These equations express the law which *always* connects the co-ordinates of *any* point on the tangent with
those of the point of contact.   Hence, if a point through
which to draw a tangent were *given*, and the point of
contact were *required*, equation (1) would still express
the relation between the co-ordinates of these points,
only $x'y'$ would then denote the given point through
which the tangent would pass, and xy the required point
of contact.   That is, the equation which when $x'y'$ is on
the curve represents the tangent at $x'y'$, when $x'y'$ is situated elsewhere denotes a right line on which will be
found the point of contact of the tangent drawn through
$x'y'$.   Now this line, in common with every other, meets
the Conic in two points: hence, *From any given point,
there can be drawn to the Conic two tangents, real, coincident, or imaginary.*

We thus learn that our curve is of the Second *class* as
well as of the Second *order*.

649. Chord of Contact in the Conic.—From what
has just been stated, it follows that

$$Ax'x + H(x'y + y'x) + By'y$$
$$+ G(x + x') + F(y + y') + C = 0,$$

or its equivalent form (2) above, is the equation to the
chord of contact of the two tangents drawn through $x'y'$.

**650. Locus of the Intersection of Tangents whose
Chord of Contact revolves about a Fixed Point.**—Let
$x'y'$ be the fixed point, and x_1y_1 the intersection of the
two tangents corresponding to the chord.   Then, as $x'y'$
is by supposition always on the revolving chord of contact, we shall have the condition

$$Ax_1x' + H(x_1y' + y_1x') + By_1y'$$
$$+ G(x_1 + x') + F(y_1 + y') + C = 0,$$

irrespective of the direction of the chord. In other words, the intersection of the tangents will always be found upon a right line whose equation is

$$Ax'x + H(x'y + y'x) + By'y$$
$$+ G(x + x') + F(y + y') + C = 0.$$

651. The Point and the Right Line, Reciprocals with respect to the Conic.—The result of the preceding article may be stated as follows : *If through a fixed point chords be drawn to the Conic, and tangents be formed at the extremities of each chord, the intersections of the several pairs of tangents will lie on one right line.*

Also, we may write the equation to the chord of contact of two tangents drawn through $x'y'$ (Art. 649)

$$(Ax' + Hy' + G)x + (Hx' + By' + F)y$$
$$+ Gx' + Fy' + C = 0 :$$

so that (Art. 117), if we suppose $x'y'$ to move along a given right line, the chord of contact will revolve about a fixed point. In other words : *If from different points lying on one right line pairs of tangents be drawn to the Conic, their several chords of contact will meet in one point.*

Combining these two properties, we see that our curve imparts to every point in its plane, the power of determining a right line ; and to every right line, the power of determining a point. That is, it renders the Point and the Right Line *reciprocal forms.*

652. The Polar and its Equation.—We perceive, then, that the relation between a tangent and its point of contact, and the relation between the chord of contact and the intersection of the corresponding tangents, are only particular cases of a general law which, with respect to the Conic, connects *any* fixed point with a corresponding right line. From the result of the last article, moreover, it appears that we shall fitly express this law by calling the line which corresponds to any point, the *polar* (i. e. the *reciprocal*) of the point, and the point itself the *pole* of the line.

We are henceforth, then, to consider the equation

$$Ax'x + H(x'y + y'x) + By'y$$
$$+ G(x + x') + F(y + y') + C = 0$$

as in general denoting the *polar* of $x'y'$; and must regard the *tangent* at $x'y'$ as the position assumed by the polar when $x'y'$ is on the Conic.

653. The Conic referred to its Axis and Vertex.—Before taking out any additional properties of the curve, it will be best to reduce the general equation

$$Ax^2 + 2Hxy + By^2 + 2Gx + 2Fy + C = 0$$

to a simpler form. Supposing the axes of reference to be rectangular, let us revolve them through an angle θ, such that

$$\tan 2\theta = \frac{2H}{A - B}.$$

We shall thus (Art. 156) destroy the co-efficient of xy, and the equation will assume the form

$$A'x^2 + B'y^2 + 2G'x + 2F'y + C = 0.$$

If in addition we remove the origin to a point $x'y'$, we shall get (Art. 163, Th. I)

$$A'x^2 + B'y^2 + 2(A'x' + G')x + 2(B'y' + F')y$$
$$+ (A'x'^2 + B'y'^2 + 2G'x' + 2F'y' + C) = 0.$$

In order, then, that the constant term and the co-efficient of y may vanish together, we must have simultaneously

$$A'x'^2 + B'y'^2 + 2G'x' + 2F'y' + C = 0, \quad B'y' + F = 0\,;$$

that is, we must take the new origin $x'y'$ at the intersection of the curve with the right line $By + F = 0$. Making this change of origin, our equation becomes

$$A'x^2 + B'y^2 + 2G''x = 0,$$

which, by putting $2G'' : B' = -P$, and $A' : B' = -R$, and transposing, may be written

$$y^2 = Px + Rx^2.$$

Here, for every value of x, there will be two values of y, numerically equal with opposite signs: the curve is therefore symmetric to the new axis of x, which for that reason shall be called an *axis of the curve.* If we seek the intersections of the curve with the new axis of y, by making $x = 0$ in the equation, we get $y = \pm 0$; so that the new axis of y meets the curve in two coincident points at the origin; or, in other words, is *tangent* to the curve at the origin. Besides, the new axes are rectangular: hence, combining this fact with those just established, the new origin is the extreme point, or *vertex*, of the curve; and we learn that our axes of reference are the *principal axis of the curve* and the *tangent at its vertex.* And, in fact, our new equation is identical with that obtained in Art. 634.

654. Focus of the Conic, and its Polar.—Taking up our equation in its new form

$$y^2 = Px + Rx^2 \qquad (1),$$

let e be such a quantity that

$$c^2 = 1 + R \tag{2},$$

and let that point whose co-ordinates are

$$x' = \frac{P}{2(1+e)}, \qquad y' = 0 \tag{3},$$

be called the *focus* of the Conic.

The equation to the polar of any point, referred to our present axes, is at once found from the general equation of Art. 652, by putting for A, H, B, G, F, C their values as given by (1). It is therefore

$$2(y'y - Rx'x) = P(x + x') \tag{4}.$$

Hence, substituting for x' and y' the values given in (3), we get, for the *polar of the focus*,

$$2(1 + e + R)x + P = 0 \tag{5},$$

or, after replacing R by its value $e^2 - 1$ from (2),

$$x = -\frac{P}{2(1+e)e} \tag{6}.$$

Equation (6) shows that the polar of the focus is perpendicular to the axis of the Conic, and cuts it on the opposite side of the vertex from the focus, at a distance = an eth part of the distance of the focus. And we shall see, in a moment, that the ratio thus found between the distances of the *vertex* from the focus and from its polar, subsists between the distances of *any* point on the curve from those two limits.

From (3) we have (Art. 51, I, Cor. 1), for the distance ρ of any point xy from the focus,

$$\rho = \sqrt{\left\{x - \frac{P}{2(1+e)}\right\}^2 + y^2} = \sqrt{\left\{x - \frac{P}{2(1+e)}\right\}^2 + Px + Rx^2},$$

since xy is on the Conic. Replacing R by its value $e^2 - 1$, and reducing, we get

$$\rho = \frac{2(1+e)\,ex + P}{2(1+e)} \qquad (7).$$

Also, for the distance from xy to the polar of the focus, we have, from (6) by Art. 105, Cor. 2,

$$\delta = \frac{2(1+e)\,ex + P}{2(1+e)\,e} \qquad (8).$$

Hence, dividing (7) by (8), we get

$$\frac{\rho}{\delta} = e \qquad (9).$$

That is, *The distance of any point on the Conic from the focus, is in a constant ratio to its distance from the polar of the focus.*

The ratio e, we will call the *eccentricity* of the Conic.

655. The Species of the Conic, and their Figures.—The preceding investigation leads directly to the resolution of the vague and general Conic into three specific curves, and the generic property just developed will enable us at once to determine the figures of these.

For since $e^2 = 1 + R$, we shall evidently have

$$e < 1, \quad e = 1, \quad e > 1$$

according as R is negative, equal to zero, or positive. Therefore, by embodying the property of (9) in the mechanical contrivances described in the corollaries to Arts. 439, 532, 618, we can generate three distinct curves, depending on the value of e; as follows:

In the first, e will *fall short* of 1: whence the curve may be called an *ellipse.*

In the second, e will *equal* 1 : whence the curve may be called a *parabola*.

In the third, e will *exceed* 1 : whence the curve may be called an *hyperbola*.

The figures of these curves are therefore such as the methods of generation give, and need not be drawn here, as they are already familiar.

If we put $x = P : 2\,(1 + e)$ in equation (1) of the preceding article, we get, for the ordinate erected at the focus,

$$y = \pm \frac{P}{2} :$$

whence, calling the double ordinate through the focus the *latus rectum*,

$$\text{latus rectum} = P \qquad (1).$$

We thus obtain a significant interpretation for the parameter P of our equation; but we do more. For, by the generic property of the preceding article, the distance of the focus from its polar must equal an eth part of its distance from the extremity of the latus rectum, and therefore can now be expressed by

$$\delta' = \frac{P}{2e} \qquad (2).$$

Hence, when $e < 1$, the semi-latus rectum will be *less* than the distance of the focus from its polar; when $e = 1$, it will be *equal* to that distance; and when $e > 1$, it will be *greater* than that distance. In other words, the classification reached above, is identical with that of Art. 644: as may be further shown by the fact, that, if $R = -1$, $e = 0$; and if $R = +1$, $e = \sqrt{2}$.

Also, when $e = 0$, equation (7) of the preceding article gives

$$\rho = \frac{P}{2}.$$

That is, when $e = 0$, the curve is such that all its points are *equally* distant from the focus, or its figure is that of the Circle. Hence, as e increases from 0 toward ∞, the figure of the curve may be supposed to deviate more and more from the circular form, and we see the propriety of calling e the *eccentricity*.

DIAMETERS AND THE CENTER.

656. A very significant question in regard to any curve is, What is the form of its *diameters*, that is, of the lines that bisect systems of parallel chords in it? Let us, then, settle this question for the Conic.

657. Equation to any Diameter.—If we suppose θ' to be the common inclination of any system of parallel chords, $x'y'$ the intersection of any member of the system with the Conic, and xy its middle point, we shall have (Art. 102)

$$x' = x - l \cos \theta', \quad y' = y - l \sin \theta',$$

where l is the distance from xy to $x'y'$. But since $x'y'$ is on the Conic, we get, by equation (1) of Art. 646,

$$A (x - l \cos \theta')^2 + 2H (x - l \cos \theta')(y - l \sin \theta') + B (y - l \sin \theta')^2$$
$$+ 2G (x - l \cos \theta') + 2F (y - l \sin \theta') + C = 0.$$

Expanding, collecting terms, and putting S for the first member of the general equation of the second degree, we get

$$(A \cos^2 \theta' + 2H \cos \theta' \sin \theta' + B \sin^2 \theta') \, l^2$$
$$- 2 [(Ax + Hy + G) \cos \theta' + (Hx + By + F) \sin \theta'] \, l + S = 0.$$

An. Ge. 45.

Now xy being the *middle* point of a chord, the two values of l given by this quadratic must be numerically equal with opposite signs. Hence, (Alg., 234, Prop. 3d,) the co-efficient of l vanishes, and we obtain, as the equation to any diameter,

$$(Ax + Hy + G) + (Hx + By + F) \tan \theta' = 0,$$

in which θ' is the inclination of the chords which the diameter bisects.

658. Form and Position of Diameters.—Comparing the equation just obtained with that of Art. 108, we learn that every diameter of the Conic is a right line, and passes through the intersection of the two lines

$$Ax + Hy + G = 0, \quad Hx + By + F = 0;$$

that is, through the point whose co-ordinates (Art. 106) are

$$x = \frac{BG - HF}{H^2 - AB}, \quad y = \frac{AF - HG}{H^2 - AB} \quad (1).$$

Moreover, putting $\theta =$ the inclination of any diameter, we have (Art. 108)

$$\tan \theta = -\frac{A + H \tan \theta'}{H + B \tan \theta'} \quad (2):$$

so that, as θ' is arbitrary, a diameter may have any inclination whatever to the axis of x; or, *every* right line that passes through the point (1) is a diameter. Hence, as (1) is in turn upon every diameter, it is the *middle* point of every chord drawn through it, and may therefore be called the *center* of the Conic.

For the form and position of conic diameters in general, we therefore have the two theorems: *Every diameter is a right line passing through the center;* and, *Every right line that passes through the center is a diameter.*

Hence, the lines $Ax + Hy + G = 0$, $Hx + By + F = 0$ are both diameters; and, by making $\theta' = 0$ in the final equation of Art. 657, we learn that the former bisects chords parallel to the axis of x; while, by making $\theta' = 90°$, we see that the latter bisects chords parallel to the axis of y.

From (1) we see that the center of the Conic will be at a finite distance from the origin, so long as $H^2 - AB$ is not equal to zero; but will recede to infinity, if $H^2 = AB$. Now, by putting for A, B, F, G, H the values they have when the equation to the Conic takes the form

$$y^2 = Px + Rx^2,$$

we get the *co-ordinates of the center, referred to the principal axis and its vertical tangent*, namely,

$$x = -\frac{P}{2R}, \quad y = 0 \qquad (3),$$

which show that the center is situated on the principal axis, at a distance from the vertex $= -P : 2R$. This distance, then, will be finite if R is either positive or negative, but infinite if $R = 0$. Hence, (Art. 655,) the diameters of the two curves which we have named the Ellipse and the Hyperbola, meet in a finite point, and are inclined to each other; but the diameters of the curve called the Parabola meet only at infinity, or, in other words, are all parallel: a result corroborated by the fact, that, if in (2) we replace A by the value $H^2 : B$, which it will have if $H^2 = AB$, we get

$$\tan \theta = -\frac{H}{B} \qquad (4),$$

showing that all the diameters of any given parabola are equally inclined to the axis of x.

659. Farther Classification of the Conic.—It thus appears that the Ellipse and the Hyperbola may be classed together as *central* conics, while the Parabola may be styled the *non-central* conic. Adding this prior subdivision, the table of Art. 644 will appear thus:

CONJUGATE DIAMETERS AND THE AXES.

660. Relative inclination of Diameters and their Ordinates.—The halves of the chords which a diameter bisects may be called its *ordinates*. If, then, θ = the inclination of any diameter, and θ' = that of its ordinates, we have, from (2) of Art. 658,

$$B \tan \theta \tan \theta' + H(\tan \theta + \tan \theta') + A = 0$$

as the relation always connecting the inclinations of a diameter and its ordinates.

Now this may either be read as the condition that the diameter having the inclination θ may bisect chords having the inclination θ', or *vice versâ*. Hence, *If a diameter bisect chords parallel to a second, the second will bisect chords parallel to the first.*

This property is however restricted to the central conics; for it is impossible that the ordinates of any parabolic diameter should be parallel to a second, since all parabolic diameters are parallel to each other.

661. Condition that two Diameters be Conjugate.—We indicate that two diameters of a central conic are in the relation above-mentioned, by calling them *conjugate* diameters. Since, then, the conjugate of any diameter is parallel to its ordinates, by interpreting θ and θ' as the inclinations of two diameters,

$$B \tan \theta \tan \theta' + H (\tan \theta + \tan \theta') + A = 0$$

becomes the condition that the two diameters may be conjugate.

662. The Axes, and their Equation.—The condition just established may be referred to the principal axis and vertex of the Conic by putting for A, B, H the values they have in the equation $y^2 = Px + Rx^2$. It thus becomes

$$\tan \theta \tan \theta' = R \qquad (1).$$

In this, if we suppose $\theta = 0$, but not otherwise, we get

$$\tan \theta' = \infty,$$

and learn that the conjugate of the principal axis is perpendicular to it. Hence, *In a central conic there is one, and but one, pair of rectangular conjugates.**

We will call these rectangular conjugates the *axes of the conic*. The one hitherto named the *principal*, shall now be termed the *transverse* axis; and the other, the *conjugate* axis. Their respective equations, referred to the same system as the conic $y^2 = Px + Rx^2$, will be

$$y = 0, \quad 2Rx + P = 0 \qquad (2).$$

---

* Unless the conic is a *circle :* when R will $= -1$, and the condition (1) will become $1 + \tan \theta \tan \theta' = 0$; so that (Art. 96, Cor. 1) *all* the conjugates will be at right angles.

For the first equation represents the axis of x; and the second, a perpendicular to it passing through the center. Hence,

$$(2Rx + P)\, y = 0 \qquad (3)$$

is the equation to both axes, in the same system of reference.

663. Equation to the Conic, in its Simplest Forms.—If in the equation

$$y^2 = Px + Rx^2 \qquad (1),$$

we suppose $R = 0$ ∴ $e = 1$, the quantity $P : 2\,(1 + e)$, which by (3) of Art. 654 denotes the distance of the focus from the vertex, will become $= \frac{1}{4} P$: showing that in the Parabola the latus rectum P is equal to four times the focal distance of the vertex. Putting this latter distance $= p$, and giving to R in (1) its corresponding value $= 0$, the equation to the Parabola will be

$$y^2 = 4px \qquad (2).$$

But if R be positive or negative, or (1) denote the Ellipse or the Hyperbola, this simplification is impossible. If, however, we transform (1) to the *center and axes*, by putting $[x - (P : 2R)]$ for x, we get

$$y^2 = P\left\{ x - \frac{P}{2R} \right\} + R \left\{ x - \frac{P}{2R} \right\}^2 ;$$

or, after obvious reductions,

$$4R^2x^2 - 4Ry^2 = P^2.$$

Here, making y and x successively $= 0$, we obtain, for the *lengths of the semi-axes*,

$$x = \pm \frac{P}{2R}, \qquad y = \pm \frac{P}{2}\sqrt{-\frac{1}{R}}\,.$$

Putting a to denote the first of these lengths, and b to denote the second, we get

$$4R^2 = \frac{P^2}{a^2}, \qquad 4R = -\frac{P^2}{b^2} \qquad \text{(A)},$$

and the central equation becomes

$$\frac{x^2}{a^2} \pm \frac{y^2}{b^2} = 1 \qquad (3),$$

the upper sign corresponding to $-R$, and the lower to $+R$.

Equations (2) and (3) are the simplest forms of the equation to the Conic. In them, we have reached the same forms with which we set out upon the separate investigation of the Parabola, the Ellipse, and the Hyperbola. Of course, then, we can now develop all the properties derived from them in the preceding Chapters, and the reader will be convinced of the adequacy of the purely analytic method without proceeding farther. We will therefore present but a single topic more, whose treatment from the generic point of view has an especial interest.

THE ASYMPTOTES.

664. We have shown (Art. 646) that every right line meets the Conic in two points, real, coincident, or imaginary. A particular case of the *real* intersections deserves notice.

The quadratic by which we determine the intersections of a right line with the Conic, may sometimes take the form of a simple equation, by reason of the absence of the co-efficient A or B in the equation to the Conic. Thus

$$\text{-} \qquad Hxy + By^2 + 2Gx + 2Fy + C = 0 \qquad (1)$$

gives, on making $y = 0$, only the simple equation

$$2Gx + C = 0 \qquad (2)$$

to determine the intersections of the curve with the axis of x, apparently indicating but a single intersection. In fact, it does indicate a single *finite* intersection; but it is a settled principle of analysis (Alg., 238) that an equation arising in the manner (2) does, shall be regarded as a quadratic of the form

$$0x^2 + 2Hx + B = 0,$$

one of whose roots is *finite*, and the other *infinite*. Hence, the consistent interpretation of such an equation as (2), will be that the corresponding line meets the Conic in one finite point and in one point infinitely distant from the origin.

665. Transforming the general equation to polar co-ordinates, we obtain

$$(A \cos^2 \theta + 2H \cos \theta \sin \theta + B \sin^2 \theta) \, \rho^2$$
$$+ 2 \, (G \cos \theta + F \sin \theta) \, \rho + C = 0 \qquad (1).$$

The condition, then, that the radius vector may meet the Conic at infinity is

$$A \cos^2 \theta + 2H \cos \theta \sin \theta + B \sin^2 \theta = 0 \qquad (2):$$

a quadratic in θ, and therefore satisfied by two values of the vectorial angle, which evidently will be real, equal, or imaginary, according as $H^2 - AB$ is greater than, equal to, or less than zero. Hence, as the origin may be taken at any point, *Through any given point there can be drawn two real, coincident, or imaginary lines which will meet the Conic at infinity.*

Moreover, since a change of origin (Art. 163, Th. I) does not affect the co-efficients A, H, B, the directions of these lines for any given conic will in all cases be determined by the *same* quadratic (2). That is, *All lines that meet the Conic at infinity are parallel.*

It is to be noted, that in general each of the radii vectores determined by (2) also meets the curve in one *finite* point, whose position is given by the *finite* terms of (1), namely, by

$$2 \left(G \cos \theta + F \sin \theta \right) \rho + C = 0 \qquad (3).$$

A convenient method of finding the equation to the two lines which pass through the *origin* and meet the curve at infinity, will be to multiply (2) throughout by ρ^2, and then put x for $\rho \cos \theta$, and y for $\rho \sin \theta$. We thus obtain

$$Ax^2 + 2Hxy + By^2 = 0 \qquad (4),$$

the equation of Art. 127. The two lines, then, in case the conic is an *ellipse*, will be *imaginary;* in case it is a *parabola*, they will be *coincident;* and in case it is an *hyperbola*, they will be *real.*

666. If we now suppose the general equation to be transformed to the center, the co-efficients G and F (Art. 163, Th. III) will vanish. For that origin, then, the condition (2) of the preceding article will occur simultaneously with the disappearance of the co-efficient of ρ in (1), and the roots of the latter equation will therefore be simultaneously infinite and equal. Hence, *Through the center there can be drawn two lines, each of which will meet the Conic in two* coincident *points at infinity.*

These tangents at infinity may appropriately be called *asymptotes;* since the curve must converge to them as it recedes to infinity, but can not merge into them except at infinity. Since, then, all the lines that meet the curve at infinity are parallel, equation (4) of the preceding article will in general denote a pair of parallels to the asymptotes, passing through the origin. Hence, supposing the *center* to be the origin, we have, for the *equation to the asymptotes,*

$$Ax^2 + 2Hxy + By^2 = 0 \qquad (1),$$

since the co-efficients A, H, B remain the same for every origin. This is the same as saying, that, given any *central* equation to a conic, *the asymptotes are found by equating to zero its terms of the second degree.* The form of (1) shows that these lines are real in the Hyperbola, coincident in the Parabola, and imaginary in the Ellipse.

If now, in the condition of Art. 661, we make $\tan \theta' = -\cot \theta$, the corresponding conjugates will be at right angles; that is, they will be the *axes.* But then

$$H \tan^2 \theta + (A - B) \tan \theta - H = 0.$$

Multiply this by ρ^2, put x for $\rho \cos \theta$, and y for $\rho \sin \theta$: then

$$Hx^2 - (A - B) xy - Hy^2 = 0 \qquad (2),$$

the *equation to the axes,* if the *center* is origin.

Now (2) is the equation of Art. 129, and therefore denotes two right lines bisecting the angles between the lines represented by (1). Hence, *The axes bisect the angles between the asymptotes, and are real whether the asymptotes are real or imaginary.*

CONDITIONS DETERMINING A CONIC.

667. The general equation of the second degree,

$$Ax^2 + 2Hxy + By^2 + 2Gx + 2Fy + C = 0,$$

may of course be divided through by any one of its co-
efficients, and therefore contains five, and only five,
arbitrary constants. Hence, *Five conditions are necessary and sufficient to determine a conic.*

Thus, a conic may be made to pass through five given
points ; or, to pass through four points and touch a given
line ; or, to pass through three points and touch two given
lines ; etc. And in case the equation to a conic contains
less than five constants, we must understand that the
curve has already been subjected to a series of conditions, equal in number to the difference between five
and the number of constants in its equation. Thus, the
conic

$$y^2 = Px + Rx^2$$

has already been subjected to *three* conditions; namely,
passing through a given point (the vertex), touching a
given line (the axis of y), and having the focus on a
given line (the axis of x).

668. The solution of two general problems which are
often of use in connection with conics, may conveniently
be presented here.

I. *To determine the relation between the parameters* P
and R *in the vertical equation to the Conic.*—From the
first of the equations at (A) in Art. 663, we have

$$P^2 = 4a^2R^2 \quad \therefore \quad P = 2aR \qquad (1).$$

Also, by combining both of the equations at (A),

$$4a^2R^2 = -4b^2R \quad \therefore \quad R = -\frac{b^2}{a^2} \qquad (2).$$

II. *To determine the axes and eccentricity of a conic
given by the general equation.*—Comparing (3) of Art.

663 with (3), (d), and (e) of Art. 156, and taking the radicals in (d) and (e) as negative, we get

$$\frac{1}{a^2} = \frac{A'}{C''} = \frac{A + B - Q}{2C''}, \quad \frac{1}{b^2} = \frac{B'}{C''} = \frac{A + B + Q}{2C''},$$

where $Q^2 = (A - B)^2 + (2H)^2$, and C' [Art. 155, (b)] $= - \varDelta : (H^2 - AB)$. Hence,

$$a^2 = - \frac{2\varDelta}{(H^2 - AB)(A + B - Q)}$$
$$b^2 = - \frac{2\varDelta}{(H^2 - AB)(A + B + Q)} \qquad (3).$$

These equations give the semi-axes in terms of the general co-efficients. For the eccentricity, we have, by putting $e^2 - 1$ for R in (2) above,

$$c^2 = 1 - \frac{b^2}{a^2} = \frac{a^2 - b^2}{a^2}.$$

Substituting for a^2 and b^2 from (3), we therefore get

$$e^2 = \frac{2Q}{A + B + Q}.$$

669. Two conics that have the same eccentricity, are said to be *similar*. It follows, then, that all circles, all parabolas, and (Art. 540) all hyperbolas included within equally inclined asymptotes, are similar.

Moreover, since e (Art. 668) is a function of A, B, H, these co-efficients must be the same for all similar conics.

THE CONIC IN THE ABRIDGED NOTATION.

670. The Anharmonic Ratio.—With respect to this ratio, we shall only develop the fundamental property of the Conic. The reader who desires to follow this property through its manifold consequences, may consult the writings of Salmon and Chasles.

Let A, B, C, D be four fixed points on any conic, and O the variable point of the curve. Then, if a, β, γ, δ be the *equations* to the four chords which connect the fixed points, the equation to the curve, referred to this inscribed quadrilateral (see paragraph 2d, p. 236) will be

$$a\gamma = k\beta\delta \qquad (1).$$

Now, if a, b, c, d denote the *lengths* of the four chords, we have, for the lengths of the perpendiculars let fall upon the chords from O,

$$a = \frac{OA.OB \sin AOB}{a}, \qquad \beta = \frac{OB.OC \sin BOC}{b},$$

$$\gamma = \frac{OC.OD \sin COD}{c}, \qquad \delta = \frac{OD.OA \sin DOA}{d}.$$

Substituting in (1), and reducing,

$$\frac{\sin AOB \sin COD}{\sin BOC \sin DOA} = k \frac{a.c}{b.d} \qquad (2).$$

But (Art. 285) the first member of (2) is the anharmonic of the pencil O - $ABCD$, and the second is constant. Hence, *The anharmonic of a pencil radiating from any point of a conic to four fixed points of the curve is constant.*

671. Definitions.—In any hexagon, two vertices are said to be *opposite*, when they are separated by two others. Thus, if A, B, C, D, E, F are the six successive vertices, A and D, B and E, C and F are opposite.

Two *sides* are also called opposite when separated by

two others. Thus, *AB* and *DE*, *BC* and *EF*, *CD* and *FA* are opposite sides.

Opposite *diagonals* are those which join opposite vertices, and are therefore three in number; namely, *AD*, *BE*, *CF*.

672. Pascal's Theorem.—Let a, β, γ, λ, μ, ν be the successive *sides* of a hexagon *inscribed* in any conic. Then, if ∂ be the diagonal joining the opposite vertices νa and $\gamma\lambda$, the equations

$$a\gamma - k\beta\partial = 0, \quad \lambda\nu - l\mu\partial = 0 \qquad (1)$$

will each represent the conic. We may therefore suppose the constants k and l to be so taken that $a\gamma - k\beta\partial$ is identically equal to $\lambda\nu - l\mu\partial$; that is, in such a manner that

$$a\gamma - \lambda\nu = (k\beta - l\mu)\,\partial \qquad (2).$$

Hence, all the conditions that will cause $a\gamma - \lambda\nu$ to vanish identically, are included in

$$\partial = 0, \quad k\beta - l\mu = 0 \qquad (3).$$

Now the points νa, $\gamma\lambda$ evidently satisfy $a\gamma - \lambda\nu = 0$, and these by hypothesis are on the line $\partial = 0$: so that the points $a\lambda$, $\gamma\nu$, which also satisfy $a\gamma - \lambda\nu = 0$, but which by hypothesis are *not* on the line $\partial = 0$, must lie upon the line $k\beta - l\mu = 0$. But, by the form of its equation, this line contains the point $\beta\mu$. In short, $a\lambda$, $\beta\mu$, $\gamma\nu$, which are the intersections of the opposite sides of the hexagon, are all on the same line. Hence, *The opposite sides of any hexagon inscribed in a conic intersect in three points which lie on one right line.*

This is known as *Pascal's Theorem.* From this single property, its discoverer BLAISE PASCAL is said to have developed the entire doctrine of the Conic, in a system of four hundred theorems, when he was but sixteen years old; but his treatise was never published, and has unfortunately been lost. Leibnitz, however, has given a sketch of it, in a letter written in 1676 to Pascal's nephew Périer.

By joining six points on a conic in every possible way, we can form sixty different figures, each of which may be called an inscribed hexagon, and in each of which the intersections of the opposite sides will lie on one right line. Consequently there are sixty such lines for every six points on the curve, which are called the *Pascal lines*, or simply the *Pascals*, of the corresponding conic.

673. Brianchon's Theorem.—If we take the symbols of the preceding article as tangentials, $a, \beta, \gamma, \lambda, \mu, \nu$ will be the *vertices* of a hexagon *circumscribed* about a conic, and ∂ will denote the *intersection* of the opposite *sides* $\nu a, \gamma \lambda$. The equations at (1) will then be tangential equations to the conic, and the relation (2) will show that the three *lines* $a\lambda, \beta\mu, \gamma\nu$ intersect in the same *point* $k\beta - l\mu = 0$. That is, *The three opposite diagonals of any hexagon circumscribed about a conic meet in one point.*

This is known as *Brianchon's Theorem*, having been discovered in the early part of the present century by BRIANCHON, a pupil of the Polytechnic School of Paris. It was one of the fruits of Poncelet's *Method of Reciprocal Polars.*

By producing six tangents to a conic till they meet in every possible way, we can form sixty different figures, each of which may be called a circumscribed hexagon. Consequently, for every six points of a conic, there are sixty different *Brianchon points*, determined by the system of six *tangents;* just as there are sixty different *Pascal lines*, determined by the system of six *chords.*

EXAMPLES ON THE CONIC IN GENERAL.

1. If two chords at right angles to each other be drawn through a fixed point to meet any conic, to prove that

$$\frac{1}{S.\,s} + \frac{1}{S'.\,s'} = \text{constant},$$

where S, s are the segments of one chord, and S', s' the segments of the other.

2. If through a fixed point O there be drawn two chords to any conic, and if their extremities be joined both directly and transversely, to prove that the line PQ which joins the intersection of the direct lines of union to the intersection of the transverse ones is the polar of O.

3. Prove that any right line drawn through a given point to meet a conic, is cut harmonically by the point, the curve, and the polar of the point; also, that the chord through any given point, and the line which joins that point to the pole of the chord, are harmonically conjugate to the two tangents drawn from the point.

4. A conic touches two given right lines: to prove that the locus of its center is the right line which joins the intersection of the tangents with the middle point of their chord of contact.

5. Prove that in any quadrilateral *inscribed* in a conic, as $ABCD$, either of the three points E, F, O is the pole of the line which joins the other two. By means of this property, show how to draw a tangent to any conic from a given point outside, with the help of the ruler only.

[This graphic problem is only one of a series resulting from the method of transversals and anharmonics, all of which are solvable with the ruler alone: for which reason, the doctrine of the solutions is sometimes called *Lineal Geometry.*]

6. Prove that in any quadrilateral *circumscribed* about a conic, each diagonal is the polar of the intersection of the other two.

BOOK SECOND:
CO-ORDINATES IN SPACE.

CO·ORDINATES IN SPACE.

674. In removing, at this point in our investigations, the restriction which has confined loci to a given plane, we shall only enter upon the consideration of the most elementary parts of the Geometry of Three Dimensions. That is, we shall only undertake to give the student a clear *general* outline of the principles by which we represent and discuss the surfaces of the First and Second orders. In order to accomplish this, we must begin, as in the case of the Geometry of Two Dimensions, by explaining the conventions for representing a point in space.

CHAPTER FIRST.

THE POINT.

675. About a century after the publication of Descartes' method of representing and discussing plane curves, CLAIRAUT extended the method to lines and surfaces in space, by the following contrivance for representing the position of any conceivable point in space.

Let XY, YZ, ZX be three planes of indefinite extent, intersecting each other two and two in the lines $X'X$, $Y'Y$, $Z'Z$. [The point Y' is supposed to be concealed behind the plane ZX in the diagram.] Then, if P be any point whatever in the surrounding space, its position will be known with reference to the three planes so soon as we find

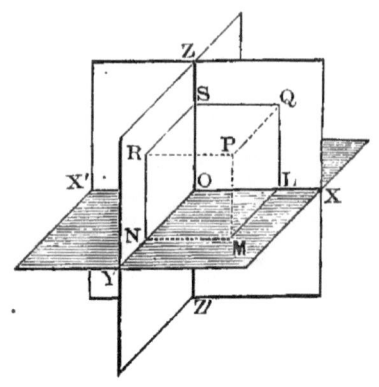

the length of PM drawn parallel to OZ, and of ML, MN drawn parallel respectively to OY and OX; or, which is obviously the same thing, so soon as we find the lengths of OL, LM, MP. In the diagram, the three planes are represented at right angles to each other: a restriction which has the advantage of simplifying the whole subject, and which can always be secured by a proper transformation, if the planes are in fact inclined at any other angle. We shall therefore suppose, in our investigations, that these *reference-planes* are always rectangular, unless the contrary is stated.

The distances OL, LM, MP, or their equals OL, ON, OS, are called the *rectangular co-ordinates* of P, and are respectively represented by x, y, z. The lines OX, OY, OZ, of indefinite extent, are termed the *axes*: OX is the axis of x, OY the axis of y, and OZ the axis of z. The point O, in which the three axes intersect, and which is therefore common to the three reference-planes, is named the *origin*.

The reference-planes evidently divide the surrounding space into eight solid angles, which are numbered as follows: $Z - XOY$ is the *first* angle; $Z - YOX'$, the *second;* $Z - X'OY'$, the *third;* and $Z - Y'OX$, the *fourth.* Similarly, $Z' - XOY$ is the *fifth* angle; $Z' - YOX'$, the *sixth;* $Z' - X'OY'$, the *seventh;* and $Z' - Y'OX$, the *eighth.*

By affecting the co-ordinates x, y, z with the proper sign, we represent a point in either of the eight angles. Thus,

First angle : $x = + a$, $y = + b$, $z = + c$;
Second " $x = - a$, $y = + b$, $z = + c$;
Third " $x = - a$, $y = - b$, $z = + c$;
Fourth " $x = + a$, $y = - b$, $z = + c$;
Fifth " $x = + a$, $y = + b$, $z = - c$;
Sixth " $x = - a$, $y = + b$, $z = - c$;
Seventh " $x = - a$, $y = - b$, $z = - c$;
Eighth " $x = + a$, $y = - b$, $z = - c$.

The student will observe that the positive x lies to the *right* of the first vertical plane YZ, and the negative x to the *left* of that plane; the positive y, in *front* of the second vertical plane ZX, and the negative y in the *rear* of that plane; the positive z, *above* the horizontal plane XY, and the negative z *below* that plane.

Corollary 1.—For any point in the plane YZ, we shall evidently have

$$x = 0 \qquad (1),$$

while y and z are indeterminate. Equation (1) is therefore the *equation to the first vertical reference-plane.*

For any point in the plane ZX, we shall have

$$y = 0 \qquad (2),$$

while z and x are indeterminate. Hence, (2) is the *equation to the second vertical reference-plane.*

Finally, for any point in the plane XY, we shall have

$$z = 0 \qquad (3),$$

while x and y are indeterminate. Hence, (3) is the *equation to the horizontal reference-plane.*

Corollary 2.—If a point is on the axis $X'X$, we shall have $y = 0$, $z = 0$ simultaneously, while x is indeterminate. If the point is on the axis $Y'Y$, $z = 0$, $x = 0$ simultaneously, while y is indeterminate. If the point is on the axis $Z'Z$, $x = 0$, $y = 0$ simultaneously, while z is indeterminate. Hence, the pairs

$$\left. \begin{array}{l} y = 0 \\ z = 0 \end{array} \right\}, \qquad \left. \begin{array}{l} z = 0 \\ x = 0 \end{array} \right\}, \qquad \left. \begin{array}{l} x = 0 \\ y = 0 \end{array} \right\},$$

are respectively the *equations to the axis of* x, *the axis of* y, *and the axis of* z.

Corollary 3.—At the point O, where the axes intersect each other, we shall evidently have, simultaneously,

$$x = y = z = 0,$$

and these three equations are *the symbol of the origin.*

POLAR CO-ORDINATES IN SPACE.

676. If MN be a fixed plane, OX a fixed line in it, and O a fixed point in that line, then, if any point P in the surrounding space be joined with O, and a plane be passed through OP perpendicular to MN, so as to intersect the latter in the line OR, the distance OP, and the angles POR, ROX,

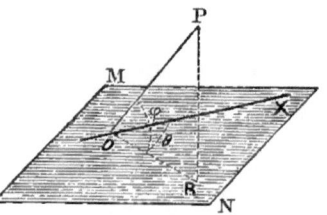

are called the *polar co-ordinates* of the point P. The distance OP is called the *radius vector*, and is represented by the letter ρ; the angles POR, ROX are termed the *vectorial angles*, and are designated respectively by φ and θ, as in the diagram.

MN is called the *initial plane*, OX the *initial line*, and O the *pole*. Instead of the angle φ, its complement is sometimes used, designated by γ.

By inspecting the diagram, it will be evident that we may use

$$\varphi = 0 \qquad\qquad (1)$$

as the *equation to the initial plane*,

$$\varphi = \theta = 0 \qquad\qquad (2)$$

as the *equations to the initial line*, and

$$\rho = 0 \qquad\qquad (3)$$

as the *equation to the pole*.

THE DOCTRINE OF PROJECTIONS.

677. Definitions.—The point in which a line in space pierces a given plane, is called the *trace* of the line upon the plane. Similarly, the line in which a surface cuts a given plane, is termed the *trace* of the surface upon the plane. In particular, the trace of one plane upon another, is the right line in which the former intersects the latter.

If a perpendicular be let fall from any point to a given plane, the trace of the perpendicular upon the plane is called the *orthogonal projection* of the point on the plane. When we use the term *projection* in what follows, we shall always intend an *orthogonal* projection. Thus, in particular, the projections of a point P on the three reference-planes, are respectively M, R, S, the traces of its three co-ordinates.

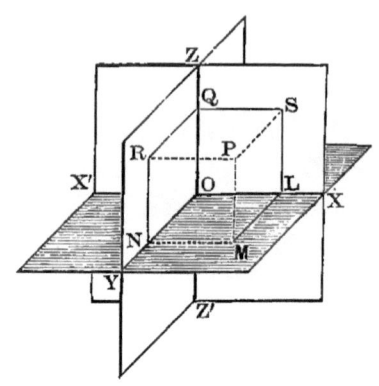

The projection of any *curve* upon a given plane, is the curve formed by projecting all of its points. The perpendiculars let fall in forming such a projection will of course form a *surface*, which is called the *projecting cylinder* of the curve.

When the curve projected is a right line, it is obvious that the projecting cylinder will become a plane. Hence, the projection of any right line upon a given plane is the right line in which the projecting plane cuts the given plane. For example, the projection of the radius vector OP upon the initial plane MN, is the line OR.

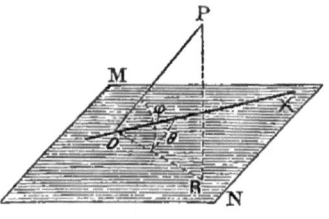

The projection of a point upon a given *line*, is the trace of that line upon the plane which passes through the given point and is perpendicular to the given line. Thus, L, N, Q are the projections of a point P upon the three co-ordinate axes.

The projection of a *right line* upon a given one, is the portion of the latter included between the projections of the extremities of the former. For example, OL is the projection of RP on the axis of x.

The angle which any right line makes with a given *plane*, is the angle included between the line and its projection on the plane; the angle which it makes with a given *line*, is the angle included between it and an intersecting parallel to that line.

678. Theorem.—*The projection of a finite right line upon any plane is equal in length to the length of the line multiplied by the cosine of the angle between the line and the plane.*

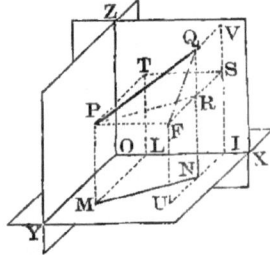

Let XY be the given plane, and PQ the given line. Then, if M and N be the projections of P and Q, the projection of PQ will be MN. Now, drawing PR parallel to MN, we get (Trig., 858)

$$MN = PR = PQ \cos QPR:$$

which proves our proposition.

679. Theorem.—*The projection of one finite right line upon another is equal in length to the length of the first multiplied by the cosine of the angle between the two.*

Let the first line be PQ, and the second OX. Then, if we pass through P and Q the planes PTL, QIF, perpendicular to OX, the projection of PQ upon OX will be IL. Let PF now be drawn parallel to OX: it will be perpendicular to the plane QIF at F. Then, in the right-angled triangle QPF, $PF = PQ \cos QPF$. But, by the construction, $PF = IL$. Hence,

$$IL = PQ \cos QPF:$$

which proves the proposition.

An. Ge. 47.

680. Theorem.—*In any series of points, the projection (on a given line) of the line which joins the first and last, is equal to the sum of the projections of the lines which join the points two and two.*

The points may be so situated that their projections on the given line advance successively from the first to the last: in which case the theorem is an obvious consequence of the sixth definition in Art. 677. Or they may be so placed that the projections of some fall on the given line *behind* those of the next preceding points: in which case we still obtain the theorem, if we consider the line which joins such a point to its predecessor as forming a *negative* projection, and understand the sum mentioned above as algebraic.

Corollary.—*The projection of the radius vector of any point, is equal to the sum of the projections of the co-ordinates of the point.*

For the points *O, L, M, P* (see diagram, Art. 675) may of course be considered as a series coming under the above theorem.

DISTANCE BETWEEN TWO POINTS IN SPACE.

681. Let P and Q be the two points, projected re- spectively at M, R, S and N, T, V. Through P, pass a plane RPF, parallel to the ref- erence-plane XY; and let PML be the projecting plane of MP, and QNH of NQ. Then, in the right-angled triangle QPF, we shall have

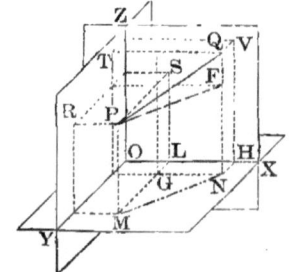

$$PQ^2 = QF^2 + PF^2 \qquad (1).$$

But, by the construction of the figure, $PF = MN$; hence, from the right-angled triangle MGN,

$$PF^2 = NG^2 + MG^2 \qquad (2).$$

Substituting in (1), we obtain

$$PQ^2 = NG^2 + MG^2 + QF^2.$$

Hence, if the co-ordinates of P be x', y', z', and those of Q be x'', y'', z'', while δ represents the distance PQ, we have

$$\delta^2 = (x'' - x')^2 + (y'' - y')^2 + (z'' - z')^2.$$

Corollary.—For the distance from the origin to any point xy in space, we therefore have (Art. 675, Cor. 3)

$$\delta^2 = x^2 + y^2 + z^2.$$

682. The last result may be interpreted thus: *The square on the radius vector of any point is equal to the sum of the squares on the co-ordinates of the point.*

This theorem leads to a remarkable relation among the so-called *direction-cosines of a right line*, that is, the cosines of the three angles which the line makes with the three co-ordinate axes. Let the angle made with the axis of x be α, that made with the axis of y be β, and that made with the axis of z be γ. Then, supposing a parallel to the given line to be drawn through the origin, the co-ordinates of any point xyz on this parallel will be the projections of its radius vector on the axes, and we shall have (Art. 679)

$$x = \rho \cos \alpha, \quad y = \rho \cos \beta, \quad z = \rho \cos \gamma.$$

Squaring and adding these equations, and observing that $\rho^2 = x^2 + y^2 + z^2$, we get, for the relation mentioned,

$$\cos^2 \alpha + \cos^2 \beta + \cos^2 \gamma = 1.$$

POINT DIVIDING THE DISTANCE BETWEEN TWO OTHERS
IN A GIVEN RATIO.

683. By an investigation analogous to that of Art. 52, the details of which the student can easily supply, the co-ordinates of such a point are found to be

$$x = \frac{mx_2 + nx_1}{m + n}, \quad y = \frac{my_2 + ny_1}{m + n}, \quad z = \frac{mz_2 + nz_1}{m + n}.$$

TRANSFORMATION OF CO-ORDINATES.

684. *To transform to parallel reference-planes passing through a new origin.*

Let x', y', z' be the co-or-dinates of the new origin, x, y, z the primitive co-ordinates of any point P, and X, Y, Z its co-ordinates in the new system. Then, as is evident upon inspecting the diagram, the formulæ of transformation will be

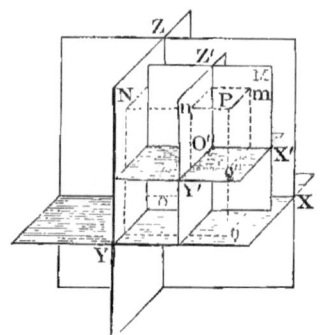

$$x = x' + X, \quad y = y' + Y, \quad z = z' + Z.$$

685. *To transform from a given rectangular system to a system having its planes at any inclination.*

Let the direction-angles of the new axis of x be α, β, γ; those of the new axis of y, α', β', γ'; and those of the new axis of z, α'', β'', γ''. Then, if we suppose each of the new co-ordinates $N'P$, $M'P$, $Q'P$ to be projected

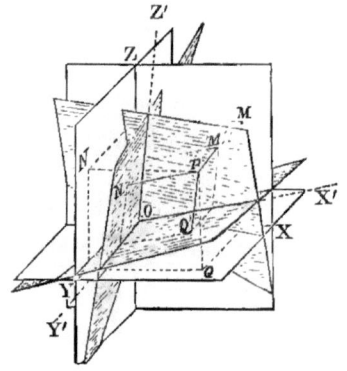

on one of the old axes, the sum of the three projections (Art. 680, Cor.) will in each case be equal to the projection of the radius vector OP. But the projection of OP on OX will be equal to the old x of P; its projection on OY, to the old y; and its projection on OZ, to the old z. Hence, (Art. 679,)

$$x = X \cos a + Y \cos a' + Z \cos a'',$$
$$y = X \cos \beta + Y \cos \beta' + Z \cos \beta'',$$
$$z = X \cos \gamma + Y \cos \gamma' + Z \cos \gamma'',$$

are the required formulæ of transformation.

Remark.—It must be borne in mind, in using these formulæ, that the direction-cosines of the new axes are subject to the conditions (Art. 682)

$$\cos^2 a \; + \cos^2 \beta \; + \cos^2 \gamma \; = 1,$$
$$\cos^2 a' \; + \cos^2 \beta' \; + \cos^2 \gamma' \; = 1,$$
$$\cos^2 a'' + \cos^2 \beta'' + \cos^2 \gamma'' = 1.$$

686. *To transform from a planar to a polar system in space.*

Let the planar system be rectangular. Then, the co-ordinates of any point P in the two systems being related as in the diagram, it is evident that we shall have

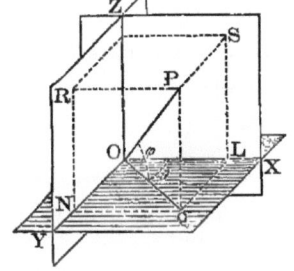

$$x = \rho \cos \varphi \cos \theta,$$
$$y = \rho \cos \varphi \sin \theta,$$
$$z = \rho \sin \varphi.$$

From these equations we can evidently also find ρ, φ, θ in terms of x, y, z.

Remark.—To combine a change of origin with this transformation or that of the preceding article, we have merely to add the co-ordinates x', y', z' of the new origin to the values found for x, y, and z.

GENERAL PRINCIPLES OF INTERPRETATION.

687. These follow from the convention of co-ordinates in space in much the same manner as the principles of Plane Analytic Geometry followed from the convention of plane co-ordinates. We may therefore state them without further argument, as follows:

I. *Any single equation in space-coördinates represents a surface.*

To hold with full generality, this statement must be understood to include (in addition to surfaces in the ordinary sense) imaginary surfaces, surfaces at infinity, surfaces that have degenerated into lines or points, and surfaces combined in groups.

II. *Two simultaneous equations in space-coördinates represent a line of section between two surfaces.*

This principle is also to be taken with restrictions corresponding to those above stated.

III. *Three simultaneous equations in space-coördinates represent* mnp *determinate points.*

These are the *points* of intersection of three surfaces, supposed to be of the mth, n^{th}, and p^{th} order respectively.

IV. *An equation which lacks the absolute term, represents a surface passing through the origin.*

V. *Transformation of co-ordinates in space does not alter the degree of a given equation, nor affect the form of its locus in any way.*

CHAPTER SECOND.

LOCUS OF THE FIRST ORDER IN SPACE.

688. Form of the Locus.—The general equation of the first degree in three variables, may be written

$$Ax + By + Cz + D = 0 \qquad (1),$$

where A, B, C, D are any four constants whatever.

Transforming (1) to parallel axes passing through a new origin $x'y'$, we get (Art. 684)

$$Ax + By + Cz + (Ax' + By' + Cz' + D) = 0.$$

Hence, if we suppose the new origin to be any fixed point in the locus of (1), the new absolute term will vanish, and our equation will take the form

$$Ax + By + Cz = 0 \qquad (2).$$

If we now change the directions of the reference-planes (Art. 685), we shall get, after expanding and collecting terms,

$$\left. \begin{aligned} &(A\cos\alpha + B\cos\beta + C\cos\gamma\,)\,x \\ +\,&(A\cos\alpha' + B\cos\beta' + C\cos\gamma'\,)\,y \\ +\,&(A\cos\alpha'' + B\cos\beta'' + C\cos\gamma'')\,z \end{aligned} \right\} = 0.$$

Hence, if we can take the new reference-planes so as to give the new axis of x and the new axis of y such directions that

$$A\cos\alpha + B\cos\beta + C\cos\gamma = 0$$
$$= A\cos\alpha' + B\cos\beta' + C\cos\gamma' \qquad (\text{Q}),$$

we shall reduce our equation to the simple form

$$z = 0 \qquad (3).$$

Now, obviously, the transformation from (1) to (2) is always possible; and that we can always effect the transformation from (2) to (3) will readily appear. For we can leave the primitive vertical reference-planes unchanged, obtaining our new system by merely revolving the primitive horizontal plane about the origin: in which case, we shall have

$$a = 90° - \gamma, \quad \beta = 0; \quad a' = 0, \quad \beta' = 90° - \gamma';$$

and the conditions at (Q), upon which the transformation we are now considering depends, will become

$$A \sin \gamma + C \cos \gamma = 0 \quad i.\,e. \quad \tan \gamma = -\frac{C}{A},$$

$$B \sin \gamma' + C \cos \gamma' = 0 \quad i.\,e. \quad \tan \gamma' = -\frac{C}{B}:$$

suppositions compatible with *any* real values of A, B, C.

We conclude, then, that by a proper transformation of co-ordinates we can always reduce the general equation of the first degree to the form

$$z = 0.$$

But this [Art. 675, Cor. 1, (3)] denotes the new reference-plane XY. Hence, (Art. 687, V,) *The locus of the First order in space is the Plane;* or, as we may otherwise state our result, *Every equation of the first degree in space represents a plane.*

THE PLANE UNDER GENERAL CONDITIONS.

689. General Form of the Equation to the Plane.—From what has just been shown, we learn that

$$Ax + By + Cz + D = 0$$

is the *Equation to any Plane.*

690. The Plane in terms of its Intercepts on the Axes.—Let the plane ABC, making upon the co-ordinate axes the intercepts $OA = a$, $OB = b$, $OC = c$, represent any plane

$$Ax + By + Cz + D = 0.$$

Making y and z, z and x, x and y simultaneously $= 0$ in succession, we obtain from this equation (Art. 675, Cor. 2)

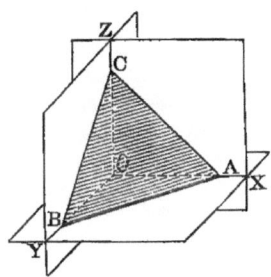

$$x = a = -\frac{D}{A} \quad \therefore \quad A = -\frac{D}{a} \qquad (1),$$

$$y = b = -\frac{D}{B} \quad \therefore \quad B = -\frac{D}{b} \qquad (2),$$

$$z = c = -\frac{D}{C} \quad \therefore \quad C = -\frac{D}{c} \qquad (3).$$

Substituting these values of A, B, C in the general equation, we obtain the equation to the Plane in terms of its intercepts, namely,

$$\frac{x}{a} + \frac{y}{b} + \frac{z}{c} = 1.$$

691. The Plane in terms of the Direction-cosines of its Perpendicular.—Let the perpendicular from the origin upon any plane be $= p$, and let its direction-angles be α, β, γ. Then, a, b, c being the intercepts of the plane, we shall have

$$a = \frac{p}{\cos \alpha}, \quad b = \frac{p}{\cos \beta}, \quad c = \frac{p}{\cos \gamma}.$$

Substituting these values in the equation of the preceding article, we obtain

$$x \cos \alpha + y \cos \beta + z \cos \gamma = p,$$

the equation to the Plane in the terms now required.

692. Reduction of the General Equation to the form last found.—We may suppose the reduction to be effected by dividing the general equation

$$Ax + By + Cz + D = 0$$

throughout by some quantity Q. If so, we shall have $A = Q \cos \alpha$, $B = Q \cos \beta$, $C = Q \cos \gamma$: whence

$$Q^2 (\cos^2 \alpha + \cos^2 \beta + \cos^2 \gamma) = A^2 + B^2 + C^2.$$

Now (Art. 682), $\cos^2 \alpha + \cos^2 \beta + \cos^2 \gamma = 1$. Hence, $Q = \sqrt{A^2 + B^2 + C^2}$; and we learn that

$$\cos \alpha = \frac{A}{\sqrt{(A^2 + B^2 + C^2)}},$$

$$\cos \beta = \frac{B}{\sqrt{(A^2 + B^2 + C^2)}},$$

$$\cos \gamma = \frac{C}{\sqrt{(A^2 + B^2 + C^2)}},$$

and that, for the perpendicular from the origin upon a plane given by the general equation, we have

$$p = - \frac{D}{\sqrt{(A^2 + B^2 + C^2)}}.$$

By always taking the radical Q with that sign which will render p positive, the resulting signs of $\cos \alpha$, $\cos \beta$, cos γ will indicate whether the direction-angles of the perpendicular are acute or obtuse.

THE PLANE UNDER SPECIAL CONDITIONS.

693. Equation to a Plane passing through Three Fixed Points.—By a process exactly analogous to that of Art. 95, this is found to be

$$\left. \begin{aligned} [y'(z''-z''') + y''(z'''-z') + y'''(z'-z'')] \, x \\ + [z'(x''-x''') + z''(x'''-x') + z'''(x'-x'')] \, y \\ + [x'(y''-y''') + x''(y'''-y') + x'''(y'-y'')] \, z \end{aligned} \right\} = \left\{ \begin{aligned} (y'' \, z''' - y''' z'') \, x' \\ + (y''' z' \ -y' \ z'') \, x'' \\ + (y' \ \ z'' - y'' z' \ ) \, x''', \end{aligned} \right.$$

in which $x'y'z'$, $x''y''z''$, $x'''y'''z'''$ are the three points which determine the plane.

694. Angle between two Planes.—This is evidently equal, or else supplemental, to the angle between the perpendiculars thrown upon the planes from the origin. Now, if p, p' be the lengths of these perpendiculars, α, β, γ and α', β', γ' their direction-angles, and δ the distance between the points xyz, $x'y'z'$ in which they pierce their respective planes, we shall have (Art. 681)

$$p^2 + p'^2 - 2pp' \cos \varphi = \delta^2 = (x - x')^2 + (y - y')^2 + (z - z')^2,$$

where φ = the angle between the planes or their perpendiculars. But (Art. 681, Cor.) $p^2 = x^2 + y^2 + z^2$, and $p'^2 = x'^2 + y'^2 + z'^2$. Hence, after obvious reductions,

$$pp' \cos \varphi = xx' + yy' + zz';$$

or, since $x = p \cos \alpha$, $y = p \cos \beta$, $z = p \cos \gamma$; $x' = p' \cos \alpha'$, $y' = p' \cos \beta'$, $z' = p' \cos \gamma'$,

$$\cos \varphi = \cos \alpha \cos \alpha' + \cos \beta \cos \beta' + \cos \gamma \cos \gamma' \quad \text{(A)}.$$

Here it becomes evident, upon a moment's reflection, that the direction-angles γ and γ' are respectively equal to the angles which the two planes make with the *hori-zontal* reference-plane; that a and a' are respectively equal to those made with the *first* vertical plane; and that β and β' are respectively equal to those made with the *second* vertical plane. Calling these new angles ξ and ξ', υ and υ', ζ and ζ', we have, then, as the expression for the *angle between two planes in terms of their inclinations to the reference-planes,*

$$\cos \varphi = \cos \xi \cos \xi' + \cos \upsilon \cos \upsilon' + \cos \zeta \cos \zeta' \quad (1).$$

Replacing the cosines in (1) by their values from Art. 692, namely,

$$\cos \xi = \frac{C}{\sqrt{(A^2 + B^2 + C^2)}}, \qquad \cos \xi' = \frac{C'}{\sqrt{(A'^2 + B'^2 + C'^2)}},$$

$$\cos \upsilon = \frac{A}{\sqrt{(A^2 + B^2 + C^2)}}, \qquad \cos \upsilon' = \frac{A'}{\sqrt{(A'^2 + B'^2 + C'^2)}},$$

$$\cos \zeta = \frac{B}{\sqrt{(A^2 + B^2 + C^2)}}, \qquad \cos \zeta' = \frac{B'}{\sqrt{(A'^2 + B'^2 + C'^2)}},$$

we obtain, as the expression for the *angle between two planes in terms of the co-efficients of their equations,*

$$\cos \varphi = \frac{AA' + BB' + CC'}{\sqrt{(A^2 + B^2 + C^2)(A'^2 + B'^2 + C'^2)}} \quad (2).$$

Corollary 1.—The two planes will be parallel if $\varphi = 0$; that is, if $\cos \varphi = 1$. As the condition of parallelism, then, the terms of the second member of (2) must be equal; or, after squaring and transposing,

$$(AB' - A'B)^2 + (BC' - B'C)^2 + (CA' - C'A)^2 = 0:$$

a condition which can only be satisfied by having simultaneously

$$\frac{A'}{B'} = \frac{A}{B}, \quad \frac{B'}{C'} = \frac{B}{C}, \quad \frac{C'}{A'} = \frac{C}{A}.$$

Corollary 2.—If the two planes are perpendicular to each other, we shall have $\cos \varphi = 0$: whence, as the condition of perpendicularity,

$$AA' + BB' + CC' = 0.$$

695. Equation to a Plane parallel to a given one.—From the condition reached in the first corollary to the preceding article, it is evident that this can finally be written in the form

$$Ax + By + Cz + D' = 0,$$

A, B, C being the co-efficients of x, y, z in the equation to the given plane. We learn, then, that *the equations to parallel planes differ only in their constant terms.*

Corollary.—The equations to planes parallel respectively to the three reference-planes, will be

$$z = \text{constant}, \quad x = \text{constant}, \quad y = \text{constant}.$$

696. Equation to a Plane perpendicular to a given one.—If $A'x + B'y + C'z + D' = 0$ be the given plane, we may write the required equation in either of the forms

$$Px - By - Cz - D = 0 \qquad (1),$$
$$Ax - Qy + Cz + D = 0 \qquad (2),$$
$$Ax + By - Rz + D = 0 \qquad (3),$$

by merely making, in accordance with Art. 694, Cor. 2,

$$P = \frac{BB' + CC'}{A'}, \quad Q = \frac{CC'' + AA'}{B'}, \quad R = \frac{AA' + BB'}{C'}.$$

Corollary.—In particular, the equations to planes perpendicular to the reference-planes will assume the forms

$$Ax + By + D = 0, \quad By + Cz + D = 0, \quad Cz + Ax + D = 0.$$

For (Art. 675, Cor. 1) R must vanish for the horizontal reference-plane, P for the first vertical, and Q for the second.

697. Length of the Perpendicular from a Fixed Point to a Given Plane.—Let the fixed point be xyz, and the given plane $x \cos a + y \cos \beta + z \cos \gamma - p = 0$. If we produce the perpendicular p, and then project upon it the radius vector of xyz, it is evident that the *required* perpendicular will be equal to the difference between this projection and p. Hence, (Art. 680, Cor.,) we have

$$P = \pm \, (x \cos a + y \cos \beta + z \cos \gamma - p),$$

the upper or lower sign being used according as the given point and the origin lie on *opposite* sides of the given plane, or on the *same* side.

Corollary 1.—For the perpendicular from xyz to the plane $Ax + By + Cz + D = 0$, we have (Art. 692)

$$P = \frac{Ax + By + Cz + D}{\sqrt{(A^2 + B^2 + C^2)}} \, .$$

Corollary 2.—Since we have agreed to consider the perpendicular from the origin upon any plane as positive in all cases, consistency requires that perpendiculars dropped upon a plane from any point on the *same* side of it as the origin, shall be reckoned *positive;* and those dropped from the *opposite* side, *negative.*

698. Equation to a Plane passing through the Common Section of two given ones.—By reasoning similar to that of Arts. 107, 108, it is evident that this may be written

$$(Ax + By + Cz + D) + k \, (A'x + B'y + C'z + D') = 0;$$

or, by adopting abridgments similar to those used in Plane Geometry,

$$P + kP' = 0.$$

Corollary.—Analogy leads at once to the conclusion, that an equation of the form

$$lP + mP' + nP'' = 0,$$

in which l, m, n are arbitrary constants, denotes a plane passing through the *point* in which the *three* planes P, P', P'' intersect.

699. Equation to the Plane bisecting the angle between two given ones.—The reasoning of Art. 109 applies here, and the required equation (Art. 692) is

$$Q'P \pm QP' = 0 \qquad (1),$$

or, if the equations to the given planes are already reduced to terms of their direction-cosines,

$$\alpha \pm \beta = 0 \qquad (2),$$

the upper sign denoting the *external* bisector, and the lower the *internal* one.

700. Condition that Four Points shall lie on one Plane.—The fourth point must of course satisfy the equation to the plane of the other three, and the required condition is therefore obtained by putting $x^{\text{iv}} y^{\text{iv}} z^{\text{iv}}$ instead of xyz in the equation of Art. 693.

701. Condition that Three Planes shall pass through one Right Line.—The equation to the third plane must take the form (Art. 698) of the equation to a plane passing through the common section of the other two. There must, then, be some constant $-n$, such that

$$-nP'' = lP + mP'.$$

Hence, the required condition is

$$lP + mP' + nP'' = 0.$$

In other words, *Three planes pass through one right line whenever their equations, upon being multiplied by three suitable constants and added together, vanish identically.*

702. Condition that Four Planes shall meet in One Point.—By applying the reasoning of the preceding article to the result of the corollary to Art. 698, we learn that this condition may be written

$$lP + mP' + nP'' + rP''' = 0;$$

or, if the equations to the planes be in terms of their direction-cosines,

$$l\alpha + m\beta + n\gamma + r\delta = 0.$$

Hence, *Four planes pass through one point whenever their equations, upon being multiplied by four suitable constants and added together, vanish identically.*

QUADRIPLANAR CO-ORDINATES.

703. The condition of the preceding article subjects its constants l, m, n, r to certain restrictions, consistent with the *identical* vanishing of the function

$$l\alpha + m\beta + n\gamma + r\delta.$$

But if we now free these constants from this condition for the converging of four planes, making them absolutely arbitrary, we learn, by reasoning entirely analogous to that of Arts. 208—217, that if $a = 0$, $\beta = 0$, $\gamma = 0$, $\delta = 0$ be the equations to any four planes forming a tetrahedron, the equation

$$la + m\beta + n\gamma + r\delta = 0$$

is a general symbol for any plane in space.

We thus arrive at what may be called a system of *quadriplanar co-ordinates*, analogous to the trilinear system of Plane Geometry.

LINEAR LOCI IN SPACE.

704. By II of Art. 687, it appears that all lines in space, whether right or curved, are to be solved as the *common sections of two surfaces*, and hence must be represented by *two simultaneous equations in three variables*. In particular, the Right Line in Space, which is the only line we shall have room to consider, must be treated as the common section of two *planes*.

705. Equations to the Right Line in Space.—We might represent this line by the two general equations

$$Ax + By + Cz + D = 0, \quad A'x + B'y + C'z + D' = 0,$$

but it is far more convenient to denote it by the simultaneous equations of its two *projecting* planes (Art. 677), in accordance with the method by which all curves in space are usually represented by means of their " projecting cylinders."

In pursuance of this method, then, the equations to the Right Line projected upon the two vertical reference-planes, will be of the form (Art. 696, Cor.)

$$By + Cz + D = 0, \quad Nz + Mx + L = 0.$$

An. Ge. 48.

Now it is noticeable, that, while these equations taken *together* involve *three* variables, each of them taken *separately* involves but *two*. The first, interpreted as an equation in two variables, denotes a right line in the first vertical plane; the second, similarly interpreted, denotes a right line in the second vertical plane. But these lines, by the principle of the corollary to Art. 696, must also lie in the two planes which the equations denote when interpreted in space: hence, they are the common sections of these planes and the vertical planes of reference; or, in other words, they are the *projections* of the right line represented by the *simultaneous* equations $By + Cz + D = 0$, $Nz + Mx + L = 0$, upon the two vertical reference-planes.

We see, then, that we may either regard the two determining equations of the Right Line as the space-equations to its two projecting planes, or as the plane-equations to its two projections. It is customary to interpret them in the latter way, and as each involves two, and only two, arbitrary constants, to write them

$$x = mz + a, \quad y = nz + b.$$

Thus the axis of z is made their common axis of abscissas, and the constants m, a, n, b take meaning as follows:

$m =$ the tangent of the angle which the projection on the second vertical plane makes with the axis of z.

$a =$ the intercept which the same projection forms on the axis of x.

$n =$ the tangent of the angle which the projection on the first vertical plane makes with the axis of z.

$b =$ the intercept which this projection forms on the axis of y.

We learn, then, that *the position and direction of a right line in space, depend upon the magnitudes and signs of four arbitrary constants.*

706. Symmetrical Equations to the Right Line in Space.—Let the line pass through an arbitrary point $x'y'z'$, and let its direction-angles be a, β, γ.

Then, if $l =$ the distance from $x'y'z'$ to any point xyz of the line, the projections of l upon the three co-ordinate axes (Art. 679) will be $l \cos a$, $l \cos \beta$, $l \cos \gamma$. But by definition (Art. 677) these projections are respectively equal to $x - x'$, $y - y'$, $z - z'$. Hence,

$$l \cos a = x - x', \quad l \cos \beta = y - y', \quad l \cos \gamma = z - z':$$

whence, solving for l and equating the three results,

$$\frac{x - x'}{\cos a} = \frac{y - y'}{\cos \beta} = \frac{z - z'}{\cos \gamma} :$$

which are the symmetrical equations sought.

707. To find the Direction-cosines of a Right Line given by its Projections.—The direction-cosines of any right line are of course the same as those of its parallel through the origin. Let the projections of such a parallel be

$$\frac{x}{l} = \frac{y}{m} = \frac{z}{n}.$$

Then, if ρ be the radius vector of any point xyz on the parallel, we shall have (Art. 681, Cor.)

$$\rho^2 = x^2 + y^2 + z^2,$$

and, from the above equations of projection,

$$y = \frac{mx}{l}, \quad z = \frac{nx}{l}.$$

Solving the last three equations for x, y, z, we obtain

$$x = \frac{l\rho}{\sqrt{(l^2 + m^2 + n^2)}}, \quad y = \frac{m\rho}{\sqrt{(l^2 + m^2 + n^2)}},$$

$$z = \frac{n\rho}{\sqrt{(l^2 + m^2 + n^2)}}.$$

But, by the doctrine of projections,

$$x = \rho \cos \alpha, \quad y = \rho \cos \beta, \quad z = \rho \cos \gamma.$$

Substituting, and dividing through by ρ.

$$\cos \alpha = \frac{l}{\sqrt{(l^2 + m^2 + n^2)}}, \quad \cos \beta = \frac{m}{\sqrt{(l^2 + m^2 + n^2)}},$$

$$\cos \gamma = \frac{n}{\sqrt{(l^2 + m^2 + n^2)}}$$

Corollary.—To find the direction-cosines of a line whose projections are given in any form whatever, throw its equations into the form

$$\frac{x - x'}{l} = \frac{y - y'}{m} = \frac{z - z'}{n}:$$

when the required functions will be l, m, n, each divided by $\sqrt{l^2 + m^2 + n^2}$.

708. Angle between two Lines in Space.—The angle θ between two right lines in space is obviously equal to that between their respective parallels through the origin. Hence, by formula (A) of Art. 694,

$$\cos \theta = \cos \alpha \cos \alpha' + \cos \beta \cos \beta' + \cos \gamma \cos \gamma' \quad (1):$$

which expresses the *angle between two right lines in terms of their direction-cosines.*

Substituting for cos *a*, cos *a'*, etc., from the preceding article, we get

$$\cos \theta = \frac{ll' + mm' + nn'}{\sqrt{(l^2 + m^2 + n^2)(l'^2 + m'^2 + n'^2)}} \quad (2):$$

which expresses the *angle between two right lines in terms of their projections.*

Corollary 1.—The condition that two right lines in space shall be parallel, derived from (1), is

$$\cos a \cos a' + \cos \beta \cos \beta' + \cos \gamma \cos \gamma' = 1 \quad (1),$$

or, derived from (2) by steps analogous to those in the first corollary of Art. 694,

$$\frac{l'}{m'} = \frac{l}{m}; \quad \frac{m'}{n'} = \frac{m}{n}; \quad \frac{n'}{l'} = \frac{n}{l} \quad (2).$$

Corollary 2.—The condition that two right lines in space shall be perpendicular to each other, derived from (1), is

$$\cos a \cos a' + \cos \beta \cos \beta' + \cos \gamma \cos \gamma' = 0 \quad (1),$$

or, derived from (2),

$$ll' + mm' + nn' = 0 \quad (2).$$

709. Equation to a Right Line perpendicular to a given Plane.—If $Ax + By + Cz + D = 0$ be the given plane, the required equation may be written

$$\frac{x - a}{A} = \frac{y - b}{B} = \frac{z - c}{C}.$$

For we may suppose the perpendicular to pass through any fixed point *abc*; and, by Art. 692, its direction-cosines must be proportional to A, B, C.

710. Angle contained between a Right Line and a Plane.—This being the complement of the angle contained between the given line and a perpendicular to the plane, if the given line be

$$\frac{x-x'}{l} = \frac{y-y'}{m} = \frac{z-z'}{n},$$

we have, by comparing Arts. 708, 709,

$$\sin \theta = \frac{Al + Bm + Cn}{\sqrt{(A^2 + B^2 + C'^2)(l^2 + m^2 + n^2)}}.$$

Corollary.—The condition that a right line shall be parallel to a given plane, is

$$Al + Bm + Cn = 0.$$

711. Condition that a Right Line shall lie wholly in a given Plane.—If a right line lies wholly in a given plane, the z co-ordinate resulting from an elimination between the equation to the plane and those of the line must of course be indeterminate. Hence, if the plane be $Ax + By + Cz + D = 0$, and the line $(x = mz + a, y = nz + b)$, so that we have by elimination $A(mz + a) + B(nz + b) + Cz + D = 0$, or

$$z = -\frac{Aa + Bb + D}{Am + Bn + C},$$

we must have, as the condition required, the simultaneous relations

$$Am + Bn + C = Aa + Bb + D = 0.$$

Remark.—This result is corroborated by the fact, that the vanishing of the numerator of z indicates that the point $(a, b, 0)$, in which the line pierces the horizontal reference-plane, is in the given plane; while the vanishing of the denominator shows, by the corollary to the previous article, that the line is parallel to the given plane: two conditions which obviously place the line wholly in that plane.

712. Condition that two Right Lines in Space shall intersect.—Two right lines in space will not in general intersect, because the *four* equations

$$x = mz + a, \qquad y = nz + b,$$
$$x = m'z + a', \qquad y = n'z + b',$$

being in general independent, are not compatible with simultaneous values of the *three* variables x, y, z. If, then, the two lines represented by these four equations do intersect, one of the equations must be derivable from the other three, and the condition of such a derivation will be the required condition of intersection.

We form this condition, of course, by eliminating x, y, z from the four equations. To do this, solve the first and third, and also the second and fourth, for z, and equate the two values thus found. The result is

$$\frac{m - m'}{n - n'} = \frac{a - a'}{b - b'}.$$

EXAMPLES INVOLVING EQUATIONS OF THE FIRST DEGREE.

1. Show that, if L, M and N, R be the equations to two *intersecting* right lines, they will be connected by some identical relation

$$lL + mM + nN + rR = 0,$$

and that the plane of the two intersecting lines may be represented by either of the equations

$$lL + mM = 0, \quad nN + rR = 0.$$

2. Find the equation to the plane which passes through the lines

$$\frac{x - a}{l} = \frac{y - b}{m} = \frac{z - c}{n}, \quad \frac{x - a}{l'} = \frac{y - b}{m'} = \frac{z - c}{n'}.$$

3. Find the equations to the traces of any given plane upon the three reference-planes, and prove that if a right line be perpendicular to a given plane, its projections will be perpendicular to the traces of the plane.

4. Find the equations to the three planes which pass through the traces of a given plane upon the reference-planes, and are each perpendicular to the plane.

5. Find the equation to the plane which passes through a given right line and makes a given angle with a given plane.

6. If $(\alpha', \beta', \gamma')$, $(\alpha'', \beta'', \gamma'')$ be the direction-angles of two right lines, prove that the direction-cosines of the *external* bisector of the angle between them, are proportional to

$$\cos \alpha' + \cos \alpha'', \quad \cos \beta' + \cos \beta'', \quad \cos \gamma' + \cos \gamma'',$$

and that those of the *internal* bisector are proportional to

$$\cos \alpha' - \cos \alpha'', \quad \cos \beta' - \cos \beta'', \quad \cos \gamma' - \cos \gamma''.$$

7. Three planes meet in one point, and through the common section of each pair a plane is drawn perpendicular to the third · prove that in general the planes thus drawn pass through one right line.

8. Find the equation to a plane parallel to two given right lines, and thence determine the shortest distance between the lines.

9. A plane passes through the origin: find the bisector of the angle between its traces on two of the reference-planes.

10. Prove that the locus of the middle points of all right lines parallel to a given plane, and terminated by two fixed right lines which do not intersect, is a right line.

CHAPTER THIRD.

LOCUS OF THE SECOND ORDER IN SPACE.

713. The general equation of the second degree in three variables, which is the symbol of the space-locus of the Second order, may be written

$$Ax^2 + 2Hxy + By^2 + 2Kyz + Ez^2 + 2Lzx$$
$$+ 2Gx + 2Fy + 2Dz + C = 0 \quad (1),$$

where A, B, E; H, K, L; C, D, F, G are any ten constants whatever.

Since we can divide this equation throughout by C, it appears that the number of *independent* constants is *nine*. Hence, nine conditions are necessary and sufficient to determine the locus. Thus we learn that, for example, *the space-locus of the Second order is a surface of such a form that one, and but one, such surface can be passed through any nine points which do not lie in the same plane.*

714. The most general and complete criterion of the form of any surface, is afforded by its curves of section with different planes. Let us apply this criterion to test the figure of the surface denoted by (1).

If in (1) we make $z = 0$, that is, if we combine (1) with the equation to the horizontal reference-plane, we get

$$Ax^2 + 2Hxy + By^2 + 2Gx + 2Fy + C = 0,$$

the general equation to the Conic. Hence, as we can transform the reference-plane of XY to any plane that we please, and as such a transformation will not affect the degree of the equation of section just found, *Every plane section of a surface of the Second order is a conic.*

Moreover, if we combine (1) with $z = k$, that is (Art. 695, Cor.), if we intersect our locus by any plane parallel to the plane of XY, we get

$$Ax^2 + 2Hxy + By^2 + 2G'x + 2F'y + C' = 0,$$

where $G' = G + kL$, $F' = F + kK$, $C' = C + 2kD + k^2E$. Hence, (Art. 669,) since the co-efficients A, H, B remain unchanged whatever be the value of k, *The sections of a surface of the Second order by parallel planes are similar conics.*

An. Ge. 49.

To denote, then, that the surface of the Second order is represented by an equation of the second degree in space, and that all its plane sections are curves of the Second order, we shall henceforth call it the *Quadric.*

THE QUADRIC IN GENERAL.

715. To increase the clearness of our conception of the quadric figure, we must now reduce the general equation (1) to its simplest forms. We can effect this reduction most rapidly, however, by taking out a few leading properties of the surface, partly from the general equation itself, and partly from the results of its first transformations.

716. Let us transform (1) to parallel axes through a new origin $x'y'z'$. Since we merely have to write $x + x'$ for x, $y + y'$ for y, and $z + z'$ for z, it is easily seen that the new equation will be

$$Ax^2 + 2Hxy + By^2 + 2Kyz + Ez^2 + Lzx$$
$$+ 2G'x + 2F'y + 2D'z + C' = 0 \qquad (2),$$

where C', the new absolute term, is the result of substituting $x'y'z'$ in (1), and in the new co-efficients of x, y, z we have

$$G^l = Ax' + Hy' + Lz' + G,$$
$$F' = Hx' + By' + Kz' + F,$$
$$D' = Lx' + Ky' + Ez' + D,$$

these quantities being *planar functions of the new origin.* It deserves especial notice, that G' (Alg., 411) is the

---

* For the discussion of Quadrics in complete detail, the reader is referred to Salmon's *Geometry of Three Dimensions,* from which the investigations of the following pages have in the main been reduced.

derived polynomial (or *derivative*, as we shall call it for brevity) of C' with respect to x; that F' is the derivative of C' with respect to y; and D', the derivative of C' with respect to z. Hence, if we write the original equation (1) in the abbreviated form $U = 0$, we may use for the four co-efficients C', G', F', D' the convenient symbols U', U_x', U_y', U_z'.

717. As we shall also find it convenient to employ the so-called *discriminant* of the equation $U = 0$, and several of its derivatives, we will determine their values before advancing farther.

The discriminant of any function may be defined as the result obtained by solving its several derivatives for its variables, and then substituting the values of these in the function itself. Accordingly, solving for x, y, z in

$$U_x = Ax + Hy + Lz + G = 0,$$
$$U_y = Hx + By + Kz + F = 0,$$
$$U_z = Lx + Ky + Ez + D = 0,$$

and then substituting in U, we get

$$ABCE + 2ADFK + 2BDGL + 2CHKL + 2EFGH$$
$$- ABD^2 - ACK^2 - AEF^2 - BCL^2 - BEG^2 - CEH^2$$
$$+ K^2G^2 + L^2F^2 + H^2D^2 - 2DFHL - 2DGHK - 2FGKL.$$

This, then, is the discriminant of the given quadric $U = 0$, and may be appropriately represented by Δ.

If we now denote the several derivatives of Δ, taken with reference to G, F, D, C in succession, by $2g$, $2f$, $2d$, c, we shall have

$$g = BDL + EFH - BEG + GK^2 - DHK - FKL,$$
$$f = ADK + EGH - AEF + FL^2 - DHL - GKL,$$
$$d = AFK + BGL - ABD + DH^2 - FHL - GHK,$$
$$c = ABE + 2HKL - AK^2 - BL^2 - EH^2.$$

718. It will be convenient next to determine the condition upon which the radius vector of the Quadric will be bisected in the origin. To find this, throw the general equation into the vectorial form, by writing $\rho \cos \alpha$ for x, $\rho \cos \beta$ for y, and $\rho \cos \gamma$ for z, which we may evidently do if α, β, γ are the direction-angles of the radius vector. Equation (1) then becomes

$$(A \cos^2 \alpha + 2H \cos \alpha \cos \beta + B \cos^2 \beta$$
$$+ 2K \cos \beta \cos \gamma + E \cos^2 \gamma + 2L \cos \gamma \cos \alpha)\rho^2$$
$$+ 2(G \cos \alpha + F \cos \beta + D \cos \gamma)\rho - C = 0.$$

If the origin bisects the radius vector, this equation will have its roots numerically equal with opposite signs. Hence, the required condition of bisection is

$$G \cos \alpha + F \cos \beta + D \cos \gamma = 0;$$

or, after multiplying through by ρ, and replacing the corresponding x, y, z, we learn that all radii vectores bisected in the origin must lie in the plane

$$Gx + Fy + Dz = 0.$$

719. If, then, in the equation to the Quadric we had G, F, D all $= 0$, the condition of bisection would be satisfied for all possible values of α, β, γ; or, in other words, *every* right line drawn through the origin to meet the quadric would be bisected in the origin, and the origin would be a *center* of the quadric.

720. Resuming now our transformations of equation (1), let us suppose the new origin $x'y'z'$ to which (2) is referred, to be a center. The new G, F, D will then vanish, and we learn (Art. 716) that the center lies at the intersection of the three planes

$$U_x = 0, \quad U_y = 0, \quad U_z = 0.$$

Solving these three equations, we obtain, as the *co-ordi-nates of the center,*

$$x' = \frac{g}{c}, \quad y' = \frac{f}{c}, \quad z' = \frac{d}{c},$$

where g, f, d, c have the values given in the table of Art. 717.

The center, then, is a single determinate point, and will be a finite real one if c is not zero, but not otherwise. Hence, *Quadrics are either central or non-central, and central quadrics have only one center.*

721. By taking, then, the center for origin, the equation to any central quadric may be written

$$Ax^2 + 2Hxy + By^2 + 2Kyz + Ez^2 + 2Lzx + C' = 0 \;(3),$$

where (Art. 716) by substituting the co-ordinates of the center in U', we readily find

$$C' = \frac{Gg + Ff + Dd + Cc}{c} = \frac{\varDelta}{c},$$

where G, F, D, C are the co-efficients of the planar terms of (1), and $g, f, d, c, \varDelta$ have the meanings assigned in Art. 717.

722. Let us next inquire into the form of the *diametral surfaces* of the Quadric.

A diametral surface of a given surface may be defined as the locus of the middle points of chords drawn parallel to a given right line. Suppose, then, that a, β, γ are the direction-angles common to a system of chords in a quadric, and let us remove the origin of equation (1) to any point on the locus of the middle points of the system. By Art. 718, the new co-efficients G', F', D' must then fulfill the condition

$$G' \cos \alpha + F' \cos \beta + D' \cos \gamma = 0,$$

and the equation to the diametral surface of the Quadric will therefore be (Art. 716)

$$U_x \cos \alpha + U_y \cos \beta + U_z \cos \gamma = 0.$$

This (Art. 698, Cor.) denotes a plane passing through the intersection of the three planes U_x, U_y, U_z, namely (Art. 720), through the *center;* and, as the direction-angles α, β, γ are arbitrary, we have the theorem: *Every surface diametral to a quadric is a plane passing through the center, and every plane passing through the center of a quadric is a diametral plane.*

It should be observed, of course, that this theorem applies to the non-central quadrics only by regarding a point infinitely distant from the origin as their center. Such, indeed, is the fact indicated by the central co-ordinates (Art. 720) $g : e$, $f : e$, $d : e$, in which $e = 0$ for the non-central quadrics. But if the diametral planes pass through a common point at infinity, their several common sections will meet in a point at infinity; in other words, will be parallel. Hence, *The diametral planes of a non-central conic are parallel to a fixed right line.*

723. The diametral planes which bisect chords parallel to the axis of x, the axis of y, and the axis of z respectively, are found by successively supposing $\beta = \gamma = 90°$, $\gamma = \alpha = 90°$, $\alpha = \beta = 90°$ in the equation of the preceding article. They are therefore respectively

$U_x = 0$, $U_y = 0$, $U_z = 0$; or, writing the abbreviations in full,

$$Ax + Hy + Lz + G = 0,$$

$$Hx + By + Kz + F = 0,$$

$$Lx + Ky + Ez + D = 0.$$

For brevity, a diametral plane is said to be *conjugate* to the direction of the chords which it bisects. Now the condition that the plane U_x, which is conjugate to the axis of x, may be parallel to the axis of y, according to the corollary of Art. 696 is $H = 0$. But, obviously, this is also the condition that the plane U_y, which is conjugate to the axis of y, may be parallel to the axis of x. Hence, as the co-ordinate axes may have any direction, *If a diametral plane conjugate to a given direction be parallel to a given right line, the plane conjugate to this line will be parallel to the first direction.*

724. If in our general equation (1) we had H, K, L all $= 0$, the equations of the preceding article would be reduced to

$$Ax + G = 0, \quad By + F = 0, \quad Ez + D = 0.$$

The diametral planes conjugate to the three axes would thus (Art. 695, Cor.) become parallel to the reference-planes; and, by the theorem last proved, *each would be conjugate to the common section of the other two.*

Three diametral planes thus related are called *conjugate planes*, and the three right lines in which they cut each other two and two are called *conjugate diameters.* Three diameters are therefore conjugate, when each is conjugate to the plane of the other two.

We thus reach the important result, that whenever the

equation to a quadric lacks the co-efficients H, K, L, the co-ordinate axes to which it is referred are parallel to a set of conjugate diameters; and, conversely, that by employing axes parallel to a set of conjugates, we can always cause these co-efficients to vanish from the equation to a central quadric.

725. As the foregoing argument evidently does not conflict with the supposition all along made, that the co-ordinate axes are rectangular, it follows that *every central quadric has one set of conjugate planes and diameters which are at right angles to each other.*

In fact, diametral planes perpendicular to the chords which they bisect, or *principal planes*, as they are called, exist in all quadrics whether central or not; though the triconjugate groups, of course, are peculiar to central quadrics. For if we seek the condition that the plane (Art. 722)

$$U_x \cos \alpha + U_y \cos \beta + U_z \cos \gamma = 0$$

may be perpendicular to its conjugate chords, the principle (Art. 709) that the direction-cosines of the chords must be proportional to the co-efficients of the plane, gives us, if we put k = the constant ratio between these quantities,

$$A \cos \alpha + H \cos \beta + L \cos \gamma = k \cos \alpha,$$

$$H \cos \alpha + B \cos \beta + K \cos \gamma = k \cos \beta,$$

$$L \cos \alpha + K \cos \beta + E \cos \gamma = k \cos \gamma.$$

Eliminating $\cos \alpha$, $\cos \beta$, $\cos \gamma$ from these equations, the required condition is

$$k^3 - (A + B + E) k^2$$
$$+ (AB + BE + EA - H^2 - K^2 - L^2) k - c = 0,$$

where c has the same value as in Art. 717. Having thus a cubic for determining the ratio k, we learn that *a quadric has in general three, and only three, principal planes.*

In the non-central quadrics however, since in them (Art. 721) we have $c = 0$, one of the roots of this cubic must be 0, and the equation to one of the principal planes will therefore assume the form

$$0x + 0y + 0z + \text{constant} = 0.$$

In the non-central quadrics, therefore, by the analogy of Art. 110, the third principal plane is situated at infinity.

726. We are now prepared to put our general equation into its simplest forms.

First, let us suppose that the derivative c is *not zero.* From equation (3) in Art. 721, which is already referred to the center as origin, we can at once proceed by taking for new axes a set of conjugate diameters; and, as we still adhere to rectangular co-ordinates, let these new axes be conjugate to the principal planes. Then (Art. 724) the co-efficients H, K, L vanish from (3), and the equation to any *central* quadric takes the form

$$A'x^2 + B'y^2 + E'z^2 + C' = 0 \qquad (4).$$

Secondly, suppose that c is *equal to zero.* We can not then arrive at the form (3); but, going back to (1), we may first change the direction of the rectangular axes, and then remove the origin. Now it can readily be shown, that, in passing from one set of rectangular axes to another, the new co-efficients of x^2, y^2, z^2 (or A', B', E') are the three roots of the cubic *

$$u^3 - (A + B + E) u^2$$
$$+ (AB + BE + EA - H^2 - K^2 - L^2) u - c = 0.$$

---

* We append Salmon's proof of this, as it is remarkable for its brevity. See his *Geometry of Three Dimensions*, p. 50.

" Let us suppose that by using the most general transformation, which is of the form

$$x = \lambda \bar{x} + \mu \bar{y} + \nu \bar{z}, \quad y = \lambda' \bar{x} + \mu' \bar{y} + \nu' \bar{z}, \quad z = \lambda'' \bar{x} + \mu'' \bar{y} + \nu'' \bar{z},$$

that $$A x^2 + 2Hxy + By^2 + 2Kyz + Ez^2 + 2Lzx$$

becomes $$A' \bar{x}^2 + 2H' \overline{xy} + B' \bar{y}^2 + 2K' \overline{yz} + E' \bar{z}^2 + 2L' \overline{zx},$$

which we write for shortness $U = \bar{U}$. If both systems of co-ordinates be rectangular, we must have

$$x^2 + y^2 + z^2 = \bar{x}^2 + \bar{y}^2 + \bar{z}^2,$$

which we write for shortness $S = \bar{S}$. Then if k be any constant, we must have $U + kS = \bar{U} + k\bar{S}$. And if the first side be resolvable into factors, so must also the second. The discriminants of $U + kS$ and $\bar{U} + k\bar{S}$ must therefore vanish for the same values of k. But the first discriminant is

$$k^3 - (A + B + E)k^2 + (AB + BE + EA - H^2 - K^2 - L^2)k - c.$$

Equating then the co-efficients of the different powers of k to the corresponding co-efficients in the second, we learn that if the equation be transformed from one set of rectangular axes to another, we must have

$$A + B + E = A' + B' + E',$$

$$AB + BE + EA - H^2 - K^2 - L^2 = A'B' + B'E' + E'A' - H'^2 - K'^2 - L'^2,$$

$$ABE + 2HKL - AK^2 - BL^2 - EH^2 =$$
$$A'B'E' + 2H'K'L' - A'K'^2 - B'L'^2 - E'H'^2.''$$

By solving these three equations for either A', B', or E', we obtain the cubic in the text above, where u is merely a symbol for the unknown co-efficient.

Hence, as we now have $c = 0$, one of the roots of this cubic, and therefore one of the new co-efficients A', B', E', must vanish whatever be the directions of the new rectangular axes. By taking these axes parallel to the principal planes, thus causing the new H', K', L' to disappear, we can therefore reduce the original equation to the form

$$B'y^2 + E'z^2 + 2G'x + 2F'y + 2D'z + C = 0,$$

as this transformation (Art. 685) does not affect the absolute term. And now we can remove the origin to the point in which the line $F' = 0$, $D' = 0$ pierces the Quadric, thus destroying the absolute term as well as the co-efficients of y and z. The equation to a *non-central* quadric will then take the form

$$B'y^2 + E'z^2 + 2G''x = 0 ;$$

or, as it may be more symmetrically written,

$$y^2 + Qz^2 = Px \qquad\qquad (5).$$

Equations (4) and (5) are the simplest forms of the space-equation of the second degree.

CLASSIFICATION OF QUADRICS.

727. By means of equations (4) and (5), we can now ascertain the several varieties of quadric surfaces, and the peculiar figure of each.

728. We begin with the CENTRAL QUADRICS, represented by the equation

$$A'x^2 + B'y^2 + E'z^2 + C' = 0.$$

I. Let A', B', E' all be *positive*. Then, if C' is *negative*, the equation can at once be put into the form

$$\frac{x^2}{a^2} + \frac{y^2}{b^2} + \frac{z^2}{c^2} = 1,$$

where a, b, c are the lengths of the intercepts cut off upon the axes of x, y, and z respectively. The sections of the surface with any planes of the form $z = k$, $x = l$, $y = m$, are the ellipses

$$\frac{x^2}{a^2} + \frac{y^2}{b^2} = 1 - \frac{k^2}{c^2}, \quad \frac{y^2}{b^2} + \frac{z^2}{c^2} = 1 - \frac{l^2}{a^2}, \quad \frac{z^2}{c^2} + \frac{x^2}{a^2} = 1 - \frac{m^2}{b^2}.$$

These are real for every positive or negative value of k, l, m that is not greater respectively than a, b, c, but are imaginary for all greater values. The quadric, therefore, lies wholly inside the rectangular parallelopiped formed by the six planes $z = \pm c$, $x = \pm a$, $y = \pm b$, but is continuous within those limits, and, having elliptic sections with the reference-planes and all planes parallel to them, is properly called the *Ellipsoid*.

Its semi-axes are of course respectively equal to a, b, c, and we suppose that the reference-planes are so taken that a is in general greater than b, and b greater than c. But the following particular cases must be considered:

1. If $b = c$, the section with any plane $x = l$ becomes $y^2 + z^2 = constant$, and is therefore a circle. The surface may then be generated by revolving an ellipse upon its axis major, and is called an ellipsoid of major revolution; or, with greater exactness, the *Prolate Spheroid*.

2. If $b = a$, the section with any plane $z = k$ becomes a circle, the surface may be generated by revolving an ellipse upon its minor axis, and is therefore called an ellipsoid of minor revolution; or, the *Oblate Spheroid*.

3. If $a = b = c = r$, all the plane sections of the surface are circles, and the equation becomes

$$x^2 + y^2 + z^2 = r^2,$$

which is therefore the equation to the *Sphere*.

Next, if C' is *zero*, our general central equation, taking the form

$$A'x^2 + B'y^2 + E'z^2 = 0,$$

can only be satisfied by the simultaneous values

$$x = 0, \quad y = 0, \quad z = 0,$$

and thus denotes the *Point*, which may therefore be regarded as an *infinitely small ellipsoid*.

Finally, if C' is *positive*, the general equation becomes

$$\frac{x^2}{a^2} + \frac{y^2}{b^2} + \frac{z^2}{c^2} = -1,$$

which having the ellipsoidal form in its first member, but involving an impossible relation, may be said to denote an *imaginary ellipsoid*.

II. Let A' and B' be *positive*, but E' be *negative*. Then, if C' be also *negative*, we can write the equation, in terms of the intercepts,

$$\frac{x^2}{a^2} + \frac{y^2}{b^2} - \frac{z^2}{c^2} = 1.$$

Here, the sections formed by the planes $x = l$, $y = m$, are the hyperbolas

$$\frac{y^2}{b^2} - \frac{z^2}{c^2} = 1 - \frac{l^2}{a^2}, \quad \frac{x^2}{a^2} - \frac{z^2}{c^2} = 1 - \frac{m^2}{b^2},$$

which are real for all values of l and m, but whose branches cease to lie on the right and left of the center, and are found above and below it, when $l > a$ and $m > b$. The section by any plane $z = k$ is an ellipse

$$\frac{x^2}{a^2} + \frac{y^2}{b^2} = 1 + \frac{k^2}{c^2},$$

which being real for every value of k, the surface is continuous to infinity. This being so, and the sections by the vertical reference-planes and all their parallels being hyperbolas, the surface is called the *Hyperboloid of One Nappe.*

The quantities a, b, c are called its semi-axes, though it is evident, by making $x = y = 0$ in the equation, that the axis of z does not meet the surface. The real meaning of c will soon appear. In general, a is supposed greater than b. In case we have $a = b$, the sections parallel to the plane XY become circles, the surface may be generated by revolving an hyperbola upon its conjugate axis, and is called an *hyperboloid of revolution of one nappe.*

Next, when $C' = 0$, the equation assumes the form

$$A'x^2 + B'y^2 - E'z^2 = 0.$$

The section made by the reference-plane $z = 0$, is the point $A'x^2 + B'y^2 = 0$, while that by any parallel plane $z = k$, is the ellipse $A'x^2 + By^2 = E'k^2$. The sections formed by the planes $x = 0$, $y = 0$, are pairs of intersecting right lines, $B'y^2 - E'z^2 = 0$ and $A'x^2 - E'z^2 = 0$; as also the section by *any* vertical plane $y = mx$, is the pair of lines $(A' + m^2B') x^2 - E'z^2 = 0$. The surface is therefore a *Cone*, whose vertex is the origin. If we have

$$A' = \frac{1}{a^2}, \quad B' = \frac{1}{b^2}, \quad E' = \frac{1}{c^2},$$

the cone is said to be *asymptotic* to the preceding hyper-boloid. If $a = b$, or $A' = B'$, the section by the plane $z = k$ is the circle $x^2 + y^2 = constant$, and the cone is a *circular* one.

Finally, when C' is *positive*, by changing the signs throughout we may write the equation

$$\frac{z^2}{c^2} - \frac{x^2}{a^2} - \frac{y^2}{b^2} = 1.$$

The plane $z = k$ now evidently cuts the surface in imag-inary ellipses so long as $k < c$, but in real ones when k passes the limit c whether positively or negatively. The surface therefore consists of two portions, separated by a distance $= 2c$, and extending to infinity in opposite directions. The sections by the planes $x = l$, $y = m$, are hyperbolas. The surface is therefore called the *Hyperboloid of Two Nappes.*

By making $x = y = 0$, we find that the intercept of this surface on the axis of z is $= c$; while the intercepts upon the axes of x and y, found by putting $y = z = 0$ and $z = x = 0$, are the imaginary quantities $a\sqrt{-1}$, $b\sqrt{-1}$. Moreover, the sections by the planes $x = 0$, $y = 0$, being

$$\frac{z^2}{c^2} - \frac{y^2}{b^2} = 1, \quad \frac{z^2}{c^2} - \frac{x^2}{a^2} = 1,$$

are hyperbolas conjugate to those in which the same planes cut the Hyperboloid of One Nappe. We there-fore perceive that the present hyperboloid is conjugate to the former, and that the *real* meaning of c in the equation to the former is, the semi-axis of its conjugate surface.

When $a = b$, this hyperboloid also becomes one of revolution, and is called an *hyperboloid of revolution of two nappes.*

III. Let A' be *positive*, and B' and E' both *negative*. The general equation may then be written

$$\frac{x^2}{a^2} - \frac{y^2}{b^2} - \frac{z^2}{c^2} = 1,$$

when C' is negative ; or

$$\frac{y^2}{b^2} + \frac{z^2}{c^2} - \frac{x^2}{a^2} = 0,$$

when C' is equal to zero ; or

$$\frac{y^2}{b^2} + \frac{z^2}{c^2} - \frac{x^2}{a^2} = 1,$$

when C' is positive. The present hypothesis therefore presents no new forms, but merely puts those of the preceding supposition into a different order. It is usual, however, to write the equation to the Hyperboloid of Two Nappes in the form

$$\frac{x^2}{a^2} - \frac{y^2}{b^2} - \frac{z^2}{c^2} = 1,$$

rather than in that obtained by the final supposition of II. The advantage of doing so appears in connection with the hyperboloids of revolution ; for, if we make $b = c$ in the above equation, we learn that an hyperboloid of revolution of *two* nappes may be generated by revolving an hyperbola upon its transverse axis. Under the present hypothesis, therefore, the hyperboloids of two nappes and of one may be generated by revolving the same hyperbola, first upon its transverse, and then upon its conjugate axis. Under the hypothesis of II, on the contrary, the two surfaces would be generated by a pair of conjugate hyperbolas revolving upon the same axis.

729. Let us, secondly, consider the NON-CENTRAL QUADRICS, represented by the equation

$$y^2 + Qz^2 = Px.$$

I. Suppose Q to be *positive.* Any plane $x = l$ cuts the surface in an ellipse $y^2 + Qz^2 = Pl$, which, if P is positive, will be real only on condition that l is not negative; or, if P is negative, only on condition that l is not positive. The surface, then, consists of a single shell, extending to infinity on one side of the plane YZ, and in fact *touched* by that plane in the point $y^2 + Qz^2 = 0$, that is, in the origin. The sections of this shell by the planes $y = 0$, $z = 0$, are the parabolas $z^2 = (P : Q) x$, $y^2 = Px$. Hence, all sections by planes parallel to these are also parabolas, and the surface is the *Elliptic Paraboloid.*

1. If $Q = 1$, the section by the plane $x = l$ is the circle $y^2 + z^2 = Pl$, and the surface may be generated by revolving a parabola upon its principal axis. It is then called a *paraboloid of revolution.*

2. We have thus far not made $P = 0$, because the transformation to $y^2 + Qz^2 = Px$ in general excludes that supposition, being made upon the assumption that we can not, in the equation (Art. 726)

$$B'y^2 + E'z^2 + 2G'x + 2F'y + 2D'z + C = 0,$$

destroy all the three co-efficients G', F', D' together. But if G' were itself $= 0$, then *any* point on the line $F' = 0$, $D' = 0$ would be a center of the quadric; and as this line would thus pierce the surface only at infinity, we could not, by placing the origin at the piercing-point, cause the absolute term to disappear. However, by taking

An. Ge. 50.

the origin upon the line $F' = 0$, $D' = 0$, the equation would be reduced to the form

$$y^2 + Qz^2 = R.$$

This equation, at first sight, appears to represent an ellipse in the plane YZ. But, obviously, it is true not only for points whose $x = 0$, but for points answering to *any* value of x that corresponds to a given y and z. It denotes, then, a cylinder whose base is the ellipse $y^2 + Qz^2 = R$, and whose axis is the axis of x. Hence we learn that a particular case of an elliptic paraboloid is the *Elliptic Cylinder*.

When $R = 0$, this cylinder breaks up into two imaginary planes, whose common section, however, being projected in the real point $y^2 + Qz^2 = 0$, is a *real right line*, perpendicular to the plane YZ.

II. Suppose $Q = 0$. The equation $y^2 + Qz^2 = Px$ then becomes

$$y^2 = Px,$$

and therefore denotes the *Parabolic Cylinder*.

By shifting the origin along the axis of x, the equation to this cylinder takes the more general form

$$y^2 = Px + N.$$

When, therefore, $P = 0$, this cylinder breaks up into the *two parallel planes*

$$y = \pm \sqrt{N},$$

which are real, coincident, or imaginary, according as N is positive, equal to zero, or negative.

III. Suppose Q to be *negative*. Then the surface

$$y^2 + Qz^2 = Px$$

will meet all planes parallel to $y = 0$ and $z = 0$ in parabolas; but, being met by any plane $x = l$ in the real hyperbola $y^2 + Qz^2 = Pl$, is called the *Hyperbolic Paraboloid*.

It evidently meets the plane $x = 0$ in the two intersecting lines $y^2 + Qz^2 = 0$, and extends to infinity on both sides of that plane. As a particular case,

1. Corresponding to the negative Q, we have the cylinder

$$y^2 + Qz^2 = R.$$

As this meets the plane $x = 0$ in the hyperbola $y^2 + Qz^2 = R$, it is called the *Hyperbolic Cylinder*.

When $R = 0$, this evidently breaks up into *two intersecting planes*.

730. The foregoing include all the varieties of the Quadric. By means of the several equations contained in the two preceding articles, we could now proceed to develop all the known properties of these surfaces. The student, however, can hardly have failed to observe the remarkable analogy, not only in respect to the preceding classification and its corresponding equations but in respect to such properties as have already been developed, which subsists between these surfaces and the several varieties of the Conic. He will therefore anticipate that by applying to the equations we have just obtained, the methods with which the discussion of conics has now thoroughly familiarized him, he can obtain the analogous properties of quadrics for himself. Accordingly, several of the more important ones have been presented in the examples at the close of this Chapter.

It may deserve mention, however, that in these quadric analogies to conics, lines generally take the place of points, and surfaces the place of lines; also, that where conic elements go by twos, the quadric elements generally go by threes. Thus, the conception of two conjugate diameters is replaced by that of three conjugate planes.

Often, in connection with this principle of substitution, the result of the analogy is altogether unexpected. For instance, the quadric analogue of the conic *focus*, is itself a conic, known as the *focal conic*, which lies in either of the principal planes; so that, in general, *every quadric has three infinitudes of foci.*

In fact, the subject of Foci and Confocal Surfaces is one of the most recent as well as the most intricate in connection with quadrics. It was originally investigated by CHASLES and MACCULLAGH independently; of whose discoveries Salmon has given a brilliant account in the Abridged Notation.*

SURFACES OF REVOLUTION OF THE SECOND ORDER.

731. Any surface that can be generated by revolving any curve about a fixed right line is called a *surface of revolution*. The revolving curve is named the *generatrix*, and the fixed line around which it moves is termed the *axis*.

We came upon the Quadrics of Revolution and their equations, in the preceding investigations. But we shall here give some account of them from another point of view, for the sake of putting the student in possession of the general *method of revolutions*.

732. Let the shaded surface in the diagram represent

---

* See his *Geometry of Three Dimensions*, p. 101.

the surface generated by any curve, revolving about an axis OC. Let the equations to the "projecting cylinders" of the generatrix be

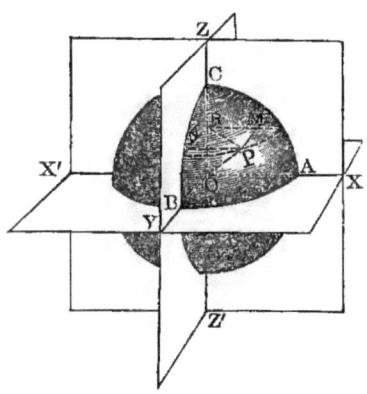

$$x = f(z), \quad y = \varphi(z).$$

Then, from the definition of a surface of revolution, the co-ordinates of any point P on the surface must at once satisfy the conditions for being in the generatrix and in a circle perpendicular to the axis. Hence, if $r =$ the distance PR of any point on the surface from the axis $Z'Z$, we shall have simultaneously

$$x^2 + y^2 = r^2, \quad \overline{f(z)}^2 + \overline{\varphi(z)}^2 = r^2.$$

Eliminating the indeterminate r from these equations, we get

$$x^2 + y^2 = \overline{f(z)}^2 + \overline{\varphi(z)}^2,$$

which expresses the uniform relation among the variable co-ordinates of the surface, and is therefore the *general equation to any surface of revolution.*

This becomes the equation to the surface generated by a *given* curve, when we expand $f(z)$ and $\varphi(z)$ in accordance with the equations to the generatrix. Were we to suppose the generatrix in one of the reference-planes, say the plane ZX, which we may always do when the generatrix intersects its axis, we should have $y = \varphi(z) = 0$, and the general equation would assume the simpler form

$$x^2 + y^2 = \overline{f(z)}^2.$$

In this, we suppose the axis of z to be the axis of revolution; but, obviously, analogous equations of revolution about the axis of x or of y are

$$y^2 + z^2 = \overline{f(x)}^2, \qquad z^2 + x^2 = \overline{f(y)}^2.$$

733. Equation to the Right Circular Cone.—Let the axis of the cone be the axis of z, and its base the plane XY. Then, if the co-ordinates of the vertex be $x' = 0$, $z' = c$, the equation to the generatrix (Art. 101, Cor. 1) will be $x = m(z - c)$, where (Art. 705) $m = \cot\varphi$, if $\varphi =$ the inclination of the side to the base of the cone. Hence

$$f(z) = \frac{z - c}{\tan\varphi};$$

and, substituting in the equation of revolution, we get

$$(x^2 + y^2)\tan^2\varphi = (z - c)^2$$

as the required equation to a right circular cone.

734. Section of a Right Circular Cone by any Plane.—Since the sections formed by parallel planes are similar (Art. 714), it will be sufficient to consider the section formed by any plane NBL passing through the axis of y. As this plane is projected upon the plane ZX in the line OL, its equation may be written

$$z = x\tan\theta,$$

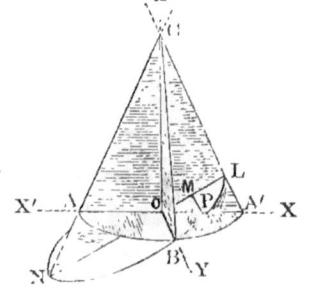

where $x = OQ$, the abscissa of any point P in the plane; $z = QM$, the corresponding ordinate; and $\theta =$ the angle $A'OL$, which measures the inclination of the plane to the plane XY. Substituting for z in the above equation to the cone, we obtain

$$(x^2 + y^2)\tan^2\varphi = (x\tan\theta - c)^2,$$

the equation to the *curve* of section NBL. Transforming this to its own plane, we shall have, in the formulæ of Art. 685, $a = \theta$, $\beta = 90°$; $a' = 90°$, $\beta' = 0$; $a'' = 90°$, $\beta'' = 90°$. We therefore replace x by $x\cos\theta$, and leave y unchanged, thus obtaining, as the equation to any plane section of a right circular cone,

$$x^2(\tan^2\varphi - \tan^2\theta)\cos^2\theta + y^2\tan^2\varphi + 2cx\sin\theta = c^2,$$

in which $\varphi =$ the angle OAC, and $\theta =$ the angle $A'OL$.

735. The Curves of the Second Order are Conics.—The proof of this theorem, promised in Art. 633, is furnished by the equation just obtained. For this evidently conforms to the type

$$Ax^2 + 2Hxy + By^2 + 2Gx + 2Fy + C = 0,$$

giving to the general co-efficients the particular values

$$A = (\tan^2\varphi - \tan^2\theta)\cos^2\theta, \quad H = 0, \quad B = \tan^2\varphi,$$
$$G = c\sin\theta, \qquad\qquad\qquad F = 0, \quad C = -c^2.$$

It therefore denotes a curve of the Second order, whose species, depending on the sign of $H^2 - AB$, in fact here depends on the sign of A; since $H = 0$, and B is necessarily positive.

I. Let A be *positive*. The function $H^2 - AB$ will then be negative, and (Art. 158) the section will be an ellipse. But if A is positive, $\tan^2\theta$ must be less than $\tan^2\varphi$; or, we shall have

$$\theta < \varphi.$$

That is, *If the inclination of the secant plane be less than that of the side of the cone, the section will be an* ELLIPSE.

1. One form of the condition $\theta < \varphi$ is $\theta = 0$. But under this supposition, the equation to the section assumes the form

$$x^2 + y^2 = c^2 \cot^2 \varphi,$$

and we obtain a *circle*, as a particular case of the Ellipse.

2. If under the condition $\theta < \varphi$, we suppose $c = 0$, or that the secant plane passes through the vertex of the cone, the section becomes

$$x^2 (\tan^2 \varphi - \tan^2 \theta) \cos^2 \theta + y^2 \tan^2 \varphi = 0,$$

and we have a *point*, as the limiting case of the Ellipse.

II. Let $A = 0$. The function $H^2 - AB$ will then also equal 0, and (Art. 191) the section will be a parabola. When $A = 0$, however, $\tan^2 \theta = \tan^2 \varphi$; or,

$$\theta = \varphi.$$

That is, *If the secant plane be parallel to the side of the cone, the section will be a* PARABOLA.

1. If we suppose $\theta = \varphi = 90°$, and $c = \infty$, the equation to the section can readily be put into the form $y^2 = constant$; and we learn that when the vertex of the cone recedes to infinity, the Parabola breaks up into *two parallels*.

2. If $\theta = \varphi$, and $c = 0$, the equation to the section becomes $y^2 = 0$, showing that the limiting case of the Parabola is a *right line*.

III. Let A be *negative*. The function $H^2 - AB$ will then also be negative, and (Art. 174) the section will be an hyperbola. But if A is negative, $\tan^2 \theta > \tan^2 \varphi$; or,

$$\theta > \varphi.$$

That is, *If the inclination of the secant plane be greater than that of the side of the cone, the section will be an* HYPERBOLA.

1. If $\theta > \varphi$, and be at the same time of such a value that $(\tan^2\varphi - \tan^2\theta)\cos^2\theta + \tan^2\varphi = 0$, the equation to the section will satisfy the condition $A + B = 0$, and the section (Art. 177) will be a *rectangular hyperbola.*

2. If $\theta > \varphi$, and $c = 0$, so that the secant plane passes through the vertex, the section becomes

$$x^2(\tan^2\varphi - \tan^2\theta)\cos^2\theta - y^2\tan^2\varphi = 0,$$

and the limiting case of the Hyperbola appears as a *pair of intersecting right lines.*

We have thus shown that every *real* variety of the curve of the Second order can be cut from a right circular cone by a plane. The *imaginary* varieties, of course, can not be obtained by any geometric process.

736. Equation to the Circular Cylinder.—The generatrix of this surface is a right line parallel to the axis: hence, if we take the latter for the axis of z, the generatrix will be represented by $x = a$. We have, then, $f(z) = \text{constant} = a$; and the required equation (Art. 732) is

$$x^2 + y^2 = a^2,$$

in which a is the radius of the base.

737. Equation to the Sphere.—Taking the plane of the generating circle for the plane ZX, the equation to the generatrix is $x^2 + z^2 = r^2$. Hence,

$$\overline{f(z)}^2 = r^2 - z^2,$$

which substituted in the equation of Art. 732 gives

$$x^2 + y^2 + z^2 = r^2$$

as the equation now required.

An. Ge. 51.

738. Equations to the Ellipsoids of Revolution.—Let the plane of the generating ellipse be the plane XY. The equation to the generatrix will be $b^2x^2 + a^2y^2 = a^2b^2$: whence

$$\overline{f(x)}^2 = \frac{b^2}{a^2}(a^2 - x^2).$$

Supposing then the axis major to be the axis of revolution, and therefore substituting in the equation for revolution about the axis of x, we get

$$\frac{x^2}{a^2} + \frac{y^2 + z^2}{b^2} = 1,$$

the equation to the *Prolate Spheroid.*

If the generatrix lie in the plane ZX, its equation may be written $c^2x^2 + a^2z^2 = a^2c^2$, and we have

$$\overline{f(z)}^2 = \frac{a^2}{c^2}(c^2 - z^2).$$

Then, taking the axis minor for the axis of revolution, we substitute in the equation for revolution about the axis of z, and obtain

$$\frac{x^2 + y^2}{a^2} + \frac{z^2}{c^2} = 1,$$

the equation to the *Oblate Spheroid.*

739. Equations to the Hyperboloids of Revolution.—Changing the signs of b^2 and c^2 in the expressions of the preceding article for $\overline{f(x)}^2$ and $\overline{f(z)}^2$, we find

$$\overline{f(x)}^2 = \frac{b^2}{a^2}(x^2 - a^2), \quad \overline{f(z)}^2 = \frac{a^2}{c^2}(z^2 + c^2).$$

Supposing the hyperbola to revolve about its transverse axis, we substitute in the equation for revolution about the axis of x, and obtain

$$\frac{x^2}{a^2} - \frac{y^2 + z^2}{b^2} = 1,$$

the equation to the *Hyperboloid of Revolution of Two Nappes.*

Substituting in the equation for revolution about the axis of z, which here implies that the hyperbola revolves about its conjugate axis, we get

$$\frac{x^2 + y^2}{a^2} - \frac{z^2}{c^2} = 1,$$

the equation to the *Hyperboloid of Revolution of One Nappe.*

740. The Ellipse of the Gorge.—This name is given to the curve cut from the narrowest part of the throat of an hyperboloid of one nappe by a plane perpendicular to its axis. Its equation, found by putting $z = 0$ in the equation to the hyperboloid (Art. 728, II), is

$$\frac{x^2}{a^2} + \frac{y^2}{b^2} = 1.$$

In the Hyperboloid of Revolution, this curve becomes the circle $x^2 + y^2 = a^2$, which is called the *Circle of the Gorge.*

741. Equation to the Paraboloid of Revolution.—Beginning with the generatrix in the plane XY, its equation is $y^2 = 4px$: whence

$$\overline{f(x)}^2 = 4px,$$

which, substituted in the equation for revolution about the axis of x, gives

$$y^2 + z^2 = 4px,$$

the required equation to the generated surface.

TANGENT AND NORMAL PLANES TO THE QUADRICS.

742. General equation to the Tangent Plane.—By reverting to the vectorial equation near the beginning of Art. 718, it will be seen that the radius vector will meet the Quadric in two consecutive points at the origin, if when $C = 0$ we also have

$$G \cos a + F \cos \beta + D \cos \gamma = 0.$$

That is to say, multiplying through by ρ, and then substituting the corresponding x, y, z, every right line in the plane

$$Gx + Fy + Dz = 0$$

is a tangent to the surface at the origin. Hence, the equation just written is the equation to the *tangent plane at the origin.*

Supposing the origin not to be on the surface, by transforming to any point $x'y'z'$ of the surface, the equation to the tangent plane at such point would be, after putting for the new G, F, D their values as given in Art. 716,

$$xU_x' + yU_y' + zU_z' = 0,$$

which, by re-transformation to the original axes, gives

$$(x - x')U_x' + (y - y')U_y' + (z - z')U_z' = 0,$$

as the *general equation to the tangent plane at any point*
x'y'z'.

Remark.—The generic interpretation of this equation,
as of its analogue in the Conics, is to regard it as the
symbol of the *polar plane* of the point $x'y'z'$, which in
this view is not restricted to being a point upon the
surface.

**743. Tangent Planes to the different Quad-
rics.**—The equations to these, in their simplest forms,
are found by deriving U_x', U_y', U_z' in the equations
of Arts. 728, 729, and substituting the results in the
general equation last obtained. In this way, we get

$$\frac{x'x}{a^2} + \frac{y'y}{b^2} + \frac{z'z}{c^2} = 1,$$

which represents, in terms of the semi-axes, the *tangent
plane to any ellipsoid;*

$$\frac{x'x}{a^2} + \frac{y'y}{b^2} - \frac{z'z}{c^2} = 1,$$

which represents the *tangent plane to any hyperboloid of
one nappe;*

$$\frac{x'x}{a^2} - \frac{y'y}{b^2} - \frac{z'z}{c^2} = 1,$$

which represents the *tangent plane to any hyperboloid of
two nappes;*

$$2\,(y'y + Q\,z'z) = P\,(x + x'),$$

which represents the *tangent plane to any paraboloid.*

744. Normal of a Quadric.—The right line perpendicular to any tangent plane at its point of contact

with a quadric, is called the *normal* of the surface at that point. Hence, by comparing Arts. 709, 742, we learn that

$$\frac{x - x'}{U_x'} = \frac{y - y'}{U_y'} = \frac{z - z'}{U_z'}$$

are the *general equations to the normal at any point* x′ y′ z′.

By deriving U_x', U_y', U_z' in the equations of Art. 728, we obtain the equations to the normals of the central quadrics, namely,

$$\frac{a^2 (x - x')}{x'} = \pm \frac{b^2 (y - y')}{y'} = \pm \frac{c^2 (z - z')}{z'} ,$$

in which we must use the upper or lower signs in accordance with the variety of the surface. If we derive U_x', U_y', U_z' in the equation of Art. 729, we obtain

$$\frac{x - x'}{P} = \frac{y' - y}{2y'} = \frac{z' - z}{2 \langle \! \rangle z'} ,$$

as the equations to the normal of a paraboloid. These can of course be thrown into other forms when convenient.

745. Normal Planes. — Any plane that passes through the normal of a quadric at any point, is called a *normal* plane to the quadric. Comparing Arts. 698 and 744, we learn that the general equation to a normal plane will be of the form

$$\frac{l (x - x')}{U_x'} + \frac{m (y - y')}{U_y'} = \frac{(l + m) (z - z')}{U_z'} ,$$

where the arbitrary k of Art. 698 is for the sake of symmetry replaced by the ratio $m : l$.

By deriving U_x', U_y', U_z' from any specific equation to either of the quadrics, and substituting the results in the preceding formula, we can obtain the equation to a normal plane for any given quadric in any given system of reference.

It is noticeable that the above equation involves the *indeterminate* ratio $m : l$. This is as it should be; for there is obviously an infinite number of normal planes corresponding to any point on a quadric. When the normal plane, however, satisfies such conditions as determine it, we can readily find the corresponding value of $m : l$.

EXAMPLES ON THE QUADRICS.

1. Determine, by means of their discriminating cubics (see Art. 726), whether the quadrics

$$7x^2 + 6y^2 + 5z^2 - 4yz - 4xy = 6,$$

$$7x^2 - 13y^2 + 6z^2 + 24xy + 12yz - 12zx = \pm 84,$$

$$2x^2 + 3y^2 + 4z^2 + 6xy + 4yz + 8zx = 8,$$

are ellipsoids, hyperboloids, or paraboloids.

2. In any central quadric, the sum of the squares on three conjugate semi-diameters is constant.

3. The parallelopiped whose edges are three conjugate semi-diameters, is of constant volume.

4. Tangent planes at the extremities of a diameter, are parallel.

5. The length of the central perpendicular upon a tangent plane is given by the equation

$$\frac{1}{p^2} = \frac{x'^2}{a^4} + \frac{y'^2}{b^4} + \frac{z'^2}{c^4}.$$

6. The length of the same perpendicular, in terms of its direction-cosines, is

$$p^2 = a^2 \cos^2 \alpha + b^2 \cos^2 \beta + c^2 \cos^2 \gamma.$$

7. The sum of the squares on the perpendiculars to any three tangent planes is constant.

8. The locus of the intersection of three tangent planes which are mutually perpendicular, is the sphere

$$x^2 + y^2 + z^2 = a^2 + b^2 + c^2.$$

9. Find the equation to a diametral plane conjugate to a fixed point $x'y'z'$; and prove that if two diameters are conjugate, their direction-cosines fulfill the condition

$$\frac{\cos a \cos a'}{a^2} + \frac{\cos \beta \cos \beta'}{b^2} + \frac{\cos \gamma \cos \gamma'}{c^2} = 0.$$

10 The locus of the intersection of three tangent planes at the extremities of three conjugate diameters is the central quadric

$$\frac{x^2}{a^2} + \frac{y^2}{b^2} + \frac{z^2}{c^2} = 3.$$